Springer

Berlin
Heidelberg
New York
Hong Kong
London
Milan
Paris
Tokyo

Rainer Friedrich · Stefan Reis (Eds.)

Emissions of
Air Pollutants

Measurements, Calculations and Uncertainties

With 116 Figures

Springer

Editors

PROFESSOR DR.-ING. RAINER FRIEDRICH
DIPL. WIRTSCH.-ING. STEFAN REIS
University of Stuttgart
Institute of Energy Economics and
the Rational Use of Energy
Heßbrühlstraße 49a
70565 Stuttgart
Germany

http://genemis.ier.uni-stuttgart.de/

ISBN 3-540-00840-3   Springer-Verlag Berlin Heidelberg New York

Library of Congress Cataloging-in-Publication Data Applied For

A catalog record for this book is available from the Library of Congress.
Bibliographic information published by Die Deutsche Bibliothek
Die Deutsche Bibliothek lists this publication in die Deutsche Nationalbibliographie; detailed
bibliographic data is available in the Internet at <http://dnb.ddb.de>.

Springer-Verlag Berlin Heidelberg New York
Springer-Verlag is a part of Springer Science+Business Media

springeronline.com

© Springer-Verlag Berlin Heidelberg 2004
Printed in Germany

Cover Design: oh! Holleschek Grafik Design, Garmisch Partenkirchen; *design & production*,
Heidelberg; Erich Kirchner, Heidelberg

Typesetting: Camera-ready by the editors

Printed on acid free paper   30/3141 – 5 4 3 2 1 0

Preface

The scientific work described in this book is the result of an international cooperation of over 60 researchers from many different European countries over more than 10 years. Considerable advances have been made in the development, application and improvement of methods and models for the calculation of emissions. Work on the quantification of uncertainties of emission data could be enhanced and carried out and uncertainty assessments conducted. The developed methods present the current state of the art in this field. Collaboration of teams has taken place under the EUROTRAC, the EUREKA project on the transport and chemical transformation of trace constituents in the troposphere over Europe extensive and networks of joint research work could be established across Europe.

The editors wish to express their gratitude to all contributing authors and their teams for their cooperation, without which this book could never have been compiled. Furthermore, authors and editors alike are much obliged to the International Scientific Secretariate (ISS) of EUROTRAC, in particular Dr. Pauline Midgley, for outstanding support in the preparation of this book.

Rainer Friedrich   Stefan Reis

Contents

List of Authors

Andersen, J. M.
Danish Farmers Unions
Vesterbrogade 4A, 1620 Copenhagen V
Denmark

Asman, W. A.H.
Danish Institute of Agricultural Sciences (DIAS)
P.O. Box 50, 8830 Tjele
Denmark

Bauerle P.
University of Stuttgart, Institute of Process Engineering and Power Plant
Technology, Department of Air Pollution Prevention
Pfaffenwaldring 23, 70569 Stuttgart
Germany

Baumbach, G.
University of Stuttgart, Institute of Process Engineering and Power Plant
Technology, Department of Air Pollution Prevention
Pfaffenwaldring 23, 70569 Stuttgart
Germany

Bayram, M.
Dokuz Eylul University, Faculty of Engineering, Department of Environmental Engineering, Kaynaklar Campus
35160 Buca/Izmir
Turkey

Blank, P.
University of Stuttgart, Institute of Energy Economics and the Rational Use of
Energy (IER)
Hessbruehlstrasse 49 a, D-70565 Stuttgart
Germany

Cellier, P.
Institut de Recherche Agronomique (INRA), Unité de Recherches "Environ-
nement et Grandes Cultures"
78850 Thiverval-Grignon
France

Cetin, E.
Dokuz Eylul University, Faculty of Engineering, Department of Environmental Engineering, Kaynaklar Campus
35160 Buca/Izmir
Turkey

Colles, A.
    Centre for Remote Sensing and Atmospheric Processes, Flemish Institute for
    Technological Research (Vito)
    Boeretang 200, B-2400 Mol
    Belgium

Dinçer, F.
    Dokuz Eylul University, Faculty of Engineering, Department of Environ-
    mental Engineering, Kaynaklar Campus
    35160 Buca/Izmir
    Turkey

Ebel, A.
    University of Cologne, Institute for Geophysics and Meteorology
    Aachener Str. 201-209, D-50931 Köln
    Germany

Elbir, T.
    Dokuz Eylul University, Faculty of Engineering, Department of Environ-
    mental Engineering, Kaynaklar Campus
    35160 Buca/Izmir
    Turkey

Emeis, S.
    Forschungszentrum Karlsruhe GmbH, Institute for Meteorology and Climate
    Research, Atmospheric Environmental Research (IMK-IFU)
    Kreuzeckbahnstraße 19, D-82467 Garmisch-Partenkirchen
    Germany

Engewald, W.
    University of Leipzig, Department of Analytical Chemistry
    Linnéstr. 3, 04103 Leipzig
    Germany

Falkenberg, G.
    Hamburger Synchrotronstrahlungslabor (HASYLAB) at Deutsches Elek-
    tronen-Synchrotron (DESY)
    Notkestr. 85, D-22607 Hamburg
    Germany

Friedrich, R.
    Institute of Energy Economics and the Rational Use of Energy, University of
    Stuttgart
    Heßbrühlstraße 49 a, D-70565 Stuttgart
    Germany

Génermont, S.
    Institut de Recherche Agronomique (INRA), Unité de Recherches "Environ-
    nement et Grandes Cultures"
    78850 Thiverval-Grignon, France

Glaser, K.
University of Stuttgart, Institute of Process Engineering and Power Plant
Technology, Department of Air Pollution Prevention
Pfaffenwaldring 23, 70569 Stuttgart

González, R.M.
Department of Meteorology and Geophysics, Faculty of Physics, Complutense
University of Madrid
Madrid
Spain

Hassel, D.
TÜV Rheinland Sicherheit und Umweltschutz GmbH, Testzentrum Ener-
gietechnik
Am Grauen Stein, 51105 Köln
Germany

Hausberger, S.
Institute for Internal Combustion Engines, Graz University of Technology
Inffeldgasse 25, A-8010 Graz
Austria

Hoffmann, H.
Forschungszentrum Karlsruhe GmbH, Institute for Meteorology and Climate
Research, Atmospheric Environmental Research (IMK-IFU)
Kreuzeckbahnstraße 19, D-82467 Garmisch-Partenkirchen
Germany

Hutchings, N. J.
Danish Institute of Agricultural Sciences (DIAS)
P.O. Box 50, 8830 Tjele
Denmark

Jahn, C.
Forschungszentrum Karlsruhe GmbH, Institute for Meteorology and Climate
Research, Atmospheric Environmental Research (IMK-IFU)
Kreuzeckbahnstraße 19, D-82467 Garmisch-Partenkirchen
Germany

Janssen, L.
Centre for Remote Sensing and Atmospheric Processes, Flemish Institute for
Technological Research (Vito)
Boeretang 200, B-2400 Mol
Belgium

Klemp, Dieter
FZ Jülich, Institut für Chemie und Dynamik der Geosphäre:Institut II: Tropo-
sphäre
D-52425 Jülich
Germany

Kühlwein, J.
Institut für Energiewirtschaft und Rationelle Energieanwendung (IER), Universität Stuttgart
Heßbrühlstraße 49 a, D-70565 Stuttgart
Germany

Lewyckyj, N.
Centre for Remote Sensing and Atmospheric Processes, Flemish Institute for Technological Research (Vito)
Boeretang 200, B-2400 Mol
Belgium

Mannschreck , K.
UFS Schneefernerhaus
c/o Meteorologisches Observatorium Hohenpeissenberg, 82383 Hohenpeissenberg
Germany

Memmesheimer, M.
University of Cologne, Institute for Geophysics and Meteorology
Aachener Str. 201-209, D-50931 Köln
Germany

Mensink, C.
Centre for Remote Sensing and Atmospheric Processes, Flemish Institute for Technological Research (Vito)
Boeretang 200, B-2400 Mol
Belgium

Michael, H.
University of Leipzig, Department of Analytical Chemistry
Linnéstr. 3, 04103 Leipzig
Germany

Mittermaier, B.
Forschungszentrum Jülich, Institut für Chemie und Dynamik der Geosphäre, ICG-II: Troposphäre
52425 Jülich
Germany

Möllmann-Coers, M.
FZ Jülich Geschäftsbereich S-UE, Forschungszentrum Jülich
D-52425 Jülich
Germany

Müezzinoğlu, A.
Dokuz Eylul University, Faculty of Engineering, Department of Environmental Engineering, Kaynaklar Campus
35160 Buca/Izmir
Turkey

Münier, B.
National Environmental Research Institute (NERI)
P.O. Box 358, 4000 Roskilde
Denmark

Odabasi, A.
Dokuz Eylul University, Faculty of Engineering, Department of Environmental Engineering, Kaynaklar Campus
35160 Buca/Izmir
Turkey

Osán, J.
KFKI Atomic Energy Research Institute
P.O. Box 49, H-1525 Budapest
Hungary

Peña, J.I.
Environmental Software and modelling Group, Computer Science School,
Technical University of Madrid (UPM). Campus de Montegancedo
Boadilla del Monte, 28660 Madrid
Spain

Pérez, J.L.
Environmental Software and modelling Group, Computer Science School,
Technical University of Madrid (UPM). Campus de Montegancedo
Boadilla del Monte, 28660 Madrid
Spain

Pregger, T.
Institute of Energy Economics and the Rational Use of Energy (IER), University of Stuttgart
Heßbrühlstraße 49 a, D-70565 Stuttgart
Germany

Reis, S.
University of Stuttgart, Institute of Energy Economics and the Rational Use of Energy (IER)
Hessbruehlstrasse 49 a, D-70565 Stuttgart
Germany

San José, R.
Environmental Software and modelling Group, Computer Science School,
Technical University of Madrid (UPM). Campus de Montegancedo
Boadilla del Monte, 28660 Madrid
Spain

Schäfer K.
Forschungszentrum Karlsruhe GmbH, Institute for Meteorology and Climate Research, Atmospheric Environmental Research (IMK-IFU)
Kreuzeckbahnstraße 19, D-82467 Garmisch-Partenkirchen
Germany

Schmitz,T.
   Engelhard Technologies
   D-30173 Hannover
   Germany

Seyfioglu, R.
   Dokuz Eylul University, Faculty of Engineering, Department of Environ-
   mental Engineering
   Kaynaklar Campus, 35160 Buca/Izmir
   Turkey

Sjödin, Å.
   IVL Swedish Environmental Research Institute
   P.O. Box 47086, SE-402 58 Göteborg
   Sweden

Slemr, F
   Max-Planck-Institut für Chemie, Abt. Chemie der Atmosphäre
   Postfach 3060 Mainz, D-55020 Mainz
   Germany

Smiatek, G.
   Atmospheric Environmental Research (IMK-IFU),
   Kreuzeckbahnstraße 19, D-82467 Garmisch-Partenkirchen
   Germany

Sommer, S. G.
   Danish Institute of Agricultural Sciences (DIAS)
   P.O. Box 536, 8700 Horsens
   Denmark

Sonnborn, K.-S.
   TÜV Rheinland Sicherheit und Umweltschutz GmbH, Testzentrum Ener-
   gietechnik
   Am Grauen Stein, 51105 Köln
   Germany

Staehelin, J.
   Institut für Atmosphäre und Klima, ETH Hönggerberg
   HPP J 7, CH-8093 Zürich
   Switzerland

Steinbrecher, R.
   Forschungszentrum Karlsruhe, Institute for Meteorology and Climate Re-
   search
   Kreuzeckbahnstraße 19, D-82467 Garmisch-Partenkirchen
   Germany

Sturm, P. .J.
   Institute for Internal Combustion Engines, Graz University of Technology
   Inffeldgasse 25, A-8010 Graz
   Austria

Theloke, J.
Institute of Energy Economics and the Rational Use of Energy, University of Stuttgart
Heßbrühlstraße 49 a, D-70565 Stuttgart
Germany

Török, S.
KFKI Atomic Energy Research Institute
P.O. Box 49, H-1525 Budapest
Hungary

Vogt, U.
University of Stuttgart, Institute of Process Engineering and Power Plant Technology, Department of Air Pollution Prevention
Pfaffenwaldring 23, 70569 Stuttgart
Germany

Weber, F.-J.
TÜV Rheinland Sicherheit und Umweltschutz GmbH, Testzentrum Energietechnik
Am Grauen Stein, 51105 Köln
Germany

Wickert, B.
Institut für Energiewirtschaft und Rationelle Energieanwendung (IER), Universität Stuttgart
Heßbrühlstraße 49 a, D-70565 Stuttgart
Germany

Winiwarter, W.
ARC Seibersdorf research
A-2444 Seibersdorf
Austria

1 Introduction and Summary

R. Friedrich

1.1 Aims of this book

Understanding and quantifying the emissions of air pollutants from human activities and from natural sources is the first, essential step in understanding, controlling and mitigating air pollution. For this, it is necessary to describe the sources of emissions and their geographical distribution, describe the strength of the emissions from a particular source under particular conditions (commonly called emission factors), collect or calculate the volume of activity, that leads to emissions, calculate emission rates by connecting activities with emission factors, quantify the uncertainties of emission rates and identify emission reduction measures and their potential and costs. In this book, advances in all the above mentioned areas are described. These advances have been achieved with research carried out in the frame of the EUROTRAC-2[1] subproject on the Generation and Evaluation of Emission Data (GENEMIS). The pollutants covered include anthropogenic and biogenic precursors of ozone, aerosols and acidifying substances. As methods for developing emission inventories for CO, SO_2 and NO_x are more advanced and the resultant uncertainties in their emission estimates are lower than those for VOC, PM and NH_3, the focus lies on the latter pollutants.

The general aim of the results described here is to support the generation of validated emission data that can be used for the development of air pollution abatement strategies in Europe. This includes

- the improvement of methods, models and emission factors for the generation of emission data (*described in Chap. 2 of this book*),
- the assessment of the accuracy and the validation of emission data (*Chap. 3*) and
- the development and improvement of tools that generate emission data for atmospheric models *(Chap. 4)*.

The following sections of this chapter summarize the findings, that are explained in more detail in chapters 2 – 4 of this book. References are given, so that for each result the reader is referred to the corresponding section in the book, where more details are presented.

---

[1] The EUREKA project on the transport and chemical transformation of trace constituents in the troposphere over Europe; second phase (http://www.gsf.de/eurotrac)

1.2 Summary

1.2.1 Anthropogenic emissions of volatile organic compounds

The main sources of anthropogenic non-methane volatile organic compounds (NMVOCs) are road transport and solvent use. Thus, in the following progress on generating NMVOC emission data for these source categories is described first. Then some other anthropogenic sources are analysed Next, progress in splitting information about total NMVOC into single species and classes of species is described. A very important contribution to NMVOC emissions stem from trees, this is addressed in the last section.

1.2.1.1 Road transport

Although the majority of vehicle kilometres are driven under highway conditions at normal operating temperatures, a significant fraction of journeys in urban areas are driven at sub-optimal engine (and catalyst) temperatures. Such cold start periods of passenger cars with the current emission control equipment are characterized by inefficient operation of the catalyst. It takes a few minutes for the catalyst to reach operating temperature (cf Sect. 2.2.2.3: Hassel et al.). In this period a substantial fraction of transport-related VOC emissions takes place and these cold start emissions are gaining in relative importance as other emission pathways are being reduced. Until now, cold start emissions have been estimated by using results of dynamometric tests based on the test driving cycle. However, recent research has shown that real world driving patterns e.g. in Germany are very different from those of the test cycle – the engine takes longer to reach its normal operating temperature – and that cold start emissions of VOC are up to 50% higher than previously assumed.

Another important contribution to the NMVOC emissions is that made by poorly maintained vehicles and those with defective catalysts, as illustrated by Sjodin et al. in Sect. 2.2.2.4. Real-world emission measurements on road vehicles have been carried out by means of optical remote sensing (FEAT) at two different sites in Gothenburg, Sweden, during 2001. The measurements indicate that, for petrol passenger cars, vehicle age is the most important factor affecting the tailpipe exhaust concentrations and the mass emission of CO, HC and NO per litre fuel burned. The largest differences can of course be observed between non-catalyst cars and catalyst cars, however, there are also significant differences between catalyst cars of more recent model years and earlier model years. Results also show that less than 10% of the passenger cars cause ca. 50% of the HC and CO emissions.

An analysis of measurements of in-use cars in Germany revealed that the exhaust emissions of VOC from cars with catalysts (current fleet) are about 9%

higher than calculated with actual emission models due to malfunctioning equipment.

1.2.1.2 Solvent use

While for road transport, methods to calculate non-methane VOC emissions are well established and harmonised throughout Europe, this has not been the case for emissions from solvent use. This leads to data sets of different quality for the different countries of Europe, and in some cases not all processes will be accounted for. Thus, a method has been developed by Theloke et al. (Sect. 2.2.3) and applied to Germany which is based on statistical data on the production of goods containing solvents and the import and export of such products.

The share of emissions from solvent use as one of the main anthropogenic NMVOC emission sources has increased during recent years relative to that of traffic, the second largest emission source for NMVOCs. In 2000, roughly 50% of all German NMVOC emissions originated from the use and application of solvents, in total ~ 945 kt. About 450 kt originated from paint application (48%), 158 kt from domestic solvent use (17%), 79 kt from the printing industry, 31 kt from the processing of synthetic materials and 33 kt from degreasing, dry cleaning and electronics production. The remaining share of emissions originates from a number of other sectors. The increase in emissions from solvent use observed from 1994 to 2000 currently prevails. Thus, it is not clear if Germany can comply with the emissions target of the recently adopted EC National Emissions Ceilings Directive for the year 2010. The implementation of additional measures to further reduce emissions of anthropogenic NMVOC emissions in Germany has to be taken into consideration.

The developed method for estimating solvent use emissions could be used throughout Europe, as it uses available statistical data. Important sources of solvent emissions include paint application, degreasing processes, domestic solvent use, printing processes, application of glues and adhesives and the preservation of wood.

1.2.1.3 Other sources

Emissions of VOC compounds have been measured in several stove types used for the combustion of brown-coal briquettes (Sect. 2.2.2.6: Michael and Engewald). Such stoves are typically used in houses in Eastern Germany. Species in the range C3-C18, mainly alkanes, alkenes, aromatics, aldehydes, phenols and thiophenols were found. Surprisingly, the share of oxygenated species was – with 45% of the total NMVOC- much larger than previously assumed.

Activities during refuelling and transport of petrol are important anthropogenic VOC emission sources. These activities include the handling of fuels within gas stations and other refuelling units respectively transport systems such as ships and road tankers, tanks and tank cleaning. To determine the emission rate from diffuse area sources and heterogeneous distributed sources measurements have been carried out, using inverse modelling to determine emission rates of fuel stations and of ships. Results for fuel stations show (Sect. 2.2.2.5: Schäfer et al.) that diffuse emissions have been seriously underestimated in emission inventories. For tank ships, it was shown that the ventilation of the tank atmosphere of river tank ships via the loading/unloading pipes into the land-site vapour recovery unit is a technical solution to reduce VOC emissions drastically.

1.2.2 VOC split

VOC is a collective name for a very large number of different chemical species that have different physico-chemical properties, and which can contribute to the formation of secondary pollutants with different efficiencies. It is therefore important to have information on the speciation of the VOCs, i.e. to split the total VOCs into species or classes of species with similar chemical and physical properties. However, emission inventories usually only provide data for total non-methane VOC only. A detailed speciation of VOCs has been developed for the different source categories by specific measurements of selected sources as well as by using data from measurements performed elsewhere (for an in-depth discussion of the VOC Split, see Chap. 2.2). Measurements included emissions from passenger cars as well as stationary sources.

To determine VOC splits of NMVOC emissions from passenger cars, measurements were made independently by several groups (e.g. Klemp and Mittermaier, Sect. 2.2.2.2) for diesel and petrol motors in different driving cycles.

Road traffic VOC splits were determined for different car types under various conditions (driving pattern, maintenance etc.). The experimental determination of VOC splits of cars shows a strong dependency of the split on the driving pattern of vehicles with three-way-catalysts. For non-catalyst and diesel cars the split differs significantly less. For diesel vehicles alkenes and carbonyls dominate, whereas in the exhaust gas of petrol cars aromatics or alkanes have the highest share of the VOC's.

Based on these measurements and other findings a detailed VOC-split data base has been developed and is now available (Sect. 2.2). Combustion processes and solvent evaporation emit very different VOC species. Solvents emit mainly alcohols and other oxygenated VOCs, whereas combustion processes tend to emit alkanes, aromatics and alkenes (Theloke et al. 2001, 2003; Friedrich and Obermeier 1999).

1.2.3 Biogenic emissions of volatile organic compounds

Recent work has shown, that – during hot sunny episodes – biogenic emissions in Europe are higher than anthropogenic emissions. On the other hand, uncertainties of estimates of biogenic emissions are much larger than those estimated for anthropogenic emissions. So, the improvement of the methodology for estimating biogenic emissions is a research topic with high priority (Sect. 2.2.1: Steinbrecher et al.).

In order to extrapolate BVOC emissions from the leaf or plant level to the ecosystem, regional or global scales, detailed information on meteorology, the area of the emitting land use type in each grid cell of the model area, leaf biomass or leaf area index (LAI), and emission factors are needed. Land use data are generally given in categories, such as agriculture, deciduous, coniferous or mixed forest. This classification is available at high resolution, e.g. in the CORINE database (www.dataservice.eea.eu.int), but this coarse land cover classification causes large systematic errors in the emission inventories, especially in landscapes where deciduous forest consists of both non-emitting plant species like beech and high emitting species like oak.

For constructing BVOC emission inventories at different spatial and temporal scales a geographical information system (GIS) and a relational data base management system (RDBMS) has been set up, managing land cover and plant species-specific data (VOC emission factors, biomass and LAI values) (Sect. 2.2.1 Steinbrecher et al. and Sect. 4.2 Smiatek et al.). GIS handles all geo-data, such as land use, administrative boundaries and other data in raster or vector form. The RDBMS stores all data on the species area and biomass as well as descriptive information. In total more than 44 different land use data sets with 14 different categories, 30 data sets with vegetation index (VI) and leaf area index (LAI) data covering Germany and large areas of Europe have been processed.

As an indication of the complexities involved in modelling BVOC emissions, it may be noted that more than 130 European plant species, 33 chemical compounds and 668 emission factors have been collated from various sources, 3600 records of information on forest area with 6 different categories and more than 400 regions have been integrated. Information on species specific emission factors for 1100 plant species is available.

A monthly-average BVOC inventory for Germany has been established for the year 1998, with hourly-average emissions estimated for a 3 days ozone episode in August 1998. Most of the biogenic emissions occur in the middle and southern part of Germany with values of more than 400 kg C km^{-2} $month^{-1}$. Compared to BVOC emission inventories which only consider general forest types (e.g. deciduous, coniferous, and mixed forest), the use of tree species distribution for the six major forest trees in Germany significantly reduced systematic errors associated with the assignment of emission factors to source categories. Isoprene emissions using plant species data are in general 30% lower compared to emission estimates considering only forest types. Furthermore, the plant specific emission inventory for Germany shows significantly different geographic distribution patterns shifting the focus from areas with large forest cover to areas with large oak stands.

The total annual emission rate of BVOC for Germany for 1998 was 547 kt C BVOC, or 20% of the total VOC emissions for the country. However, under high temperature and light conditions during August 1998 the BVOC contribution rose to 60% of the total VOC emission. Since the reactivity of BVOCs towards OH is greater than that of the anthropogenic VOCs, their role in ozone formation is likely to be greater than is apparent from just considering the mass emission rates.

1.3 Anthropogenic emissions of ammonia

Most emissions of ammonia (NH_3) originate from agricultural practices, principally the production of manure and the application of fertiliser. Emissions from animal husbandry depend on many factors (Sect. 2.4 Asmann et al.). The nitrogen excretion of animals is influenced by the nitrogen content of the feed intake and the relative share of amino acids in the feed. For animals housed indoors, the ammonia emissions depend on factors such as the type of building floor, the ventilation rate, the frequency of cleaning and the type of manure (liquid or solid) produced. Emissions during storage depend on the type of storage system used (pile, open or closed tanks) and to some extent on meteorological conditions (wind speed, temperature). Following the application of manure, ammonia emissions depend on factors such as the properties of the manure and the meteorological and soil conditions. Emissions resulting from different manure handling systems for the same kind of animal can be up to a factor of two different. Recently control techniques have been applied in some areas to reduce agricultural ammonia emissions.

Experimental work has shown that ammonia emissions after manure application in the field increases by 2% when the temperature increases by 1° C; it increases by 4% when the wind speed increases by 1 m s^{-1} and it is 10% higher for wet soil than for dry soil (Hutchings et al. 2002).

In many countries, no statistical information is available on different manure handling systems. Even when information is available it is not usually geographical disaggregated within countries.

Mechanistic models have been developed for ammonia emissions following application of manure. These models give a better understanding of the complex interactions during the emission process and their results agree reasonably well with measurements. However, there is a need for further development of such models so that a coherent description can be given of all processes influencing ammonia emissions. The ammonia emission rate is influenced by atmospheric turbulence, but vertical mixing in the atmosphere depend also on the atmospheric turbulence. It has been demonstrated that when there is much turbulence, the emission rate is higher, but also that the emitted NH_3 is transported over longer distances before it is deposited back to the surface, either as NH_3 or as NH_4^+.

This co-dependence is not yet taken into account in most atmospheric transport models that are used to make policy decisions concerning ammonia emissions. This is important because it may change the "blame-matrices" for acidifying and eutrophying compounds that are used to develop strategies for emission reduction. Model experiments have shown that a significant fraction of emitted NH_3 is lost from the atmosphere by dry deposition very close to the source. This is important to consider as otherwise the ammonia emission rates feeding into atmospheric transport, chemistry and deposition models may be overestimated.

1.4 Emissions of particulate matter to the atmosphere

Information on emission sources and factors has been systematically collected and detailed emission inventories have been set up for Germany (Sect. 2.3.3.2: Pregger et al..) and Austria (Sect. 2.3.3.1: Winiwarter et al.). In addition to that, measurements of emission factors have been carried out, investigating particulate emissions of small wood combustion furnaces in Hungary (Sect. 2.3.2: Török et al.) by microchemical analysis methods, focussing on including elemental concentrations and aerodynamic parameter. The majority of stack-gas particles were found in the respirable size range. The completeness of PM inventories is now substantially improved.

For Austria (Sect. 2.3.3.1: Winiwarter et al.) results indicate, that the emissions of all sources except road dust suspension (base year 1995) amount to 75,000 t of TSP, 45,000 t of which are PM_{10}, and 26,000 t are $PM_{2.5}$. Most important source sectors are fugitive emissions from bulk materials handling in industry as well as building and agricultural operations. Wood combustion is the most important source of fine particle emissions.

An important, yet very uncertain source might be vehicle-induced suspension and abrasion of particles from road surfaces (Sect. 2.3.1.1, Sturm.et al.) A first model has been set up to calculate vehicle induced emissions; the model is based on a model design developed by US-EPA, but adjusted to European conditions using measurements in Berlin. It shows that suspension and abrasion processes might lead to PM_{10} emissions in the same order of magnitude as those from exhaust gases from diesel engines (Pregger and Friedrich 2003). Further results show that suspension of road dust may be the cause of up to half of the total PM_{10} emissions for Austria. The model methodology is under dispute, however, and more measurements (Sect. 2.3.1.3 Sjodin et al.) are required under a range of European conditions to confirm this result.

1.5 Assessment of the quality of emissions data

Emission data are needed to estimate future trends in the concentration and deposition of air pollutants, to identify air pollution control strategies that lead to an avoidance of exceeding ambient thresholds and to report emissions for different reporting obligations in the frame of conventions and directives, e.g. IPCC, national emissions ceilings directive, and others. Emissions data however can only be used as useful aid for decision making in the field of air pollution control if their quality is known. There are several aspects of inventory quality, basically all of which deal with the question: how well does the inventory fit reality.

An overview of potential quality control options in urban emission inventories has been prepared (Sect. 3.2.1 Sturm et al.).

To assess the quality of emission data, two options have been used and are described here; one is the analysis of input uncertainties in emission inventories and their representation in the result (Sect. 3.3.1 Winiwarter et al.), and the other is the comparison of an inventory with independently derived information on emission fluxes, especially with measurement data (Sect. 3.3.2 and 3.3.3).

Quantifying uncertainties in emissions data requires a statistical approach. The spreads of statistical errors of model input parameters can be calculated by basic statistical methods such as error propagation or Monte-Carlo analysis. Statistical evaluations provide confidence intervals of calculated emission values. However, they give no information on the true emission values, as possible systematic errors (missing sources, wrong assumptions) are not fully detected, even if attempts have been made in this direction; so in addition to a statistical analysis verification is necessary (see Sect. 3.2.2. Kühlwein et al).

The verification of emissions data requires the comparison of the calculated emission rates with emission rates derived with other independent methods (see 3.2.2). The use of an inter-comparison of different emission inventories is limited as they will, at least in part, make use of the same methods and parameters. Nevertheless, comparisons of available inventories for the same regions, but prepared by different groups, show quite large deviations, this demonstrates the necessity of quality assurance and control (Sect. 3.3.3.2, Slemr et al.). A further possibility for verification is the comparison of emission trends with ambient concentration trends (short term as well as long term). True emission rates can be estimated within a certain error margin by measurements of ambient concentrations. Methods include the measurements of upwind and downwind fluxes, e.g. for tunnels (Sect. 3.3.2.1, Staehlin and Sturm), motorways or whole cities. Furthermore, tracers can be released (see Sect. 3.3.3.4, Moellmann-Coers et al.) at known emission rates and the ratio of measured tracer concentration to the concentration of another substance then used to determine the emission rate of this substance. Other methods include the comparison of concentration ratios with emission ratios of individual substances (cf. Sect. 3.3.3.7, Mannschreck et al.) and receptor modelling (Sect. 3.3.3.6, Klemp et al.). Measurements under real conditions are restricted to particular conditions (e.g. emissions from one motorway section with a particular road gradient, fleet composition and driving pattern). All this methods have been applied and are described here.

Road tunnel studies have been used to validate road traffic emission models (see Chap. 3.3.2.1). Measurements of primary air pollutants performed in different years in the same European road tunnels impressively demonstrated the large success in the reduction of the emission factors of total Non Methane Volatile Organic Compounds (t-NMVOC), carbon monoxide (CO) and nitrogen oxides (NO_x) of petrol powered vehicles during the 1990s, largely attributable to the introduction of catalytic converters.

In general it is necessary to use compilations of data from as many experiments as possible to get a full picture of the uncertainties in emissions estimates. Results of such a metaanalysis demonstrate clearly that there is a systematic error in the calculation of emissions from trucks: the commonly used NO_x emission factors for the current truck fleet are too low by a factor of 2 (Kühlwein and Friedrich 2003).

In another experiment, emissions from fuel stations have been measured with remote sensing devices (Schäfer et al. 2000). This showed that the vapour recovery systems in use often do not work properly, with VOC emissions 3 to 5 times larger than from well maintained stations. This shows that more efforts should be made to regularly maintain emission reduction technologies and to check them to ensure effective operation.

Within the evaluation project EVA, whose aim was to evaluate emissions from all sources of a whole city, two field campaigns were successfully performed in March and October 1998 (see Sect. 3.3.3) using aircraft, tethered balloons, air ships, radio sondes, ground stations, and remote sensing techniques. For the experiment, the city of Augsburg has been chosen, as the terrain there is relatively flat and there are not too many large emission sources outside the city borders. Augsburg has on the one hand significant emissions, on the other hand it is small enough to be analysed with the limited number of measurement devices and aircrafts available. Emphasis has been laid on QA/AC during the measurement campaigns. At the first and last days of each field experiment an intercalibration with transfer standards and comparisons of measurements under field conditions has been carried out.

Three different measurements (two mass balance approaches including aircraft measurements, see Sect. 3.3.3.3 Slemr et al. and Sect. 3.3.3.8 Baumbach et al., one tracer experiment, described in Sect. 3.3.3.4, Moellmann-Coers et al.) and two different models have been used to determine the emissions. Given the anticipated standard deviation of modelled (Kühlwein et al. 2002a) as well as measured emission rates of ca. 20%, no statistically significant deviation could be detected for NO_x and CO (Kühlwein et al. 2002b). However, for some single VOC species derived from solvent use, especially from white spirits, calculated emission rates are much higher than measured values, suggesting that emissions from solvent use may be currently overestimated. Unfortunately VOC species from solvents are rather difficult to measure and so relatively few reliable data are available. This suggests that more measurements of higher alkanes and oxygenated VOC emissions in urban areas are necessary.

This overestimation of solvent use emissions was also detected in another analysis (Sect. 3.3.3.10., Staehlin et al.) for Swiss emission inventories. Based on a statistical analysis of extended monitoring measurements the conclusion was drawn, that the proportion of VOC from motor traffic emissions including fuel vaporisation amounts to 48-67% whereas the Swiss emission inventory (basis year 1990) gives lower values (33-48%).

In summary, modelled emissions data still have considerable sources of error associated with them. Future efforts to improve activity data and emission factors should reduce these uncertainties. Statistical uncertainty analyses of modelled emission rates of CO, NO_x and NMVOCs result in coefficients of variation in the order of 10% to 50%. Uncertainties in individual VOC emission rates can be much higher. Comparisons between measured and modelled emission values at individual road sections in general show lower model emission factors especially for trucks. Good agreement between modelled and measured total NO_x and CO emission rates from the city of Augsburg has been found within the 1 σ confidence intervals.

1.6 Improvement and application of emission models and tools

1.6.1 Urban modelling

In the case of urban modelling, work focused on developing specifications of emission data for emission modelling and development of methodologies for emission inventories models and especially methodologies for the estimation of traffic and biogenic emissions. A group from Turkey (Müezzinoglu et al, Sect. 4.3.4) developed a specific decision support system for urban air quality management for the İzmir metropolitan area. An emission inventory for five air pollutants of sulphur dioxide (SO_2), particulate matter (PM), nitrogen oxides (NO_x), volatile organic compounds (VOC) and carbon monoxide (CO) by using appropriate emission factors was prepared. The results show that industry is the most polluting sector in the study area contributing about 91% of total SO_2 emissions, 40% of total PM emissions, 90% of total NO_x emissions, 40% of total VOC emissions and 70% of total CO emissions per year.

New EU guidelines with the intention to restrict concentrations of harmful pollutants came into force recently. Model based pollution maps are necessary to recognize receptor points where concentration limits might be exceeded. To prepare these maps, emission data in different spatial scales need to be available. Thus, a microscale traffic emission model is being established as described by Kühlwein et al. (see Chap 3). Emission data in high spatial resolution (< 50 m) is calculated and provided as input data for high-performance chemistry transport models.

1.6.2 Regional modelling

A number of emission models (EMIMO, MIMOSA, ECM, see San Jose et al., Lewyckyi et al. and Reis et al., Chap. 4) have been developed which can be used by air quality mesoscale models for all parts of the world. The EMIMO model uses both the top-down and the bottom-up approaches together with an important set of global and European emission inventories such as EMEP, CGEIC, GEIA and EDGAR.

Due to the increasing demand of emission data for CTMs, a geographic pre-processor (GIS) has been designed and implemented into the GENEMIS emission module (ECM) by Friedrich et al. (1999 and 2003) to improve the spatial resolution of the European database and enable the conversion of data in model grids of any geographic projection. The fine spatial structure of emissions is based on land-use data sets, on digital road and railway maps and other available information.

Ozone forecasting using the ECM model

To provide emission data in high temporal and spatial resolution, a large amount of data has to be generated. It is difficult and inefficient to generate these data, e.g. for a year in advance and then to deliver it to the location of the atmospheric model. Instead, a module (named ECM) has been developed, that can be coupled to the atmospheric model generating the input data in the moment they are needed (Friedrich et al. 1999). The most important part of this module is a data base containing data and specifications for possible data manipulations. The information in the data base allows the emission model to determine emissions of different pollutants in a high spatial and temporal resolution, e.g. hourly values for a 1 km x 1 km-grid for any investigation project. ECM is in fact more a flexible modelling toolbox than just a single emission model. ECM was and is further used for preparing emission data for various investigation areas including small-scale urban areas, regions, countries and the whole of Europe. ECM is currently applied by EURAD to provide emission data for ozone forecasting (http://www.eurad.uni-koeln.de/index_e.html) and provides emission data as input for a number of chemical transport models on different scales.

1.7 Outlook

Air pollution is still a major problem in Europe, as current or planned thresholds or critical levels and loads are frequently exceeded (e. g. for Ozone, PM_{10}, acid deposition and nutrient deposition). So, further reductions of emissions are necessary. It is obvious, that a prerequisite of identifying abatement measures is a profound knowledge of emissions and the structure of emission sources. Furthermore, the increasing emission reductions necessary to improve air quality require more and more expensive emission reduction measures. To achieve acceptance of additional measures, it is increasingly important to establish proof, that the planned measures lead to a fulfilment of the environmental goals and that the emission reduction is reached efficiently, i.e. with the least costs possible. To show this, emission data with high quality are needed, as well as input for atmospheric models as base for investigating possible abatement measures and related costs.

The results described here have contributed to the improvement of the quality of emission data including the improvement of methods for calculating

- emissions with high spatial and temporal resolution;

- PM emissions for all emission sectors;
- NMVOC emissions from solvent use;
- Biogenic emissions from forests
- Emissions from road transport (cold start, evaporation)
- measurements of VOC emission factors and VOC split
  - from fuel stations and tankers;
  - from passenger cars;
  - from lignite firings.

Furthermore, new knowledge on uncertainties of emission data has been acquired by carrying out or analysing results from whole city, tunnel and open road experiments and statistical analyses of uncertainties. Uncertainties still are large, this has to be taken into account when interpreting results of atmospheric models.

A number of models to set up emission inventories for street canyons, for urban areas and for regions have been developed and improved. A large number of regional and urban emission data sets have been generated and provided to groups applying atmospheric models.

1.8 References

Friedrich R. and Obermeier A. (1999) Anthropogenic Emissions of VOCs. In: C. N. Hewitt (ed.): Reactive Hydrocarbons in the Atmosphere. Academic Press, San Diego, CA, pp. 1-39

Friedrich R, Wickert B, Schwarz U and Reis S (1999) Improvement and Application of Methodology and Models to Calculate Multiscale High Resolution Emission Data for Germany and Europe.). In: GENEMIS Annual Report 1999, International Scientific Secretariat (ISS), Munich

Friedrich R and Reis S. (eds) (2003) Emissions of Air Pollutants – Measurements, Calculation, Uncertainties –Results from the EUROTRAC Subproject GENEMIS. Springer Publishers, 2003 (in preparation)

Kühlwein J and Friedrich R (2000) Uncertainties of modelling emissions from road transport. Atmospheric Environment 34,pp. 4603-4610

Kühlwein J, Wickert B, Trukenmüller A, Theloke J, Friedrich R (2002a) Emission modelling in high spatial and temporal resolution and calculation of pollutant concentrations for comparisons with measured concentrations. Atmospheric Environment 36 (S1), pp. S7–S18.

Kühlwein J, Friedrich R, Kalthoff N, Corsmeier U, Slemr F, Habram M, Möllmann-Coers M (2002b) Comparison of modelled and measured total CO and NOx emission rates, Atmospheric Environment 36 (S1), pp. S53–S60

Kühlwein J, Friedrich R (2003) Summarizing analyses of tunnel and open road studies. In: Friedrich R, Reis S (eds.) Emissions of Air Pollutants – Measurements, Calculation, Uncertainties: Results from the EUROTRAC Subproject GENEMIS, Springer Verlag, Berlin (to be published)

Pregger T, Friedrich R (2003) Anthropogenic Particulate Matter Emissions in Germany. In: Friedrich R., Reis, S. (eds.) Emissions of Air Pollutants – Measurements, Calculation, Uncertainties: Results from the EUROTRAC Subproject GENEMIS, Springer Verlag, Berlin (to be published)

Theloke J, Obermeier A, Friedrich R (2001) Abschätzung der Lösemittelemissionen in Deutschland. Gefahrstoffe – Reinhaltung der Luft 61(3), VDI-Verlag, Düsseldorf

Theloke J (2003) Calculating detailed VOC Splits. In. Friedrich R, Reis S (eds.) Emissions of Air Pollutants – Measurements, Calculation, Uncertainties: Results from the EUROTRAC Subproject GENEMIS, Springer Verlag, Berlin (to be published)

2 Improvement of emission factors

2.1. Introduction

In the following Sections, measurements conducted to improve and validate emission factors as well as modelling work are described. The aim of this work was to lead to an improvement of emission factors and thus to an overall improvement of the accuracy of emission data and to quantify and reduce uncertainties. The pollutants covered include anthropogenic and biogenic precursors of ozone, primary and secondary aerosols and acidifying substances. As methods for the quantification of emissions and for developing emission inventories for CO, SO_2 and NO_x are more advanced and the resultant uncertainties in their emission estimates are lower than those for VOC, PM and NH_3, the following Sections focus on the latter pollutants.

2.2 Emissions of Volatile Organic Compounds (VOC)

J. Theloke

2.2.1 Introduction

Emissions of non-methane volatile organic compounds (NMVOCs) still contribute significantly to current air pollution problems. They are responsible for the formation of photo-oxidants such as groundlevel ozone – in connection with nitrogen oxides (NO_x) in the presence of sunlight – and individual NMVOCs (e.g. benzene) do have adverse health effects on their own account. NMVOCs are partly responsible for the depletion of stratospheric ozone as well as for the reinforcement of the greenhouse effect as precursors of the ozone, which is a greenhouse gas as well. The major anthropogenic NMVOC sources are road transport and solvent use. Among these, solvent use emissions currently account for about 50% of all anthropogenic NMVOC emissions in Germany. Fig. 2.1 displays the distribution of NMVOC source groups in Germany (own calculations of IER, as well as by R. Steinbrecher IMK-IFU for biogenic sources).

In 1998, biogenic sources emitted approximately the same amount of NMVOCs as road transport. Other NMVOC emissions mainly originate from production processes, combustion in industry and public power plants and from small combustion in households and commercial buildings.

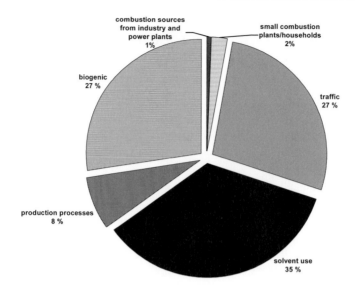

Fig. 2.1. Distribution of annual total NMVOC emissions from different source groups in Germany 1998 (total: 2436 kt)(own calculations, biogenic emissions from R. Steinbrecher, IMK-IFU, Garmisch-Partenkirchen, Germany, see Sect. 2.2.2)

2.2.2 VOC emissions from biogenic sources

R. Steinbrecher, G. Smiatek

2.2.2.1 Introduction

Biogenic volatile organic compounds (BVOC) play a prominent role in the chemistry of the atmosphere. Moreover, they may contribute together with anthropogenic NO_x and VOC emissions to regional and global changes in the HO-radical budget, ozone as well as particle distributions (Jenkin and Clemitshaw 2000; Poisson et al. 2000; Wang and Shallcross 2000; Brasseur et al. 1999). It is well known that high atmospheric pollutant levels lead to health problems and plant damage. Acidic substances further cause erosion problems on buildings (McKee 1994; Sandermann et al. 1997; Milne et al. 1999; Atkinson 2000; Hewitt 1999). The secondary products (ozone, particles) also have to be considered in climate forcing studies. Therefore, it is prerequisite to quantify the biogenic as well as anthropogenic precursors of air pollution in the most exact way possible when aiming at quantifying the environmental impacts of a changing atmosphere. Only then effective measures can be established for ensuring a good air quality. In consequence, negative effects on the environment which in a broader view may lead to socioeconomic problems are then minimised.

In total 1500 TgC of BVOC per year are emitted on global scale (Guenther et al. 1995). This is about 2% of the C-amount photosynthesised by plants (Lal 1999). In contrast to BVOC emission total anthropogenic VOC emission amounts only to 150 TgC per year (Mueller 1992; Veldt 1993; Steinbrecher 1994). On regional scale, however, the mass ratio may change completely. In densely populated, industrialised and heavily polluted areas (megacities) anthropogenic VOC may contribute to more than 80% to total regional VOC emission on a yearly basis (Schnitzler et al. 2002). Considering smaller time scale (e.g. days, weeks or months) anthropogenic VOC emission may be of minor importance for air chemistry even in these regions. This is especially the case during high pressure periods in the summer with high air temperatures and high solar radiation as these weather situations strongly enhance the BVOC emission. Under these conditions the BVOC contribution rises to 60% of total BVOC emission (Schnitzler et al. 2002). Having in mind that BVOC are mainly unsaturated compounds partly containing oxygen (Helmig et al. 1998), this compound group is by far more reactive compared to the anthropogenic VOC emitted (Derwent et al. 1998). Therefore, biogenic VOC contribute significantly to total VOC emission and hence to photooxidant formation during summer smog episode even in densely populated areas.

As a result of the high reactivity towards oxidants in the atmosphere, BVOC are degraded very near to their source as a function of (1) the reactivity of the compound, (2) the local atmospheric trace gas composition and (3) the turbulent transport properties. These effects, in consequence, lead to a reduction in the potential emission. In a modelling study it is demonstrated that for a Norway spruce forest the actual canopy emission of isoprenoids is up to 30% smaller compared to the potential emission of single compounds (Forkel et al. 2001). At the same time a canopy emission of secondary compounds is very likely, and e.g. the fluxes of aldehydes and ketones may increase by a factor of 5. Currently, only for Scots pine, Norway spruce and oak forests the canopy effect on BVOC surface fluxes can be addressed. These latest findings have not yet been implemented into large scale BVOC emission inventories.

Based on this situation it becomes evident that sound definitions of effective measures for mitigating photosmog episodes require not only anthropogenic VOC emission inventories but also reliable VOC fluxes from the Biosphere with high spatial and temporal resolution. Within GENEMIS a lot of efforts have been undertaken to improve the biogenic emission inventories for Germany and Europe by (1) implementing a more detailed land use, biomass as well as leaf area classification, (2) using a updated plant species and chemical compound specific emission factors, (3) developing a generalized emission module and (4) coupling this module to a 3D-hydrostatic meteorological chemistry model (MCCM) and a GIS/RDBMS database (Steinbrecher et al. 2001). Some results of this emission inventory tool will be presented in the next chapters.

2.2.2.3 Material and methods

The BVOC model

The BVOC emissions modelling system implemented at IMK-IFU consists of three major parts (Smiatek 2001): (1) databases storing all required tabular and spatial geo-data, (2) a GIS/RDBMS interface processing the input data according to specific model domains and sub-domains defined by the user and (3) the BVOC emission module.

Two databases store and manage the required input data. A relational database is used to manage all tabular data on the plant species distribution, specific leaf weight (SLW) describing the biomass and emission factors for the chemical compounds defined in the RADM2 and RACM specification used in Chemistry Transport Models (CTM). Within the database the semantical and referential data integrity of the stored data is secured by various constrains. Stored procedures allow the calculation of emission factors related to SLW as well as related to the leaf area index (LAI). At the current stage the species list includes 128 items and the chemical compounds list consists of 33 items. There are 29 land use data sets available covering large parts of Europe as well as 7 plant species data sets for Germany. The emission factors table consists of 992 items.

The GIS/RDBMS interface is able to process general land cover as well as plant specific land cover data for specified domain and sub-domain systems. In addition it can provide vegetation parameters, such as SLW, LAI, VOC emission factors required in BVOC modelling. By using GIS functions and assessing the data via RDBMS, the interface is able to provide input data for any domains and subdomains systems according to MM5 (Mesoscale Meteorological Model Version 5) /MCCM models requirements. Plant species-specific data are usually only available from statistical surveys related to administrative units, such as forest districts. Thus, only the relative area is known but not the actual spatial distribution of the plant species. One of the strengths of the interface is its ability to disaggregate plant species – specific land cover into raster cells of the MM5/MCCM modelling system.

The implemented semi-empirical BVOC mode is based on approaches presented by Guenther et al. (1993), Simpson et al. (1995, 1999) and Steinbrecher et al. (1999). The BVOC model implemented at IMK-IFU is described in detail by Richter et al. (1998) and Stewart et al. (2003). Here, it only should be stated that the emission E of the chemical compound i by the plant species k is described as:

$$E_{ik} = A_k * EF * SLW_k * C(T,PAR) \tag{2.1}$$

In equation (2.1) A_k describes the area of the emitting plant species k in m^2, EF is the average emission factor in $\mu g\ g^{-1}h^{-1}$, SLW_k is the foliar biomass density (g/m^2) and C is a unitless environmental correction factor. It represents the effects of temperature (T) and solar radiation, given as the photosynthetically active radiation (PAR), on the emission factor.

All required meteorology data (temperature, short wave radiation, humidity and wind speed) are derived from the output of the meteorology model MCCM (Grell et al. 2000).

2.2.2.4 Model application

The BVOC model has been applied to the area of Germany in a sensitivity study for the month May 1998, aiming at the estimation of BVOC emissions from forests. The study involves three types of data input: (1) general land cover data and specific leaf weight (SLW) (see Fig. 2.2a), (2) plant species – specific data and SLW (Fig. 2.2b) and (3) general land cover data with leaf area index (LAI). All relevant input data are summarized in Table 2.1. The grid resolution is 15km x 15km according to the spatial resolution of the meteorology data input. The plant species-specific land cover has been disaggregated into grid cells proportionally to the area of deciduous and coniferous forest in each grid cell. The general land cover data have been resampled to 1km x 1km raster, combined with the LAI data and aggregated to the required 15km x 15km grid size.

Table 2.1. Geo-Data used in the BVOC Modelling

Data	Type	Content	Source
Land Cover	250m x 250m Raster	Deciduous Coniferous Mixed Forests	CORINE Land Cover
Leaf Area Index	1km x 1km Raster	Monthly mean	Preußer et. al. (1999)
Plant Species	Area in 50km x 50km grid	9 major tree species	Ministry of Agriculture
	Area in Germany	4 major tree species	
Meteorology	15km x 15km grid	Modelled with MCCM	IMK-IFU

The application study shows substantial differences in the model output resulting from different land cover and vegetation data. The isoprene and monoterpenes emissions are significantly overestimated if general land cover data are used. In that case the varying forest composition within the grid cells is not appropriate reflected by the emission rates assigned to the deciduous, coniferous and mixed forests. Plant species specific land cover overcomes these limitations. The advantage of using plant species specific land cover maps is especially obvious when comparing forest areas covered with deciduous trees. In that case two major trees species are considered, oak (*Quercus robur* L.) and beech (*Fagus sylvatica* L.). Oak trees are strong isoprene emitter but beech is classified a non emitter. Therefore, a mean isoprene emission factor has been assigned to the land-cover class "mixed forests". This means in consequence, that forested areas covered with beech only, show up in the emission inventory as isoprene emitting, but in fact these areas are emitting no isoprene. On the other hand, for areas covered completely with oak the isoprene emission rate is significantly underestimated. This example highlights the importance of using plant species land cover classification whenever possible.

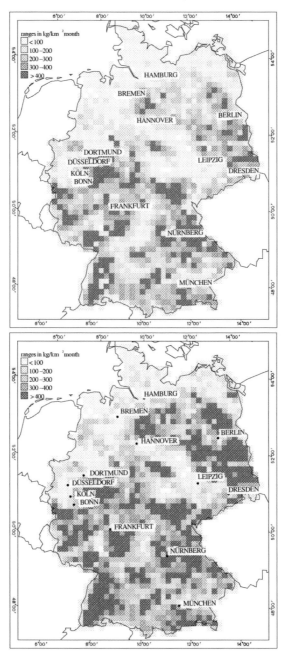

Fig. 2.2. Monthly Total BVOC estimates for Germany and May 1998. Calculated with plant species-specific land cover (top). Calculated with general land cover data (bottom).

    The effect of using LAI distributions instead of specific leaf mass (SLM) values for setting up a BVOC emission inventory is shown in Fig. 2.2. At present LAI distributions are only available for the coarse forest classes deciduous, coniferous and mixed (Preußer et al. 1999). Obviously the overall BVOC emission is drastically lower when compared to the inventories calculated with SLM values. This result is mainly due to the seasonality of the vegetation cover. In the presented study SLM based inventories do not consider the seasonal development while the LAI based inventory does. In May, especially deciduous forest have not yet developed their maximum LAI values resulting in low emissions of this forest type. In Germany these areas are located e.g. west of Frankfurt, east of Köln/Bonn, and in the southern as well as in the eastern parts. For coniferous forest, however, the effect of season on LAI is not as pronounced as for deciduous forest and landscapes, e.g. the Black Forest shows similar emission rates for the SLM and LAI based inventory type (see Fig. 2.2 and Fig. 2.3).

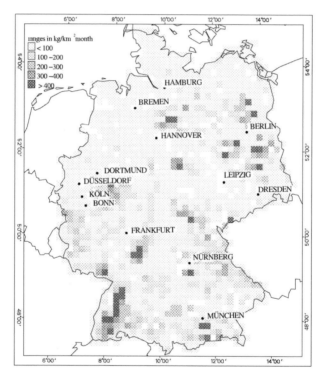

Fig. 2.3. Total BVOC emissions in May 1998 calculated with general land cover data and leaf area index data

2.2.2.5 Uncertainties

Each BVOC emission inventory is published with a certain uncertainty resulting from different sources. Four major sources can be distinguished:

- The biogenic VOC split is very complex (Helmig et al. 1998) and is not completely covered by the existing emission models (Simpson et al. 1999). Therefore often default values are used.
- The very intense changes with season and physiological activity of the plants are not sufficiently considered in current emission algorithms (Xiaoshan et al. 2000; Schaab et al. 2003)
- Synergistic effects of the environment to the plant, e.g. water availability, pests, which may enhance and change the VOC emission rates are neglected (Prieme et al. 2000; Brüggemann and Schnitzler 2002).
- The specific land use type with plant specific distributions as well as vegetation parameters (LAI, seasonality) is often not available (Steinbrecher 1998).

At present, the uncertainties associated with the items listed above can not always quantified exactly. Therefore, discussions on the reliability of BVOC emission inventories are based mostly on error estimates or on sensitivity studies.

In Table 2.2 the results of a sensitivity study are summarized. In this exercise, different land use data sets as well as different vegetation parameters were used serving as input for the emission modelling system. Total BVOC emissions of the three studies differ within a factor of 4. The general land cover data set based on LAI values, resulted in the lowest emission rates. A change from LAI to SLM by using the same land cover resulted in a 4-fold increase in total BVOC emission. This big difference is mainly due to the fact that the SLM based inventories do not consider the seasonality of the vegetation resulting in an increase in emitting surface during the year with a maximum in July/August followed by a decline in the autumn. Therefore, it is strongly recommended to use SLM and LAI values adjusted to the season in BVOC inventory calculations.

However, even when using the same vegetation parameter set (SLM based inventories), the emission estimates differ by a factor of 1.6. The difference in this study is the land cover type. The inventory, based on plant species specific land cover shows lower values compared to the general forest type inventory. As pointed out before, plant species specific emission modelling is closer to nature compared to modelling according forest types. This statement may be supported by the fact that the available information on emission factors, emission algorithms, biophysical data etc. is mostly plant specific. But it has to be noted, that the plant species specific inventory for May presented here may still be a factor of two to high as SLM values used were not corrected for the effect of season. The calculations further reveal that in May 1998 monoterpenes make up the major fraction of BVOC, followed by isoprene or the compound group "other VOC" depending whether the general land cover set or the plant species specific land cover set is used in the calculations.

Table 2.2. BVOC estimates for Germany, May 1998, calculated with different land cover data sets. The emission is related to the total area of Germany and given in 1000 kg. SLM: specific leaf weight, LAI: leaf area index

Procedure	Isoprene	Monoterp.	Other VOC	Total VOC
		[1000 kg]		
Land Cover, SLM	143.3	195.0	102.3	440.6
Land Cover LAI	35.6	52.2	26.9	114.7
Plant-Specific Land	79.2	98.6	95.6	273.4

In a very recent paper from Stewart et al. (2003) describing the set up of a BVOC emission inventory for Great Britain, a detailed error analysis was performed. The overall uncertainty of the inventory is estimated to be a factor of 4. The inventory considers only isoprene and monoterpenes.

For assessing the uncertainty of emission factors used in the calculations, three cases may be distinguished: (1) emission factors measured, (2) emission factors taxonomically assigned and (3) assigned default values. For emission factors measured the uncertainty is in the range of 39%. Taxonomically assigned emission factors contain a much larger error (200%). The largest uncertainty with a factor of 10 is associated with the default emission factors.

Smaller errors are introduced into the calculations by the environmental correction terms used in the emission algorithms (see equation (2.1)). The PAR correction used for calculating actual isoprene emission exhibits a relatively large error (36%), whereas the temperature correction seems to be reliable with an error term of 4%. For deriving actual monoterpene emission rates in the inventory, the temperature parameterisation is associated with an error of 68%. The error associated with the biomass assignments is estimated to amount to 65%. Finally, the relative error in the Great Britain study is estimated to range from 366% to 374% for isoprene and 434% to 441% for monoterpenes.

In this uncertainty analysis a recently described variability in potential emission as a result of the weather is not included. Zimmer et al. (2003) demonstrate in their study on isoprene emission from oak (*Quercus robur* L.) that as a result of cold weather periods the amount of enzymes present in the leaf synthesizing isoprene is less compared to the enzyme content in leaves of high temperature periods. By not considering this variability, maximum emission during hot days may be underestimated by up to 30%.

A detailed uncertainty assessment for the compound group "other VOC" has so far not been performed as the data base available is too small to conduct an error analysis. "Other VOC" estimates in inventories are, in general, considered to be associated with an uncertainty factor greater than 5 (Simpson et al. 1999).

Other aspects which serve attention in an uncertainty discussion of BVOC inventories are the seasonality of emission factors, the light extinction by the canopy and the influence of the terrain on the emission rates. For a Mediterranean-type landscape dominated by the broad-leaved evergreen holm oak (*Quercus ilex* L.) sensitivity studies have been performed stressing these aspects (Schaab et al. 2003). The yearly isoprenoid emission is reduced by 12% considering the seasonality of emission factors. Considering light extinction by the canopy in the calculations reduces the overall emission by 48%. The terrain effect depends on the complexity of the orography, e.g. enhanced temperatures on south faced slopes compared to north faced slopes. For this Mediterranean landscape about 3% smaller daily sums were calculated if the complexity of terrain is considered. A two fold exaggeration of the digital elevation model resulted in a 12% decrease in the daily sums.

2.2.2.6 Conclusions

BVOC emissions play an important role in regional and global air chemistry as well as in the global carbon and radiation balances that affect climate. For these reasons, substantial efforts were undertaken to understand and estimate BVOC emissions from the earth surface. In the past 10 years, significant advances have been made in our understanding about emissions from vegetation. The by far best studied compound is isoprene. Improving continuously the analytical techniques, more compounds have been detected and quantified in the BVOC emission patterns from a variety of plants and landscapes. These emissions are strongly dependent on the plant density and the plant distributions in terrestrial ecosystems. The rapidly developing remote sensing techniques have improved the detection of landcover characteristics, such as land use type and leaf area index. The combination of remote sensing and model capabilities leads to a more accurately monitoring and predicting of the meteorological and biophysical parameters that control BVOC emissions.

Although a substantial amount of information about BVOC emissions has been gathered and implemented into models, there is a need to evaluate the model results by field measurements. It is clear that further important improvements have to be gained (e.g. land cover classification) emission models reflecting bio- and air chemistry, the light situation in the terrain and canopy as well as the physiological conditions) before BVOC inventories fully reflect the complexity of nature. However, even if current BVOC emissions show a great uncertainty (factor of 4) they lead to a better understanding of the chemical and climatological processes of the earth.

2.2.3 VOC-Split of gasoline and diesel passenger cars

B. Mittermaier, T. Schmitz, D. Hassel, F.-J. Weber, D. Klemp

2.2.3.1 Introduction

The Volatile Organic Compounds (VOC) in the atmosphere comprise many different species with various properties. Some of these compounds directly influence human health due to their toxicity. VOC are also important precursors of photochemically formed secondary pollutants like ozone. Since the individual VOC react with different rates and different mechanisms, they also differ in their contribution to photochemical ozone formation. For the development of efficient abatement strategies to lower the VOC emissions in order to improve air quality in urban areas with regard to direct toxic effects or photochemical air pollution, reliable data of VOC are required.

The emission factors of the total VOC from road traffic are known (Hassel et al. 1994; De Vlieger 1996; Lenaers 1996), but for typical European in-use passenger cars there are only few data concerning the VOC composition on the basis of measurements of individual cars (Bailey et al. 1990; Hoekman 1992; Jemma et al. 1995; Rijkeboer and Hendriksen 1993; Siegl et al. 1999). Another approach to obtain emission data from road traffic is to perform measurements at roadsides or in street tunnels (Pierson et al. 1996; Staehelin et al. 1998). The main issue of these studies was the evaluation of existing emission factor models for specific traffic situations rather then the extension of those models by emission factors of individual VOC.

In recent years new methods based on chassis dynamometer studies for determining emission data of motor vehicles have been developed (Hassel et al. 1994). From these data the dependence of the emissions of VOC, CO and NO_x on the driving behaviour is known. In a recent study (Schmitz et al. 1999, 2000) the dependence of the VOC-composition on different driving situations was investigated. Also the influence of cold-/ warm-conditions was quantified. The principle results concerning VOC are presented here.

2.2.3.2 Experimental

Vehicles and driving conditions

The measuring program included 15 different passenger cars. Three engine types were investigated: petrol driven cars equipped with a fuel injection system and a closed-looped controlled three-way-catalyst (TWC), petrol driven cars without any catalyst and cars with diesel engines. Five cars of each engine type were measured. Three of the cars with diesel engines were equipped with oxidative catalytic converters, one with a direct fuel injection.

The emission measurements were performed on the chassis dynamometer of TÜV Rheinland. A detailed description of the test facility is given in Hassel et al. (1994). The measurements were carried out under the conditions of the United States Federal Test Procedure (US FTP-75) and the "Autobahn" test are presented in Fig. 2.4.

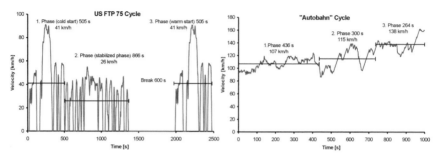

Fig. 2.4. Driving pattern of the US FTP-75 cycle and the "Autobahn"-cycle.

The FTP test is divided into three phases. The test begins with a cold start. Therefore, the cars are pre-conditioned at 20 °C over a period of 12 hours. In the first test phase an average speed of 41 km/h is driven. In the second test phase the average speed is only 26 km/h. After the second test phase the engine is shut off for 10 minutes and then the test continues with a warm start. The driving pattern of this third phase is identical with the driving pattern in the first phase. By comparing the emissions obtained in the first and in the third test phase, the effects caused by the cold start procedure can be derived from the data.

The "Autobahn"-cycle was developed by TÜV Rheinland. This cycle represents driving behaviour on German motorways. Again this cycle consists of three phases. From phase one to phase three the average velocity is increasing. In the third phase of the test cycle the cars have to accelerate from 127 km/h to the maximum speed of 162 km/h in 40 seconds. Under these conditions most of the cars tested in this program were running at their limits. The respective emissions represent high engine loads which are not covered in the US FTP-75.

2.2.3.3 Sampling and analysis

The exhaust gas is diluted by a factor of 10 – 15 with filtered ambient air using a constant volume sampler (CVS) device and sampled in glass vessels. The use of Tedlar bags is unfavourable due to substantial blank values for some hydrocarbons (Schmitz et al. 2000). For the measurement of the $C_2 - C_{10}$ hydrocarbons a GC-FID (HP 5890A) and a specially designed sampling device was used (Schmitz et al. 1997). The sampled air passes a cooling trap at –25 °C for removal of water before being sucked through a sample loop, which is kept at about –190 °C. When sampling is completed, the sample loop is heated up to 80°C and the hydrocarbons are injected on the capillary column (DB-1; 90 m x 0.32 mm x 3 µm). Temperature program: –50 °C / 2 min @ 5 °C/min → 200 °C / 15 min. A chromatogram

taken from a car with TWC in the first phase of the US FTP-75 is shown in Fig.
2.5.

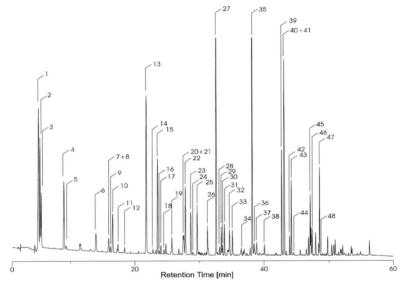

Fig. 2.5. Chromatogram taken from a TWC-vehicle: 1=ethene; 2=ethyne; 3=ethane;
4=propene; 5=propane; 6=i-butane; 7/8=i-/1-butene; 9=1,3-butadiene; 10=n-butane;
11=trans-2-butene; 12=cis-2-butene; 13=i-pentane (coelution with acetone); 14=1-pentene;
15=n-pentane; 16=isoprene; 17=trans-2-pentene; 18=cis-2-pentene;19=2,2-dimethylbutane;
20/21=cyclopentane/2,3-dimethybutane; 22=2-methylpentane; 23=3-methylpentane; 24=1-
hexene; 25=n-hexane; 26=2,4-dimethylpentane; 27=benzene; 28=cyclohexane; 29=2-
methylhexane; 30=2,3-dimethylpentane; 31=3-methylhexane; 32=i-octane; 33=n-heptane;
34=methylcyclohexane; 35=toluene; 36=2-methylheptane; 37=3-methylheptane; 38=n-
octane; 39=ethylbenzene; 40/41=m-/p-xylene; 42=styrene; 43=o-xylene; 44=n-nonane;
45=n-propylbenzene; 46=1,3,5-trimethylbenzene; 47=1,2,4-trimethylbenzene; 48=n-
decane.

2.2.3.4 Results

VOC-composition

The relative composition of the VOC measured for the different engine types is
shown in Fig. 2.6. In order to get a better overview the single compounds were
grouped into five substance classes.

The relative VOC pattern obtained from the TWC cars shows strong variations
depending on the different test phases. Under cold start conditions, the VOC pattern is dominated by aromatic compounds followed by the alkanes and the alkenes. In the corresponding warm phase (US FTP-75 phase 3) the VOC pattern is
changing completely. The more reactive alkenes and aromatic compounds are re-

duced more efficiently. Therefore, the main fraction of the VOC is given by the alkanes, a fact which also is reported in an Australian study (Duffy et al. 1999). At higher average velocity, the fraction of alkenes and aromatics is increasing continuously. In the last phase of the "Autobahn" cycle the aromatic fraction nearly reaches 60% of the VOC pattern. This again is due to the effect of rich air/fuel mixtures which are injected under high load conditions existing in the "Autobahn" cycle. Therefore, conversion efficiency of the catalyst for the hydrocarbons is diminished as described above. This effect leads to an overall increase in total hydrocarbon emission (Schmitz et. al. 2000)) and to a change in the VOC pattern especially in the third phase of the test cycle.

The VOC-patterns of cars without exhaust treatment undergo only minor changes. The pattern measured in the cold start phase is nearly identical to that obtained in the warm phase. Only at higher average speed a slight increase of the alkenes and hence a slight decrease of the alkane compounds is observed.

Alkenes are by far the largest fraction of hydrocarbons emitted from cars with diesel engines, although it has to be mentioned that diesel engines also emit considerable amounts of carbonyl compounds (Schmitz et al. 2000). A similar VOC fraction in the exhaust gas of in-use light-duty diesel vehicles was also found in other studies (Rijkeboer and Hendriksen 1993; Siegl et al. 1999). Diesel cars have the only engine type where alkynes play an important role; here their fraction is even larger than the fraction of alkanes.

The measured mass emissions in mg/km of the VOC grouped in different substance classes and the mean confidence levels (95% significance) calculated for the small number of vehicles (5 cars of each engine type) are given in Table 2.3 and Table 2.4. Detailed results for the individual compounds are given in Schmitz et al. (1999).

The mean values of the individually measured VOC show large confidence levels of about 80%. In the US FTP-75 the confidence levels are smaller than that calculated for the "Autobahn" cycle. The confidence intervals which were found for the mean values of the TWC cars and cars with diesel engines are significantly higher than that for cars without exhaust gas treatment. The large scatter is due to the small number of cars of each engine type and due to differences in the mass emissions because of vehicle age, standard of maintenance and type of manufacture.

Table 2.3. Emission rates of VOC measured in the different phases of the US FTP-75.

Substance	US FTP-75 phase 1		US FTP-75 phase 2		US FTP-75 phase 3	
	Mean emission [mg/km]	95% Confidence level [%]	Mean emission [mg/km]	95% Confidence level [%]	Mean emission [mg/km]	95% Confidence level [%]
Gasoline cars with TWC						
Alkanes	130.2	58	5.5	72	18.8	41
Alkenes	62.6	70	0.4	85	1.9	74
Alkynes	21.0	81	0.0	72	0.1	84
Aromatics	250.8	63	2.9	64	10.2	68
Gasoline cars without TWC						
Alkanes	662	29	651	51	568	49
Alkenes	344	33	341	38	292	40
Alkynes	139	30	101	34	77	24
Aromatics	1267	28	1138	35	976	34
Diesel Cars						
Alkanes	8.2	65	7.9	74	5.2	63
Alkenes	44.4	47	35.4	52	20.2	63
Alkynes	8.8	61	5.9	70	4.2	73
Aromatics	9.1	49	7.5	46	4.6	48

Table 2.4. Emission rates of VOC measured in the different phases of the Autobahn cycle.

Substance	Autobahn phase 1		Autobahn phase 2		Autobahn phase 3	
	Mean emission [mg/km]	95% Confidence level [%]	Mean emission [mg/km]	95% Confidence level [%]	Mean emission [mg/km]	95% Confidence level [%]
Gasoline cars with TWC						
Alkanes	18.0	79	11.8	63	28.8	70
Alkenes	4.2	83	7.5	83	60.7	96
Alkynes	0.1	70	0.2	119	0.9	103
Aromatics	16.7	74	18.8	81	129.5	71
Gasoline cars without TWC						
Alkanes	215	40	259	38	225	52
Alkenes	293	50	367	67	503	70
Alkynes	86	32	119	36	176	34
Aromatics	580	43	724	51	847	70
Diesel Cars						
Alkanes	1.6	77	1.5	82	1.2	103
Alkenes	6.5	103	8.1	93	9.4	104
Alkynes	1.7	109	2.3	105	4.3	110
Aromatics	1.9	73	2.2	79	3.0	116

It was found that the sum of the identified VOC and the corresponding THC measurements agree better than 20% (Schmitz et. al. 1999). Therefore, a normalisation of the VOC composition to the measured THC was performed resulting in smaller confidence intervals than for the mass emissions. In principle, the normalized VOC patterns can be adapted to existing emission factors of THC which are available for a large number of driving situations. However, it has to be carefully studied to which extend the driving patterns of the investigated test phases are representative for real word driving.

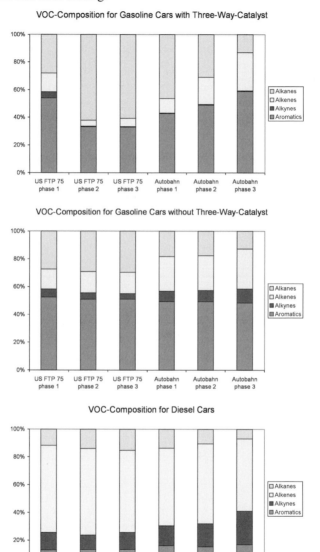

Fig. 2.6. VOC pattern obtained for the three vehicle categories for all phases of both cycles.

Ozone Formation Potential (OFP)

In order to assess the measured VOC emissions concerning the photochemical ozone production, the **O**zone **F**ormation **P**otential (OFP) using the method by Carter (1994) was calculated for the different engine types. This method is based on a model scenario in which ozone formation is calculated under optimum conditions: i.e. high solar fluxes and a base scenario with VOC/NO$_x$-ratios which yield a maximum ozone formation. For single hydrocarbons a **M**aximum **I**ncremental **R**eactivity (MIR) can be determined in terms of grams ozone formed per gram hydrocarbon added to the scenario. The OFP of a certain VOC mixture is calculated by summing up the products of measured VOC concentrations and corresponding MIR factors:

$$OFP = \sum_{i} VOC_i \cdot MIR_i \qquad (2.2)$$

It is important to note that these conditions represent typical California conditions and are significantly different from typical summertime conditions in Central Europe. Nevertheless, for just a mere ranking of VOC mixtures with regard to ozone production these MIR-factors are very useful.

When calculating the ozone forming potential for the 5 phases with warmed-up engines, a speed dependency can be derived (see Fig. 2.7). As expected, the highest ozone formation potential is observed for cars without catalyst due to the highest mass emission of THC in connection with a high share of reactive substances like alkenes and aromatics.

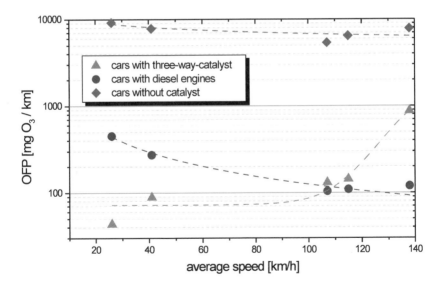

Fig. 2.7. Ozone formation potential as function of mean speed in different test phases. Note the logarithmic scale. The fitted lines only show trends between OFP and mean velocity.

The ozone formation potential for the TWC cars is found to be up to two orders of magnitude lower than for cars without catalyst and is increasing with increasing speed. The low values in the ozone formation potential at moderate speeds are caused by the high efficiency of the catalysts resulting in low mass emissions of THC and low emissions of reactive hydrocarbons. The efficiency of the catalyst decreases with higher average velocities. At an average speed of 140 km/h there is an increase of the ozone formation potential of a factor of 200 compared to the OFP values calculated for low average speed.

In contrast, the OFP of the diesel cars decreases with increasing speed. Highest OFP values are observed at moderate speed, whereas lowest OFP values are found in the Autobahn cycle. This dependency is caused by the fact that the mass emissions of diesel cars are continuously decreasing with increasing mean speed (Schmitz et al. 2000) without major changes in the reactivity of the VOC composition.

2.2.3.5 Summary and conclusions

The VOC compositions of three different engine types were measured under conditions of the US FTP-75 and the „Autobahn"-test cycle. The variations of the VOC compositions in the different phases observed for gasoline cars without catalysts and diesel cars are moderate. For the cars with three-way-catalyst there is a strong dependency of the VOC composition on the operating conditions of the engine and the temperature of the catalyst. There is a large change in the VOC-pattern from cold start phase to the phases with warmed-up engine. A significant change can be observed at high engine loads and high speeds compared to moderate driving conditions with warmed-up engines.

The resulting OFP of the VOC emissions shows a dependency with respect to the average speed. Especially for cars with three-way-catalysts the speed dependency of the OFP has to be taken into account. In this context it should be stressed that there is a lack of information with regard to traffic situations like stop and go or other relevant parameters like road inclination.

Using the emission results of this study it should be mentioned that the measurements are based on mean market fuels in Germany of the reference year 1997. It is known that VOC composition and therefore OFP will change (Decker et al. 1996) when fuel composition is significantly modified.

2.2.3.6 Acknowledgement

This work is supported by the BMBF as part of the TFS–principal subject 2: "Anthropogenic and biogenic emissions as well as deposition of trace gases" in the course of the joint project A.2:"Analysis, development and evaluation of emission models and data sets for road traffic in Germany and Europe". The authors thank the ARAL research centre for provision and detailed analysis of the test fuels.

2.2.4 Cold start emissions

D. Hassel, F.-J. Weber, K.-S. Sonnborn

2.2.4.1 Introduction

A new method for the estimation of cold start emissions from passenger cars has been developed, using data of emission measurements performed in the climatic chamber of Porsche in the framework of the German/Swiss Emission Factor Programme (Hassel et al. 1993; BUWAL 1995). Within the category 'passenger cars', a distinction was made between diesel and gasoline vehicles with and without three-way catalysts, but data for diesel operated vehicles with catalysts was too scarce to allow for a substantial and detailed analysis in that case. Emission measurements have been performed with cold and as well with hot engines, otherwise test conditions conditions were kept the same. The differences between these two measurement series makes it possible to exactly determine cold start related excess emissions. In this context, cold start is defined as a start process after more than 12 hours parking at a defined ambient temperature level.

The main intention of this investigation was to improve the existing method for estimating excess emissions by including the influence of different urban driving behaviour reflected in urban driving patterns. According to the old method, excess emissions could be calculated per (cold) start for different start temperatures as a function of distance travelled. The emission measurements of the German/Swiss Emission Factor Programme were based on the US FTP standard driving cycle. Hence, it was necessary to find a feasible way to transfer FTP based results to other urban driving patterns varying in average speed and speed profile. As an approximation, it was assumed that the ratio of excess emission and emission of a fully warmed up engine based on the FTP standard driving cycle could be transferred to all urban driving patterns. A consequence of this approximation is that different emission levels due to different mean speeds were taken into account, while the different engine operation modes were more or less neglected.

2.2.4.2 Description of the new improved method

The general idea of the new method is based on the assumption that excess emissions due to a cold start depend on the warming-up phase of the engine. The duration of the warming-up phase is mainly a function of energy provided by the engine of the vehicle. Because of the fact that vehicles are equipped with different engine sizes and thus a vehicle with a powerful engine in comparison to a vehicle with a low-performance engine is operating in the lower range of the engine map far below the maximum load curve, the energy provided by the engine cannot be used directly as parameter for an excess emission function. For general application it seemed useful to introduce a standardised parameter. The energy provided by the engine related to rated power of the engine was selected as the most appropriate parameter. The following equation reflects the principle ideas of the new method:

$$\Delta E = \Delta E(EN, T, POL, VC) = E_C - E_W \qquad (2.3)$$

where

ΔE is the excess emission for a trip expressed in g/h
E_C is the emission of the cold engine
E_W is the emission of the warm engine
T is the start temperature (°C), in this case $-10°C$, $+5°C$, $+20°C$
EN is the sum of energy provided by the engine related to the rated power of the engine in s, $EN = \int P_{rel} *dt$
$P_{rel} = (P_a + P_w + P_r) / P_{rated}$
$P_a = m*a*v$, power to overcome the inertia mass of the vehicle
$P_w = 0,5*\rho*c_w*A*v^3$, power to overcome the wind resistance
$P_r = m*g*f_r*v$, power to overcome the rolling resistance
m is the mass of the vehicle
a is the acceleration of the vehicle
v is the speed of the vehicle
ρ is the air density
c_w is the aerodynamic drag coefficient
A is the vehicle cross-sectional area
f_r is the rolling friction coefficient
g is the gravitational acceleration
P_{rated} is the rated power of the engine
POL is the specific pollutant
VC is the vehicle concept

Losses between engine and wheels of the vehicle are not taken into account. It is assumed that the engine will not cool down during driving in phases of deceleration. Therefore in phases of deceleration the energy sum provided by the engine will not decrease.

2.2.4.3 Data basis

The total vehicle sample for cold start emission measurements consisted of 12 gasoline vehicles with controlled 3-way catalysts, 8 gasoline vehicles without special measures for exhaust emission control and 8 diesel vehicles with and without oxidation catalysts. The emission measurements were based on the US FTP 75 driving cycle displayed in Fig. 2.8. The test consists of three phases and a parking time of 10 minutes between phase 2 and 3. Phase 1 begins with a cold start after a parking time of more than 12 hours and is called cold transient phase. The driving distance is 5.8 km and the cycle has a duration of 505 s. Phase 2, the stabilized phase, takes 867 s, the length is 6.3 km. Phase 3 is the hot transient phase, the cycle complies with the cycle of phase 1. The cold start was performed at three different temperatures: -10°C, 5°C and 20°C.

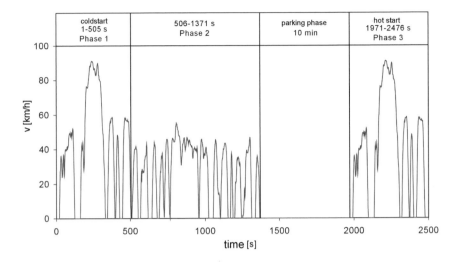

Fig. 2.8. Speed curve of the US FTP 75

The pollutants CO, HC, NO$_x$, and CO$_2$ were measured continuously in one second steps so that the emissions could be displayed as a function of driving distance. The particulate emissions of the diesel vehicles were measured for the three phases only. For the gaseous pollutants the excess emissions in dependence of start temperature could be derived from the measurements by calculating the difference of the emissions of the cold transient phase (phase 1) at the three different start temperatures and the hot transient phase at an ambient temperature of 20°C.

2.2.4.4 Statistical data processing

Based on the results of the measurement programme a correlation and regression analysis was performed between the excess emission ΔE and EN, which is the sum of energy provided by the engine related to the rated power of the engine. The statistical approach is the following:

$$\Delta E = \sum_{i=0}^{n} (EN^i * a_i) \qquad (n = 2,3,4) \qquad (2.4)$$

where a_i is the regression coefficient

This way, functions correlating the excess emissions ΔE and the sum of the standardized energy EN were derived for the different vehicle concepts and the different start temperatures. Fig. 2.9 displays an example for the HC excess emissions of gasoline vehicles with three-way catalyst.

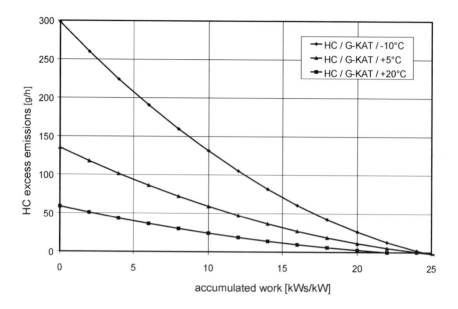

Fig. 2.9. HC excess emissions for gasoline vehicles with 3-way catalyst as a function of energy provided by the engine related to the rated power of the engine with the start temperature as parameter

These functions allow for the calculation of total excess emissions in grammes for any driving cycle. In a first step, the energy sum EN can be calculated for each second of the cycle, so time related excess emission can be allocated. Then, the time related excess emission results in the total excess emission of the specific cycle according are added up according to the following equation:

$$\Delta ET = \sum_{i=1}^{n} \Delta E_i \qquad (2.5)$$

where

ΔET is the total excess emission of the cycle in g
ΔE_i excess emission for the time i in the cycle
n is the number of seconds of the cycle

In order to determine the excess emission ΔE_i for each second of a cycle the energy sum EN provided by the engine related to the rated power of the engine has to be calculated. Therefore it is necessary to provide the technical data for the different vehicle categories compiled in Table 2.5. The vehicle categories are defined in Hassel et al. (1993) resp. in BUWAL (1995) taking into account engine concept (gasoline or diesel), engine capacity and categories of emission control technology.

Table 2.5. Average technical specifications of different vehicle categories for calculating EN (Hassel et al. 1993; BUWAL 1995)

Vehicle concept/engine capacity	m [kg]	$C_w * A$ [m²]	P_{rated} [kW]	f_r
Gasoline cars w/o cat < 1.4 l	742	0.73	35	0.0135
Gasoline cars w/o cat 1.4- 2.0 l	1006	0.72	70	0.0135
Gasoline cars w/o cat >2.0 l	1348	0.72	125	0.0135
Diesel <2.0 l	1033	0.66	45	0.0135
Diesel >2.0 l	1353	0.69	66	0.0135
3-way-cat <1.4 l	838	0.65	46	0.0135
3-way-cat 1.4 -- 2.0 l	1014	0.71	67	0.0135
3-way-cat >2.0 l	1361	0.67	108	0.0135
3-way-cat before 1987	1102	0.72	78	0.0135

2.2.4.5 Application of the new method for any urban driving pattern

In the framework of the German/Swiss Emission Factor Programme datasets with regard to driving conditions and speed have been collected. Hence, a number of driving patterns for different urban traffic situations could be identified (Hassel et al. 1993). These urban driving patterns differ for instance in driving dynamics and average speed. Fig. 2.10 displays total HC excess emissions of a typical three-way catalyst equipped vehicle with an engine capacity between 1.4 and 2.0 l as a function of the mean driving speed for three different cold start temperatures. As anticipated, the excess emissions decrease with increasing cold start temperature and increasing mean driving speed. In addition to that, it could be demonstrated that driving cycles with higher shares of deceleration phases result in higher total excess emissions.

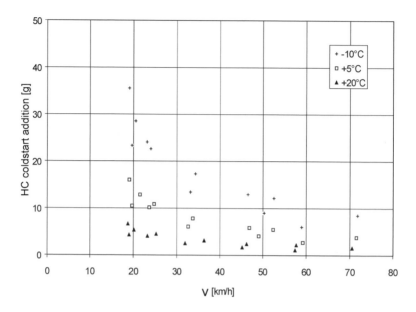

Fig. 2.10. Total HC excess emissions for the average 3-way-catalyst vehicle of engine capacity class 1.4 – 2.0 l as a function of mean driving speed with the cold start temperature as parameter

2.2.4.6 Comparison of the new method with other calculation models

The first analysis for estimating cold start related excess emissions was based on the European Urban Driving Cycle. Only a rough assessment of excess emissions in dependence of ambient temperature was possible. In contrast to this approach, Fig. 2.10 shows total HC excess emissions for an average three-way-catalyst equpped vehicle of engine capacity class 1.4 – 2.0 l as a function of mean driving speed with the cold start temperature as a parameter. The following three methods allow the calculation of total excess emissions as a function of mean speed during warming up phase, cold start temperature (equal to ambient temperature) and driving distance for different vehicle categories:

- Old Method *(developed in the framework of the German/Swiss Emission Factor Program, Hassel et al. 1993)*
- INRETS Method *(Sérié and Joumard 1997)*
- New Method *(based on the same measurement data as the Old Method)*

In Fig. 2.11 the total excess emissions calculated according to the three mentioned methods are displayed as a function of driving distance for different cold start temperatures.

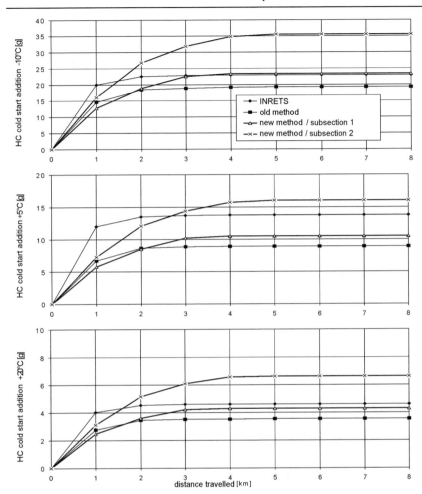

Fig. 2.11. Total HC excess emissions for the average 3-way-catalyst vehicle of engine capacity class 1.4 – 2.0 l as a function of driving distance calculated according to three different calculation models

The calculation was performed for an average three-way-catalyst equipped vehicle of the
1.4 - 2.0 l capacity class. For the warming-up phase, the urban driving pattern of area
sources was selected which can be divided into two parts of different driving dynamics.
Section 2 of the driving pattern has a higher share of deceleration phases and leads therefore in the case of the New Method to higher excess emissions. The other two methods do
not take into account the dynamics of the driving pattern and thus calculate the same value
for both sections of the cycle. The example shows that driving dynamics have a significant
influence on total excess emissions. However, is has to be mentioned that, typically, such
detailed information about driving statistics is not available for establishing emission inventories. So it seems useful to take the value of the excess emissions between section 1 and 2
of the driving pattern as the best assessment of cold start related excess emissions.

2.2.5 VOC emissions from high-emitting gasoline passenger cars

Å. Sjödin

2.2.5.1 Introduction

Late-technology catalyst cars (i.e. fuel injection cars equipped with closed-loop three-way catalysts) are rapidly penetrating the European passenger car fleet, thus leading to reduced emissions of the regulated compounds CO, HC and NOx and subsequent improvements in air quality in many European cities. For instance, the growth in the number of TWC-cars (Three Way Catalyst cars) during the 90's is believed to be the main cause behind the observed strong decrease in ambient urban levels of both NO_2 and benzene in Sweden since the early 90's (Sjödin et al. 1996; *http://www.ivl.se/en/miljo/projects/urban/trendvoc.asp*). The magnitude of further road vehicle emission reductions and related improvements in air quality will however largely depend on the real-world emission performance of the catalyst cars in the long term. Still fairly little is known about phenomena such as catalyst ageing and catalyst malfunction within the European on-road fleet. In the US, where the catalyst technology was enforced much earlier than in Europe and the fleet turn-over to TWC-cars today is more or less complete, there is still a concern about a small fraction of the passenger car on-road fleet having excess emissions, "high-emitters" or "gross-polluters", being responsible for the majority of the air pollution from road traffic.

As part of the GENEMIS-2 project, on-road emission measurements by means of optical remote sensing were carried out on a large number of gasoline passenger cars in 1995-98 in order to study the significance of the high-emitting vehicle problem within the Swedish passenger car fleet. The field measurements, conducted by means of remote sensing instruments based on the original Denver University remote sensor concept more commonly known as FEAT (Fuel Efficiency Automobile Test), measuring on-road CO and HC at tailpipe (Bishop et al. 1989), were carried out along a commuter road outside Stockholm. Chassis dynamometer measurements were made on smaller sub-selections of the measured fleets.

2.2.5.2 Abundance of high-emitting TWC-cars and their contribution to overall TWC VOC emissions

In the first year of the study (1995), emissions from more than 8,000 catalyst cars were measured by the remote sensor. Of these, ten cars representing both high- and low-emitters were recruited for US FTP (urban driving cycle) and HWFET (Highway driving cycle) measurements. On-road data and US FTP data for the ten cars are summarised in 5% in the case of HC emissions.

Table 2.6. Assuming the 8,000 catalyst cars measured by the remote sensor were representative for the whole Swedish fleet, some rough estimates of the fraction of high-emitting catalyst cars in the 1995 in-use fleet were made. These estimates were based on 1) the observed increase in both FTP and HWFET CO and HC emissions with increasing average remote sensor CO readings from repeat passes, c.f. Table 2.6, and 2) the percentile distributions for the average remote sensor CO readings for catalyst cars with repeat passes, c.f. Fig. 2.12. Based on the results presented in Table 2.6, with cutpoints of 1% and 2% CO according to the remote sensor, as averages of multiple readings, used as indicators of cars exceeding the FTP CO and HC standards, respectively, then according to Fig. 2.12 about 10% of the Swedish catalyst car fleet would have excess FTP CO emissions and be classified as CO high-emitters, whereas 3% would have excess FTP HC emissions and be classified as HC high-emitters. Acordingly, the fraction of the worst CO and HC high-emitters, i.e. the super-emitters, represented by the cars no. 1-3 in Table 2.6, was estimated to only about 0.5% for both CO and HC. However, since the latter category emitted 30-50 times more CO and HC than a normally functioning catalyst car, i.e. a low-emitter, it may still be very significant for the Swedish catalyst car fleet overall CO and HC emissions. Estimates of the contributions from the two high-emitter categories to catalyst cars' overall CO and HC emissions were made based the following assumptions:

1. The 0.5% super-emitters were assumed to emit on average 50 g CO and 2.5 g HC per km according to FTP, and 25 g CO and 1 g HC per km according to HWFET,
2. The 9.5% CO high-emitters, excluding the 0.5% super-emitters, were assumed to emit on average 5 g CO/km according to FTP and 2 g CO/km according to HWFET,
3. The 2.5% HC high-emitters, excluding the 0.5% super-emitters, were assumed to emit on average 0.3 g HC/km according to FTP and 0.1 g HC/km according to HWFET,
4. The low-emitters, 90% for CO and 97% for HC, were assumed to emit 1 g CO and 0.1 g HC per km according to FTP and 0.2 g CO and 0.05 g HC per km according to HWFET,
5. Catalyst cars' VMT (Vehicles Miles Travelled) were assumed to be equally distributed between city driving represented by the FTP and highway driving represented by HWFET.

Based on the above assumptions, the high-emitter category was estimated to account for 40-50% of the CO emissions and about 20% of the HC emissions from the overall catalyst car fleet. The corresponding contribution from the super-emitter category was estimated to 15-20% in the case of CO emissions and 10-15% in the case of HC emissions.

Table 2.6. On-road %CO according to the remote sensing measurements and US FTP emissions for the ten selected three-way catalyst cars in the 1995 measurements. Figures in italics denote exceedances of the FTP standard.

Make, model and model year	Odometer (km)	On-road %CO	US FTP (g/km)		
			CO	HC	NOx
1. Nissan Micra -90	100,400	8.9	*55.3*	*2.00*	0.18
2. Nissan Micra –89	52,800	8.1	*57.0*	*2.27*	0.10
3. Volvo 740 –92	131,200	7.4	*46.3*	*3.20*	0.52
4. Volvo 345 –88	100,800	3.1	*4.2*	*0.31*	0.07
5. Opel Ascona –87	158,500	2.7	*9.3*	*0.31*	0.51
6. Audi 100 –90	213,100	2.5	*3.1*	0.21	*1.10*
7. Suzuki Swift –89	83,900	2.2	*2.2*	*0.42*	0.14
8. Ford Scorpio –94	45,300	1.3	*3.1*	0.16	0.15
9. Peugeot 405 -90	92,300	0.7	1.1	0.08	0.12
10. Volvo 850 -94	57,300	0.5	1.1	0.08	0.10

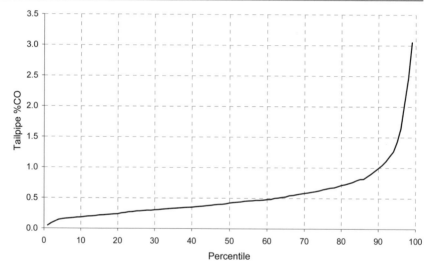

Fig. 2.12. Remote sensor CO readings distributions for the catalyst car on-road fleet for single and average readings for multiple passes (>4 passes, N=541) in the 1995 measurements.

2.2.5.3 VOC split for high- vs. low-emitting gasoline passenger cars

In 1997 the remote sensor was used to measure the emissions from about 12,000 gasoline passenger cars. A subset of nine cars, five catalyst cars and four non-catalyst cars, car no. 1-5 and car no. 6-9, respectively, as given by Table 2.7, comprising both high-emitters and low-emitters according to the remote sensing measurements, were recruited for chassis dynamometer measurements to measure the emissions of both regulated compounds CO, HC and NO_x as well as a large number of individual VOC over the US FTP cycle.

Table 2.7. Remote sensing (average %CO from repeat passes) and FTP results for CO, HC and NO_x in the 1997 measurements. Cars no. 1-5 catalyst cars, no. 6-9 non-catalyst cars.

Car no, make, model, model year	On-road % CO	US FTP g/km		
		CO	HC	NO_x
1. Saab 900, -88	6.8	55.8	2.91	0.93
2. Nissan Sunny, -92	7.8	8.7	0.39	0.44
3. Volvo 440, -94	6.7	2.4	0.15	0.10
4. Audi 80, -91	0.03	1.5	0.15	0.24
5. VW Golf, -96	0.04	0.5	0.04	0.13
6. Opel Ascona, -83	11.2	144.6	42.5	0.19
7. BMW 1602, -74	7.3	44.0	4.68	0.95
8. Volvo 740, -87	5.2	8.5	1.46	0.97
9. VW Golf, -86	0.15	6.4	1.39	0.96

The average, maximum and minimum VOC emission factors, along with the weight-% fractions of total HC, measured according to the FTP test for the sample fleet are shown in Table 2.8. As expected, the large differences between individual vehicles with regard to emissions of total HC are reflected also in large differences in emissions of individual VOC. However, these differences vary with different VOC. The largest difference was observed for acetylene, for which the ratio between the highest and lowest observation was ≈3,000. The amount of acetylene relative to the amount of ethene emitted is often used as an indicator of the performance of the emission control system of gasoline vehicles. Low ethene/acetylene-ratios, <1, in the FTP are indicative of "engine-out emissions", whereas the ratio for well-performing three-way catalysts has been reported to 3.8-6.4 (Hoekman 1992). For the TWC cars in this study the ethene/acetylene-ratio ranged from 0.3 to 11. Apart from acetylene, species such as butyne, propadiene, 1,3-butadiene, ethene, propene and ethane exhibited the largest variation in emissions, more than two orders of magnitude, between individual catalyst cars. The variation in VOC emissions was in general lower for non-catalyst cars (always less than two orders of magnitude), although even in this case acetylene exhibited the largest variation.

Thus, looking at the weight-% composition of individual VOC, high-emitting catalyst cars, represented by car no. 1, were richer in acetylene by a factor of 10-20, in butyne by a factor of 5-10, in propadiene by a factor of 3-4, in ethane by a factor of 2-3 and in ethene, propene and 1,3-butadiene by 25-50%, compared to the low-emitting catalyst cars, represented by car no. 4 and 5. Inversely, the latter were richer in hexane and acetaldehyde by a factor of 3-10, in pentane by a factor of 2-4, in toluene and o-xylene by a factor of 2-3, and in all alkenes above propene by a factor of 1.5-4 compared with the high-emitting catalyst cars. As for non-catalyst cars, high-emitters, represented by car no. 6, were richer in acetylene by a factor of 3-4 and in all pentanes, hexane and heptane by a factor of 1.5-2, compared to low-emitters, represented by car no. 8 and 9. Inversely, the low-emitters were richer in ethane, propane and isobutane by a factor of 1.5-3 and in all alkenes by a factor of 2-5. Compared to non-catalyst cars, catalyst cars in general tended to be richer in the lightest alkanes ethane and propane and in acetaldehyde (by about a factor of 5) and slightly richer in n-butane, cyclopentane, the heavier alkanes and xylenes, whereas non-catalyst cars tended to be richer in isobutane, isopentane, ethene, propene and propyne.

35 VOC species were analysed for all cars, and accounted for about 80% (TWC cars) and 60% (non-TWC cars) of the overall HC emission. Thus, for non-TWC cars a significant portion of the overall HC emissions was species not covered by the VOC analysis, mainly considered to be methane and heavier VOC ($>C_8$).

2.2.5.4 Conclusions

This study has demonstrated that high- and super-emitters (with regard to CO and HC emissions) occur within the European gasoline passenger car fleet, both among catalyst and non-catalyst cars, and may contribute significantly to overall fleet CO and HC emissions. For catalyst cars the high-emitter category was estimated to amount to 10% (CO) and 3% (HC) of the fleet and account for 40-50% of the CO and about 20% of the HC overall catalyst car emissions, respectively. The corresponding contribution from the super-emitter category, making up only 0.5% of the catalyst car fleet, was estimated to 15-20% in the case of CO emissions and 10-15% in the case of HC emissions.

Regarding the VOC-split, high-emitting cars tend to be richer in acetylene, butyne, propadiene, ethane, ethene, propene and 1,3-butadiene, whereas they tend to be less rich in hexane, acetaldehyde, pentane, toluene, o-xylene and in all alkenes above propene compared to well-maintained, low-emitting cars.

Table 2.8. Average, max and min VOC emission factors in mg/km along with weight-% fractions of THC according to the FTP test for the nine cars in the 1997 measurements.

	Catalyst cars				Non-catalyst cars			
	average	max	min	weight-%	average	max	min	weight-%
ethane	27	112	0.5	3.2±2.0	28	64	9.9	0.6±0.3
propane	2.4	5.3	0.9	1.3±1.5	11	30	3.6	0.2±0.1
iso-butane	5.4	23	0.7	0.9±0.5	99	285	21	1.5±0.6
n-butane	37	143	2.8	6.1±2.4	276	875	53	3.3±1.0
iso-pentane	35	134	2.9	5.9±1.4	865	2998	100	7.4±0.8
n-pentane	15	54	2.8	3.6±2.3	271	949	25	2.2±0.3
hexane	4.1	12	1.3	1.5±1.3	92	326	8.7	0.7±0.1
heptane		interfering peaks			68	242	5.2	0.5±0.1
octane		in chromatogram			87	195	8.1	0.5±0.3
nonane	0.6	0.9	0.2	0.5±0.6	2.1	5.8	0.3	0.03±0.02
decane	2.5	data for 1 car only		0.7	1.8	3.2	0.5	0.05±0.07
2+3-methyl-pentane	20	75	1.3	3.4±0.8	457	1595	42	3.7±0.4
cyclopentane	7.4	29	0.3	1.1±0.5	49	174	4.1	0.4±0.1
cyclohexane	2.5	5.8	0.1	0.8±0.6	49	162	6.4	0.5±0.1
ethene	56	250	1.8	5.4±2.8	313	792	95	6.2±2.9
propene	22	99	0.8	2.4±1.0	110	225	48	2.9±1.7
1-butene	2.8	12	0.3	0.4±0.2	18	36	7.1	0.5±0.3
iso-butene	10	42	0.7	1.6±0.7	60	151	17	1.2±0.6
cis-2-butene	1.4	5.5	0.2	0.3±0.2	6.5	12	3.2	0.2±0.1
trans-2-butene	1.7	7.2	0.2	0.3±0.1	9.1	16	4.8	0.3±0.2

46   Å. Sjödin

Table 2.8. (continued) Average, maximum and minimum VOC emission factors in mg/km along with weight-% fractions of THC according to the FTP test for the nine cars in the 1997 measurements.

	Catalyst cars				Non-catalyst cars			
	average	maximum	minimum	wt-%	average	maximum	minimum	wt-%
1-pentene	0.6	2.3	0.1	0.2±0.1	3.5	6.6	1.6	0.1±0.1
cis-2-pentene	0.6	2.0	0.1	0.1±0.1	2.8	4.9	0.8	0.1±0.1
trans-2-pentene	1.0	3.6	0.1	0.2±0.1	5.8	10	1.7	0.2±0.2
2-methyl-1-butene	1.8	7.4	0.1	0.2±0.1	7.3	14	2.0	0.2±0.1
3-methyl-1-butene	0.7	2.6	0.1	0.2±0.1	4.4	7.9	2.4	0.1±0.1
2-methyl-2-butene	1.3	5.2	0.1	0.2±0.1	7.0	15	2.8	0.2±0.1
propadiene	6.5	32	0.2	0.4±0.5	9.2	23	2.4	0.2±0.1
1,3-butadiene	4.6	21	0.3	0.5±0.3	22	43	9.4	0.6±0.3
acetylene	151	744	0.1	7.3±13	639	2279	34	4.5±1.5
propyne	2.6	12	0.1	0.3±0.2	18	51	4.2	0.3±0.2
1-butyne	9.8	47	2.5	0.6±0.8	8.6	17	2.3	0.2±0.1
benzene	36	141	3.6	6.5±1.0	430	1392	62	5.2±0.9
toluene	47	164	0.4	11±2.8	1012	3379	138	11±2.1
ethylbenzene	5.5	17	1.0	1.5±0.5	28	75	8.1	1.3±0.3
m+p-xylene	26	67	1.1	7.2±7.3	113	372	15	4.0±1.3
o-xylene	11	26		4.8±3.8	311	990	42	1.4±0.5
1,2,3-trimethylbenzene	16	data for 1 car only		4.9	16	16	14	0.6±0.7
1,2,4-trimethylbenzene	4.5	"		1.4	30	68	7.0	0.4±0.2
1,3,5-trimethylbenzene	1.1	"		0.3	15	35	4.0	0.2±0.1
formaldehyde	6.8	29	0.9	0.6±0.4	43	62	25	1.2±1.7
acetaldehyde	0.4	0.7	0.1	0.2±0.2	0.9	1.1	0.6	0.03±0.04

2.2.6 Emissions of fuel stations and tankers determined by the inverse method

K. Schäfer, S. Emeis, H. Hoffmann, C. Jahn

2.2.6.1 Introduction

The investigation of the formation of photo-oxidants in the troposphere needs an accurate knowledge of their precursor emission inventory as of VOC. Some of these VOC as e.g. benzene and some photo-oxidants are carcinogenic or toxic. But, emission source strengths of important VOC are not well known. Activities during refuelling and transport of Otto fuel are one of the main anthropogenic VOC emission sources. The emitting areas are fuel stations and other refuelling units, transport systems as ships and road tankers, tanks and tank cleaning..

Fuel stations are a wide spread emission source. Reduction techniques have been installed as e.g. vapour return systems (VRS) which are required in Germany at each fuel station by the 21. BImSchV (1992). Emissions from fuel stations are usually calculated by multiplying an emission factor of a compound (in mg per l Otto fuel) with the amount of Otto fuel filled into the tank of the vehicles (ECE 1990). The VOC emission factor for motor vehicle refuelling for uncontrolled re-fuelling (without VRS) is 1320 mg l^{-1} (variation from 1280 to 1512 mg l^{-1}) which results in a benzene emission factor of 12 mg l^{-1} considering a fraction of 0.9 % of benzene in the Otto fuel vapour (mean from 1984 until 1991, Obermeier 1995). For controlled devices by a VRS a reduction of 85 % can be assumed. A factor of 67 up to 92 mg l^{-1} is given for emissions which arise from spillage losses. Up to now emission data are available only from measurements at the vehicle Otto fuel storage tanks during refuelling of Otto fuel (Hassel 1997) giving a benzene emission factor of 1 to 2 mg l^{-1} with a mean of 1.8 mg l^{-1} for fuel stations with VRS.

During transport of Otto fuel by river tankers VOCs are emitted mainly by the ventilation of remaining liquid and vaporous parts of load by means of an on-board fan into ambient air during travel. This gas-free ventilation is of common practice if the tankers are operated without load or if the tankers are to be refuelled with an other product than loaded before. During a typical ventilation process 2,000 kg VOC are emitted from a tanker with a loading volume of 2,000 m³. Further sources of VOC emissions from river tankers are not measured up to now. A possible technical solution for the prevention of VOC emissions during gas-free ventilation of tanks is the use of vapour recovery units (VRU) for cleaning the exhausts. This search for alternatives is necessary because a new amendment for the execution of the 20. BImSchV (1998) came into force on 25 August 2001 which allows the ventilation of tank atmosphere of river tankers into ambient air under certain conditions only. From 01 January 2006 a general prohibition of ventilation of tank atmosphere into ambient air is valid (20. BImSchV 2001). Further VOC emissions during Otto fuel transport by river tankers are not known (Hempel et al. 2000).

To quantify the VOC emissions of these sources certain measurement techniques are necessary which determine the whole emissions source strengths. Point and in situ measurement techniques cannot be applied because the emission sources are distributed heterogeneously and are variable temporally.

2.2.6.2 Objectives

The whole VOC emissions of gas stations are measured to determine the emissions under working conditions and to control the efficiency of the VRSs.

VOC emissions of river tankers in conjunction with loading and unloading activities are investigated. The VOC emissions from ventilation of storage tanks of tankers into ambient air are measured for comparisons with these emissions and to determine the accuracy of the inverse method. The VOC emissions during tests of the ventilation of tanks of river tankers into a land-site VRU are determined.

2.2.6.3 Method

To determine the emission rate from diffuse area and heterogeneous distributed sources an inverse method was developed (Carter et al. 1993; Lehning 1994; Piccot et al. 1994, 1996; Schäfer et al. 1998). Using a numerical dispersion model the emission rates are calculated on the basis of background and downwind concentration as well as based on meteorological measurements (see Fig. 2.13).

In this work spectroscopic path-integrated concentration measurements are used: open-path Fourier Transform Infrared (FTIR) spectroscopy and Differential Optical Absorption Spectroscopy (DOAS). Mainly benzene is measured using a mono-static DOAS with up to three retro-reflectors (Opsis 1997). If local spatial conditions allow up to three open paths in different directions nearby the emission source to be installed then the determination of influences of emission sources nearby as e.g. road traffic is possible. Normally, the measurements are performed during a time frame of several weeks with a fixed measurement system installation so that upwind and downwind measurement conditions are captured if there are stable background conditions in different directions. To calculate the dispersion of air pollutants a Lagrangian particle model is used considering influences from buildings and orography upon the streaming field which is in agreement with the German VDI guideline VDI 3945/3 (1999). This model is driven by wind and turbulence field data of MISKAM, a widely used prognostic, non-hydrostatic Eulerian model in Germany (Eichhorn 1989). All single emission sources are determined as a whole by this inverse method. The accuracy of the inverse dispersion modelling was validated by tracer releases (CH_4, N_2O) and detection with FTIR spectroscopy giving a typical value of ± 30 % i.e. in the same order as the accuracy of DOAS for benzene. Several fuel stations with different types of VRS and in different terrain were investigated. Emissions of river tankers were measured in a river harbour including probe sampling on board the tanker and FID (Flame Ionisation Detector) analyses of sampled air to determine NMHC concentration, too.

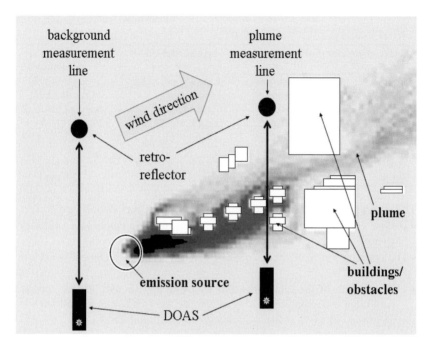

Fig. 2.13. Scheme of measurements for the inverse method. At suitable locations the measurement design can be optimised by using one DOAS rotating to both retro-reflectors.

2.2.6.4 Results

Fuel stations

The measurement results of benzene emissions at fuel stations are given in Table 2.9. The unit mg l^{-1} corresponds to mg benzene per l refuelled Otto fuel. Emissions of benzene originated from spillage losses, wet parts from leakage and ventilation pipes of storage tanks The emissions were determined with measurements during times without refuelling activities. The stock increase in the Otto fuel storage tanks of the fuel station by road tankers was also determined. The emissions from refuelling activities were calculated as difference between the emissions during operation time of the fuel station and the diffuse emissions.

Table 2.9. Emission source strengths of benzene from refuelling activities at fuel stations with VRS and during operation with non-activated VRS.

Types of emissions of benzene		Inverse method	Emission model calculations
Diffuse emissions	during closing time	15 mg l^{-1} (up to 18)	not available
	stock increase	2 mg l^{-1}	not available
	spillage losses	not available	0.8 mg l^{-1} (0.7 - 0.9)
Refuel-ling emis-sions	without VRS	9 mg l^{-1} (up to 25)	12 mg l^{-1} (11.5 - 13.6)
	with VRS	2 mg l^{-1} (1 - 9)	1.8 mg l^{-1} (1.3 - 2.3)
Total emissions	without VRS	29 mg l^{-1} (up to 40)	12.8 mg l^{-1}
	with VRS	22 mg l^{-1} (up to 24)	2.6 mg l^{-1}

The investigation results from campaigns at different fuel stations are:
- Diffuse emissions are much higher than considered in emission inventories up to now.
- The emissions at fuel stations depend strongly from the technical behaviour of the VRS.
- The reduction of the total emissions by using a VRS with higher vapour return rates and exhaust cleaning in comparison to a non-activated VRS is 40 % instead of 24 % at fuel station with conventional VRSs.

Table 2.10. Measured benzene and NMHC concentrations as well as emission source strengths during loading or unloading of river tankers and during ventilation of the atmosphere of two tanks (145 m^3 each) into ambient air and into the land-site VRU.

Activity	Measured concentration Increase		Source strength by inverse method	
	Benzene DOAS path across the harbour basin	NMHC on deck of tankers	Benzene per activity	NMHC Per activity
Loading and unloading (Otto fuel)	Not detectable	100 ppb		60 g
Pipe attachment and detachment, load Sampling	1 ppb	1 ppm	3 g	600 g
Ventilation into the land-site VRU	1 ppb	1 ppm	3 g	600 g
Ventilation into Ambient air	maximum 78 ppb	Maximum > 50 ppm	1.65 kg	330 kg

River tankers

The measurement results of VOC emissions during gas-free ventilation of the tanks of river tankers determined by the inverse method differ by about ±15 % to the values known so far. This result shows that the emissions of heterogeneous sources can be determined with sufficient accuracy. The measured benzene and NMHC concentrations together with the emission source strengths during loading or unloading of river tankers and during ventilation of two tanks (145 m³ each) are summarised in Table 2.10.

The measured emission source strengths during ventilation of the tank atmosphere into the land-site VRU are of the same amount as the measured emission source strengths during loading and unloading of Otto fuel and are caused by attachment and detachment of loading/unloading pipes only.

2.2.6.5 Conclusions

The comparison of the available emission factors used for emission modelling with the measurements made here leads to the conclusion that the emission source strengths of fuel stations with VRS during refuelling correspond well with the measured factors as long as the VRS is kept in order. Since a lot of gas stations have VRS problems, it is ensured with the new amendment of the 20. BImSchV from 2001 that each fuel station has to install a control unit in the VRS until end of 2007. In addition, there seem to be high diffuse emissions from the fuel stations; these emissions are usually neglected in emission inventories.

Also, the emission source strengths during loading and unloading activities of river tankers should be included in the emission inventories. It was shown during several measurement campaigns that the ventilation of the tank atmosphere of river tankers via the loading/unloading pipes into the land-site VRU is a technical solution to reduce the VOC drastically. Further reduction measures are closed sampling for probing the load to avoid the release of tank atmosphere, cleaning of the tanks by gas-free ventilation with a VRU instead of washing with water and chemicals to avoid destruction of washing-up water as well as the use of efficient stripping systems to reduce the rest of the load to a minimum.

2.2.6.7 Acknowledgements

The authors would like to thank the German Federal Ministry of Education and Research (BMBF), the German Federal Environment Agency (UBA) and the Environmental Ministry of Luxemburg for financial support of this research with contracts 07ATF12 and UBA-FKZ 200 44 321 as well as the staff of the investigated fuel stations, Dieter Gerstenkorn and his staff of the Magdeburger Umschlag und Tanklager KG, Dettmer GmbH & Co, Michael Bacher and Peter Sturm from the University of Graz and Burckhard Wickert and Rainer Friedrich of the University of Stuttgart for fruitful cooperation.

2.2.8 Determination of emission factors of volatile organic compounds from the residential burning of brown-coal briquettes (lignite)

H. Michael, W. Engewald

2.2.8.1 Introduction

In the measurement period from 1997 up to 2000 the decentralised heating of living rooms based on the burning of lignite was still widespread in eastern Germany (1997 in Leipzig: 62.000 single fireplaces $\cong$ 27 % of all homes) and other countries in Eastern Europe. Although there is a wide knowledge about the inorganic air pollutions arising from these sources (Ehrlich and Kalkoff 1993; Struschka et al. 1995), only relatively few data are available about the amounts of volatile and semi-volatile organic compounds (VOCs) emitted (Knobloch and Engewald 1995).

The aim of this work was the determination of real emission factors for VOCs, produced by combustion of lignite in different stove types for the heating of living rooms. Therefore, an analytical method was developed for the reliable determination of the VOCs emitted. The resultant emission factors for selected VOCs were provided to the project partners to serve as valid and representative input parameters for the compilation of an emission register.

2.2.8.2 Experimental

Sampling

With the beginning of the heating period in winter 1997/98 a measuring assembly for the systematic determination of the burning cycle of lignite was established (Fig. 2.14), equipped with a non-transportable tiled stove as well as a transportable slow-burning coal stove. The nominal heating capacity of both stove types was 5 kW.

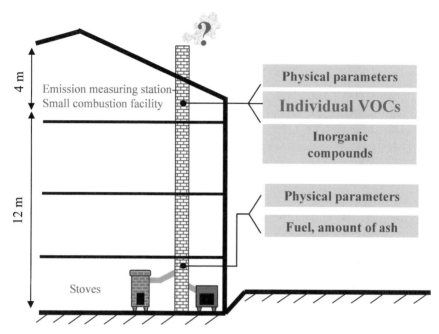

Fig. 2.14. Measuring assembly for the determination of stack gas emissions.

The investigations were carried out on both coal stoves using two different fuel types: Lusatian Briquettes and Central German Briquettes. At first the time dependent burn-up of the brown coal was investigated under real heating conditions to determine the necessary sampling time. An example of these investigations is shown in Fig. 2.15. It was found that the main part of the VOCs was emitted during the first hour after ignition of the Lusatian brown-coal briquettes. However, in case of the Central German Briquettes it was necessary to increase the sampling time up to two hours for complete sampling of the VOC emitted.

The sampling of individual VOCs was performed by adsorptive enrichment of the emitted VOCs. Therefore, a special sampling device was adapted approximately 4 m under the top of the chimney. With this device, a small part of the stack gas was taken out during the sampling time and diluted with pre-cleaned air to prevent water condensation on the adsorption material during cooling down of the hot emission gas. Considering the complex character of the samples, a multi-step enrichment of the volatile organic species on different hydrophobic carbon adsorbents, which were serially arranged in small glass tubes and connected in the order of increasing adsorption strength, was necessary. Following adsorbents were used: Carbotrap C ($10m^2/g$), Carbotrap Y ($25m^2/g$), Carbotrap X ($250m^2/g$) and Carboxen 569 ($387m^2/g$).

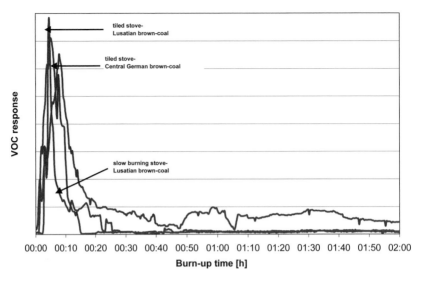

Fig. 2.15. Time dependent VOC emissions during the burn-up cycle of lignite, measured by FID-signal.

Sample analysis

After sampling the adsorption tubes were disconnected enabling a separate analytical treatment of each of the tubes. In the laboratory, the moisture was removed from the adsorption tubes by purging with dry helium. The compounds enriched were thermally desorbed and then analysed using temperature-programmed capillary gas chromatography. The fractionated enrichment of the analytes on different adsorbents fixed in two separate glass tubes led to a pre-separation into a light-volatile fraction and a fraction of semi-volatile compounds. That way, an optimum gaschromatographic separation of each fraction was possible.

The VOCs of the light-volatile fraction were separated by use of a dual-column technique. After thermal desorption the carrier gas flow containing the compounds was splitted by a Y-splitter at the end of the transfer line, whereby a simultaneous chromatographic separation on two different capillaries could be achieved. One part of the carrier gas was conveyed to a GS-Alumina PLOT column (J&W: 30m x 0.32 mm I.D.), the other to a CP-SilicaPLOT column (Chrompack 30m x 0.32 mm I.D.). This parallel column technique combines the excellent selectivity of a GS-Alumina PLOT column for separation of isomeric light hydrocarbons with the specification of polar volatile organic compounds which is achieved simultaneously using a CP-SilicaPLOT column. Most importantly it was possible to obtain 12 additional peaks for polar compounds using the CP-SilicaPLOT column in comparison with the GS-Alumina PLOT. The reason for this fact is, that the CP-SilicaPLOT shows a high degree of inertness as it also elutes polar organic compounds. By this way it was possible to identify additional polar sample constitu-

ents by means of the CP-SilicaPLOT in the same run. It would have not been possible using only the Al$_2$O$_3$-PLOT capillary. The both resulting chromatograms and
the analysis parameters for the separation of the light-volatile fraction are shown
in Fig. 2.16 and Fig. 2.17.

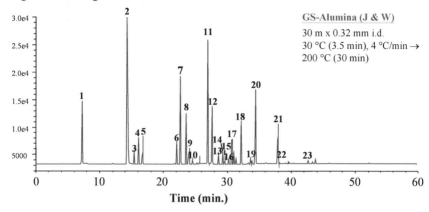

Fig. 2.16. Light-volatile fraction, separated on GS-Alumina PLOT column (resulting from
tiled stove - fired with Lusatian Briquettes).

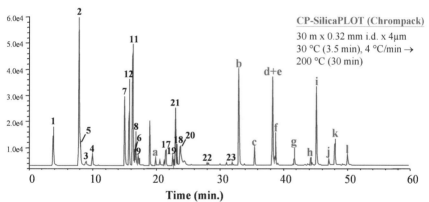

Fig. 2.17. Light-volatile fraction, separated on CP-SilicaPLOT column (resulting from tiled
stove - fired with Lusatian Briquettes).

For the separation of the compounds of the semi-volatile fraction a non-polar
thick film capillary was used (Fig. 2.18). Species in the range of C6 - C18 were
separated in order of increasing volatility. Identification of the separated organic
compounds was achieved via GC-MS-coupling and by comparison with retention
data of reference compounds. The identity of the numbered peaks is summarized
in Table 2.13. Quantification of the identified species was done by external calibration using selected reference compounds.

Additional methane and selected carbonyl compounds were determined in the
stack gas emissions. The methane emissions were measured continuously during
the burn-up cycle using a flame ionisation detector. Light-volatile carbonyl com-

pounds were collected analogous the VOCs, but using the well-established DNPH-method, followed by HPLC analysis of the derivates with UV detection.

Furthermore, the physical parameters, e.g. temperature and the volume flow were continuously registered during the completed sampling procedure. At the same time, the concentrations of the inorganic gases CO, CO_2, NO, NO_2, SO_2 and O_2 were continuously measured.

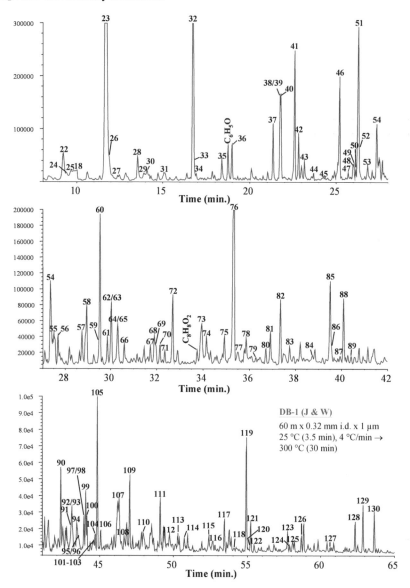

Fig. 2.18. Semi-volatile fraction.

2.2.8.3 Principal results

Composition of the VOC emissions

24 individual burn-up cycles of lignite were investigated and evaluated during the measuring periods in the winter time from 1998-2000. At the end of this project, more than 150 individual compounds emitted from the burning of Lusatian and Central German Briquettes have been reliable identified (Table 2.13). Species in the range of C1 – C18, mainly alkanes, alkenes, aromatics, aldehydes, phenols and thiophenes, were found. In Fig. 2.19, the composition of the VOC emissions is shown. The main compound was methane ($\approx$ 62 %). The differentiation of the NMOG group (non methane organic gases) resulted in a high amount of oxygenates ($\approx$ 45 %, e.g. aldehydes and ketones), followed from unsaturated hydrocarbons ($\approx$ 20 %) and aromatics ($\approx$ 19 %).

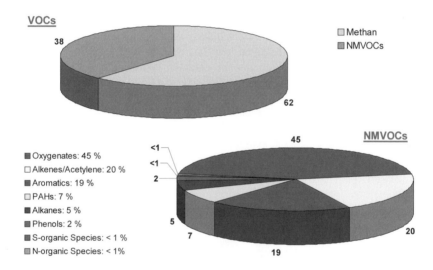

Fig. 2.19. Composition of the VOC emissions (example combination tiled stove - Central German Briquettes).

Emission factors

Based on knowledge of the mass of lignite loaded, the volume flow of the stack gas during the complete burn-up time and the amount of the enriched organic compounds, emission factors of selected important VOCs were calculated. In Table 2.11 the emission factors for the most important VOCs are shown. The values of the emission factors are average data based on 18 individual measuring campaigns. The confidence interval of the emission factors presented was calculated for the statistical certainty of these data.

The components in Table 2.12 were measured by FID (methane) and the light carbonyl components were analysed by derivatisation with 2,4-DNPH and HPLC separation. Due to the limited number of measurements for these components the ratio benzene/component "i" was calculated additionally for the assessment of these values. Thus, a correlation is possible of the mean value of the emission factor for benzene (resulting from a large number of measurements) and the emission factors of methane and the light carbonyl components (resulting from only few measurements).

2.2.8.4 Discussion

The compounds with the highest emission factors were methane, acetaldehyde, propionaldehyde, formaldehyde, benzene, propene and toluene. Further main components were naphthalene, 1,3-butadiene, phenol, styrene and 1-butene. These species were produced by thermally cracking of lignite during the burn-up cycle and non-completed combustion. Mean emission factors in the range of 400-1600 mg/kg for methane and 22-118 mg/kg for benzene respectively were found, depending on the combination of selected stove type and used briquettes.

Although the method presented here has been proved to allow a reproducible sampling and analysis of stack gas, the emission factors determined show a variation of 40-60 % reflecting the fact that a wide spectra of criteria, e.g. meteorological parameters and heterogeneity of burning material, kind of packing of the lignite in the stove used have a drastic influence on the emission characteristic. The variation of the calculated emission factors for benzene is shown in Fig. 2.20 for 9individual measurement campaigns.

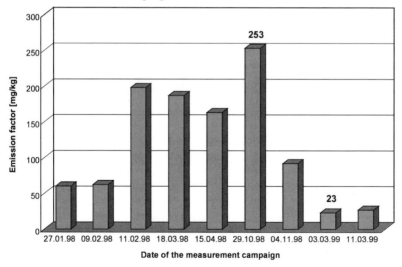

Fig. 2.20. Variations of the calculated emission factor for benzene (example combination tiled stove - Central German Briquettes).

The emission factor for benzene was ranged from 23-253 mg benzene/kg lignite (combination Lusation brown-coal briquettes and tiled stove). It was interesting that not only the benzene emissions (product of non completed combustion of the lignite loaded) were fluctuated, the same fluctuation was found for the carbon dioxide emissions (product of completed combustion), too. The course of benzene and carbon dioxide emissions was correlated inverse (Fig. 2.21).

This result shows that the variation of the determined benzene emissions (representative example for all other VOCs) is not depending on non-reproducible sampling of VOCs, but it is strongly depend on the specific of the process of combustion of the lignite.

A special aim of all investigations was the simulation of real combustion processes without any standardization. Thus, the presented emission factors represent real data suitable as valid and representative input parameters for the compilation of an emission register.

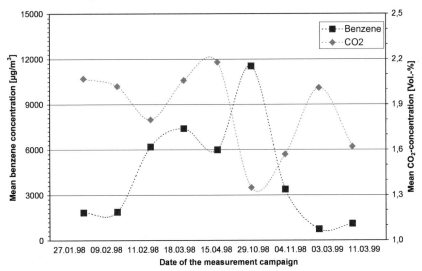

Fig. 2.21. Correlation of benzene and carbon dioxide emissions (example combination tiled stove - Central German Briquettes).

2.2.8.5 Outlook

It is significant that the ratio of the main compounds was found unaffected by the level of absolute emission factors. Knowing this ratio it could be possible to reduce the measuring expenditures by determining only a few key compounds and estimate the corresponding values for further compounds.

2.2.8.6 Acknowledgements

The authors wish to thank the German Federal Ministry for Research (BMBF) for the support of this project.

Table 2.11. Emission factors of selected NMVOCs.

Emission factors (mg/kg) of non-methane volatile organic compounds (NMVOCs)		
Component	Non transportable tiled stove		Slow burning stove
	Lusatian brown coal (n=9)	Central German brown coal (n=7)	Lusatian brown coal (n=2)
Aromatics			
Benzene	118 ± 94	56 ± 29	22
Toluene	39 ± 36	25 ± 16	8
Ethylbenzene	6 ± 5	5 ± 4	2
Styrene	14 ± 11	9 ± 4	4
o-/m-/p-Xylene	18 ± 13	15 ± 6	4
1H-Indene	10 ± 9	6 ± 3	4
Naphthalene	25 ± 19	16 ± 8	6
1,1'-Biphenyl	3 ± 3	3 ± 2	1
Acenaphthylene	5 ± 5	4 ± 3	3
Phenanthrene	4 ± 3	3 ± 2	2
Alkanes			
Propane	13 ± 14	17 ± 21	2
n-Butane	4 ± 4	6 ± 8	1
n-Hexane	7 ± 6	6 ± 4	2
Olefines			
Propene	70 ± 64	62 ± 46	14
1.3-Butadiene	23 ± 19	17 ± 9	6
1-Butene	15 ± 13	15 ± 11	4
i-Butene	9 ± 9	8 ± 6	2
1.3-Cyclopentadiene	18 ± 21	5 ± 5	2
1-Hexene	7 ± 6	4 ± 3	2
Miscellaneous			
Phenol	21 ± 21	10 ± 5	5
m-/p-Cresol	7 ± 8	4 ± 3	2
2-Nitrophenol	3 ± 2	<1	<1
Benzaldehyd	9 ± 9	6 ± 4	2
Furan	5 ± 4	3 ± 2	1
Benzofuran	7 ± 6	2 ± 1	1
Acetonitrile	3 ± 2	1 ± 2	1
Benzonitrile	2 ± 2	1 ± 2	1
Benzothiphene	3 ± 2	3 ± 3	1

Table 2.12. Emission factors of Methane and light carbonyl components.

Emission factors (mg/kg) of Methane and light carbonyl components					
Component	Non transportable tiled stove				Slow burning stove
	Lusatian brown coal		Central German brown coal		Lusatian brown coal
	EF	ratio	EF	ratio	EF	ratio
Methane	390	1:16	1570	1:26	620	1:28
Formaldehyde	57	2	100	1		
Acetaldehyde	101	4	183	2		
Propionaldehyde	52	2	44	1		
Acetone/Acroleine	27	1	33	<1		

Table 2.13. Identification of the numbered peaks in Fig. 2.16 - Fig. 2.18

Light-hydrocarbons and polar volatile organic compounds

Al$_2$O$_3$-PLOT-capillary		Additional PVOC using the CP-SilicaPLOT	
1	Propane	a	Methylenchloride
2	Propene	b	Acetaldehyde
3	i-Butane	c	Nitromethane
4	n-Butane	d	Acrolein
5	Propadiene	e	Propionaldehyde
6	trans-2-Butene	f	Acetonitrile
7	1-Butene	g	Methacrolein
8	i-Butene	h	Ethyl acid methyl ester
9	cis-2-Butene	i	Acetone
10	n-Pentane	j	2,3-Butandione
11	1.3-Butadiene	k	Methylvinylketone
12	Propyne	l	Methylethylketone
13	Cyclopentene		
14	2-Methyl-2-Butene		
15	trans-2-Pentene		
16	2-Methyl-1-Butene		
17	1-Pentene		
18	n-Hexane		
19	1-Butyne		
20	1,3-Cyclopentadiene		
21	Furan		
22	1-Hexene		
23	Benzene		

Semi-volatile compounds using the non-polar DB-1 capillary

22	1-Hexene	56	1-Decene
24	2-Methylfuran	57	1.2.3-Trimethylbenzene
25	Ethyl acid	58	2-Hydroxybenzaldehyde

Table 2.13. (continued) Identification of the numbered peaks in Fig. 2.16 - Fig. 2.18

Light-hydrocarbons and polar volatile organic compounds

Al_2O_3-PLOT-capillary		Additional PVOC using the CP-SilicaPLOT	
18	n-Hexane	59	o-Cresol
23	Benzene	60	1H-Indene
26	Thiophene	61	Acetophenone
27	Cyclohexane	62	2-Methylbenzaldehyde
28	1-Heptene	63	3-Methylbenzaldehyde
29	2.5-Dimethylfuran	64	m-Cresol
30	n-Heptane	65	p-Cresol
31	Pyridine	66	4-Methylbenzaldehyde
32	Toluene	67	x-Methylbenzofuran (1)
33	3-Methylthiophene	68	1-Undecene
34	2-Methylthiophene	69	x-Methylbenzofuran (2)
35	1-Octene	70	x-Methylbenzofuran (3)
36	2-Furanaldehyde	71	n-Undecane
37	Ethylbenzene	72	2-Nitrophenol
38	m-Xylene	73	$C_{10}H_{10}$ (x-Methylindene ?)
39	p-Xylene	74	$C_{10}H_{10}$ (x-Methylindene ?)
40	Ethinylbenzene	75	1-Phenyl-2-propin-1-one
41	Styrene	76	Naphthalene
42	o-Xylene	77	Benzo[b]thiophene
43	1-Nonene	78	1-Dodecene
44	n-Nonane	79	n-Dodecane
45	i-Propylbenzene	80	2-Nitro-m-cresol
46	Benzaldehyde	81	x-Nitro-y-cresol (1)
47	n-Propylbenzene	82	2-Nitro-p-cresol
48	Benzonitrile	83	6-Nitro-m-cresol
49	1-Ethyl-3-Methylbenzene	84	3-Hydroxybenzaldehyde
50	1-Ethyl-4-Methylbenzene	85	2-Methylnaphthalene
51	Phenol	86	1-Tridecene
52	1.3.5-Trimethylbenzene	87	n-Tridecane
53	1-Ethyl-2-Methylbenzene	88	1-Methylnaphthalene
54	Benzofurane	89	4-Hydroxybenzaldehyde
55	1.2.4-Trimethylbenzene	90	1,1'-Biphenyl
91	1-Tetradecene	111	9H-Fluorene
92	2-Ethylnaphthalene	112	n-Hexadecane
93	1-Ethylnaphthalene	113	x-Methyl-Dibenzofuran (1)
94	n-Tetradecane	114	x-Methyl-Dibenzofuran (2)
95	2,6-Dimethylnaphthalene	115	1-Heptadecene
96	2,7-Dimethylnaphthalene	116	n-Heptadecane
97	1,3-Dimethylnaphthalene	117	9H-Fluoren-9-one
98	1,6-Dimethylnaphthalene	118	Dibenzothiophene
99	1,7-Dimethylnaphthalene	119	Phenanthrene
100	x-Ethinylnaphthalene	120	1-Octadecene
101	1,4-Dimethylnaphthalene	121	Anthracene
102	2,3-Dimethylnaphthalene	122	n-Octadecane

Table 2.13. (continued) Identification of the numbered peaks in Fig. 2.16 - Fig. 2.18

Light-hydrocarbons and polar volatile organic compounds

Al_2O_3-PLOT-capillary		Additional PVOC using the CP-SilicaPLOT	
103	1,5-Dimethylnaphthalene	123	1H-Phenalen-1-one
104	1,2-Dimethylnaphthalene	124	1-Nonadecene
105	Acenaphthylene	125	n-Nonadecane
106	1,8-Dimethylnaphthalene	126	$C_{15}H_{10}$ (4H-Cyclopenta[def]phenanthene)
107	4-Nitrophenol	127	n-Eicosane
108	n-Pentadecane	128	Fluoranthene
109	Dibenzofuran	129	Acephenanthrene
110	x-Nitro-y-cresol (2)	130	Pyrene

2.2.9 NMVOC emissions from solvent use in Germany

J. Theloke, R. Friedrich

2.2.9.1 Introduction

Solvent use emissions currently account for about 50% of all anthropogenic NMVOC emissions in Germany. The examination of potential abatement strategies requires information on the sources and a substance-related compilation of the inputs of organic solvents and solvent containing products as well as the accurate determination of the emissions caused by these products.

In the frame of the international reporting requirements (e.g. for the Kyoto protocol UNFCCC (1997), for the UNECE Gothenborg protocol UNECE (1999), for the EC legislation Directive 2001/81/EC (2001) etc.) is it more and more necessary to apply a methodology for calculating emissions using consistent approaches as they are required there.

A first calculation of emissions from solvent use for the old federal states of Germany for the year 1986 was presented by Bräutigam and Kruse (1992). For more recent years, only rough estimates were carried out on the basis of this study until recently. Solvent use emissions were assessed in detail by Obermeier (1995) for the federal state of Baden-Wuerttemberg for the year 1990. More general results for a federal state (North Rhine-Westphalia) on solvent related emissions from small trade and private households were introduced by Brandt et al. (2000). With emerging new knowledge on NMVOC sources and emission factors and because of the reunification and the resulting expansion of the examination area it was necessary to develop new methods for the determination of NMVOC emissions from solvent use in Germany, as it is described in this contribution.

With the method describe here, NMVOC emissions from solvent use were assessed for Germany for the base years 1994, 1996, 1998 and 2000. The amount of consumption of solvents and solvent containing products was identified with the help of a product based approach. The domestic consumption of different organic solvent substance classes calculated thus were validated and checked by a solvent related approach.

The work described here was carried out for and financed by the German Federal Environment Agency UBA (Theloke et al. 2000, 2001).

2.2.9.2 Definitions and system boundaries

The definition of solvents and VOC as it is used by the solvent directive (1999/13/EC) as a basis of EU legislation, defines solvents as follows (Directive 1999/13/EC 1999):

"Organic solvent shall mean any VOC which is used alone or in combination with other agents, and without undergoing a chemical change, to dissolve raw materials, products or waste materials, or is used as a cleaning agent to dissolve contaminants, or as a dissolver, or as a dispersion medium, or as a viscosity adjuster, or as a surface tension adjuster, or a plasticiser, or as a preservative."

According to the "EU-VOC guideline" (Directive 1999/13/EC 1999) volatile organic compounds are defined as follows:

"Volatile organic compound shall mean any organic compound having at 293,15 K a vapour pressure of 0,01 kPa or more, or having a corresponding volatility under the particular condition of use. For the purpose of this Directive, the fraction of creosote which exceeds this value of vapour pressure at 293.15 K shall be considered as a VOC."

Based on results of work conducted at IER, the main emission sources of NMVOCs from solvent use are the production of solvent containing products and the application of solvents and solvent containing products. It has to be taken into account, though, that some products are used both as solvents and as a chemical reaction component. Toluene, for example, is used as a solvent in varnishes and adhesives and as a reaction component for the production of Toluenediisocyanat (TDI). In a similar way, Methyl Ethyl Ketone is used as a solvent in printing inks and as basic material at the synthesis of Methyl Ethyl Ketone Peroxide. Products which are used as a chemical reaction component do not count as solvents according to the solvent definition.

2.2.9.3 Method

A product based approach was applied for the calculation of solvent emissions. First, the use of solvents and solvent based products (e.g. paints, adhesives, degreasing agents) was compiled from production statistics and foreign trade statistics. Secondly, the use of these goods in specific industrial and commercial sectors and households was estimated. The emission factors were then calculated under consideration of application techniques, emission control measures and other pathways (e.g. waste, water, recycling).

For the solvents related approach, data was used as complied by solvent manufacturers about domestic solvent consumption to assess the domestic consumption of different solvents directly, i.e. without the detour via the consumption of solvent containing products. The domestic consumption of solvents was defined as the balance from production, import and export. For the solvent related approach generally accessible statistics offered only a limited basis, thus the applicability of this method depended on obtaining supplementary information mostly from solvent industry.

The product related approach is described in more detail as follows:

For the determination of the solvent inputs as well as the resulting NMVOC emissions, economic statistics were processed and evaluated as well as supplementary information obtained from industry associations and a other institutions by surveys.

The fundamental way towards determining solvent use and emissions can be subdivided into the following steps:

1. Estimate of domestic use of solvent containing products from production and foreign trade statistics (production + import - export) as well as by the way of interviews of trade associations.
2. Determination of details for the product specific solvent content as well as on the composition of the solvent speciation on the basis of interviews of the appropriate trade associations.
3. Determination of the use of solvent borne products on the basis of information of the different trade associations, e.g. The *Association of the Varnish Industry* (VDL, http://www.lackindustrie.de), *Industry Association of Adhesives* (IVK, http://www.klebstoffe.de), *Association of the Chemical Industry* (VCI, http://www.vci.de), *Association of the Mineral Printing Paint Industry* (VDMI, http://www.vdmi.de) (Gemeinsamer Abschlußbericht 1997) etc. related to single application areas as well as other information sources, e.g. environmental consultants as *Oekorecherche* (http://www.oekorecherche.de) or *Oekopol* (http://www.oekopol.de) etc., technical literature, personal communications of experts etc..
4. Estimate of business-related solvent use under consideration of data on business dependent differences regarding solvent content and solvent composition in the respective product types.
5. Estimate of solvent emissions with application data found out above under consideration of the stock and the effectiveness of secondary abatement measures as well as the further use of recycled solvents.

Finally, an assignment of the calculated emissions to the corresponding source group related classifications for the respective demands of the required emission inventory was conducted. In Fig. 2.22 the method for the determination of the solvent emissions is displayed including the plausibility check by a solvent related approach.

An essential task was to produce an assignment of goods groups of the production statistics as transparent as possible to the corresponding goods groups of the foreign trade statistics. A position of the foreign trade statistics often corresponds to several positions of the production statistics. Where this assignment was not feasible primarily due to non-disclosure cases, other proxy data had to be identified. It was, for instance, impossible to find any meaningful data about pharmaceutical production amounts in tonnes in the production statistics.

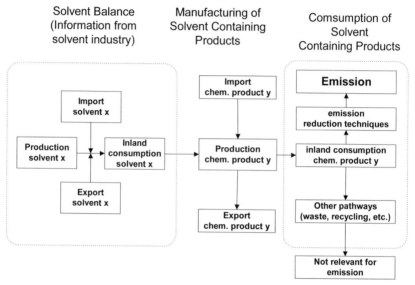

Fig. 2.22. Overview on the method for calculation of emissions from solvent use

Quantities of import and export can, however, more or less completely be found for pharmaceutical products as well as their cash values in the foreign trade statistics. In the statistics of the *German Association of the Chemical Industry* (VCI, http://www.vci.de) "Chemiewirtschaft in Zahlen" (German Association of Chemical Industry (VCI 2001) the production value of the pharmaceutical products is indicated for the respective year. For the production amount or the domestic consumption of pharmaceutical products in kilotonnes, the mean foreign trade value in € per tonnes of an exported or imported pharmaceutical product is put into relation with the production value in €. Of course this calculation is prone to uncertainties, however it helps to achieve a good estimate of the domestic produced amount of pharmaceutical products

The harmonisation of national production statistics in Europe was primarily carried out with regard to a European harmonisation of the production statistics and with the aim of a clear assignment to production and foreign trade statistics. Before this rearrangement, the assignment in many cases was not clear. Hence this method is applicable to other OECD countries as well.

Such examinations, however, are only conceivable and feasible in larger intervals. For this reason a method was developed with which intermediate projections are possible without too much effort, based upon easily available data such as e.g. production statistics or foreign trade statistics and so on.

2.2.9.4 Results

In the year 2000, emissions from solvent use in Germany have been calculated to amount to approx. 863 kilotonnes, with about 25% (222 kilotonnes) emitted by paint application. Roughly 106 kilotonnes of NMVOC emissions originated from domestic solvent use. Emissions from washing and cleaning agents as well as body and hair care products have a significant share in domestic applications. The printing industry emitted approx. 80 kilotonnes NMVOC in Germany in 2000. The remaining emissions originate from a variety of other sources. In Table 2.14 the calculated emissions from solvent use in Germany for the year 2000, split into different source groups, are presented.

Table 2.14. Emissions from solvent use in Germany 2000 dissolved to different source groups

Source group	Emissions 2000 [tons]
Paint application	
Manufacture of automobiles	13,294
Vehicle refinishing	21,135
Constructions and buildings-professional use	31,355
Construction and buildings – DIY	13,438
Coil Coating	2804
Boat building	4,783
Wood	35,761
White goods	5 840
Mechanical engineering	18,504
Car components	2,760
Metal hardware	2,618
Plate emballages	6,095
Wire varnishes	271
Foundry agents	1,038
Prefabricated construction units	3,896
Synthetics	8,532
Paper/Slide	278
Road marking varnishes	3,561
Corrosion-inhibiting primers	18,619
Other paint applications	30,368
Thinner w/o specific application	225,108
Degreasing, dry cleaning and electronics	
Metal degreasing	15,843
Dry cleaning	1,181
Electronics manufacturing	984
Precision optics	1,222
Precision mechanics	1,161
Industrial special applications	6,232
Non industrial services (e.g. brake cleaners)	7,576

Table 2.14. (continued) Emissions from solvent use in Germany 2000 dissolved to different source groups

Source group	Emissions 2000 [tons]
Chemical products manufacturing or processing	
Polyester processing	349
Polyvinylchloride processing	204
Polyurethane processing	7,973
Polystyrene foam processing	9,501
Rubber processing	3,554
Pharmaceutical products manufacturing	7,650
Paints manufacturing	3,715
Inks manufacturing	940
Glues manufacturing	576
Asphalt blowing	201
Adhesives manufacturing	362
Magnetic tapes manufacturing	227
Films/Fotographs manufacturing	6
Textile finishing	1,926
Other chemical production processes with solvent use	3,239
Other use of solvents and related activities	
Glass and mineral wool enduction	1,176
Rotogravure printing	11,802
Fat edible/non-edible oil extraction	9,167
Heatset Offset printing	17,500
Letterpress printing	462
Package printing	16,054
Screen printing	2,500
Offset printing	31,441
Application of glues/adhesives	22,414
Preservation of wood	31,497
Underseal treatment and conservation of vehicles	11,439
Domestic solvent use	105,699
Vehicles dewaxing	1,526
Domestic use of pharma products	3,295
Solvent use for application of plant protective agents	3,531
Paint remover	3,146
Concrete additives	42,389
Aircraft deicing/other deicing	8,071
Application of propellants	15,518
Universities/scientific labs	1,716

The distribution of the emissions amount and solvent inputs is represented graphically in the following illustrations. In the year 2000, 863 kilotons of NMVOC were emitted in Germany from the application of solvents and solvent containing products. Fig. 2.23 shows the distribution of the solvent emissions on the source groups. About 64 % of the solvent emissions caused by the source groups "application of paints and varnishes" and ", "domestic use of solvents ", as well as the "application of other thinners".

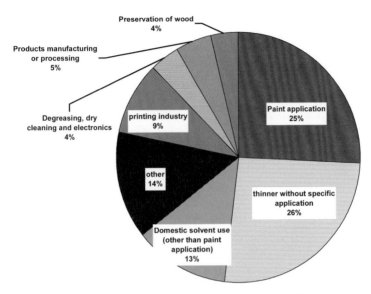

Fig. 2.23. NMVOC emissions from solvent use in Germany resolved to main source groups (in total 863 kt)

Approximately 25% (~222 kilotonnes) of solvent emissions were caused by paint application. In addition to that, other thinners without specific application profiles which are used partly also in paint application contributed significantly. Fig. 2.24 displays the speciation of emissions to different paint applications. About 10% of the emissions originated from plants that fall under the jurisdiction of the 4. regulation to the German Federal Immission Act (4. BImSchV 1999). These are required to implement emission control equipment (e.g. thermal incineration) in accordance with the regulations of the German Technical Instructions on Air (TA Luft, German Technical Instructions for air pollution control 2002). More than 50% of the emissions from paint application were caused by smaller and medium enterprises (SMEs) and skilled crafts enterprises which are not required to operate any waste control equipment. When including the areas of wood preservation, building and construction, vehicle refinishing as well as domestic paint application (Do-It-Yourself), more than 90% of NMVOC emissions from the source group "paint application" originate from SMEs and skilled crafts enterprises as well as the Do-It Yourself area.

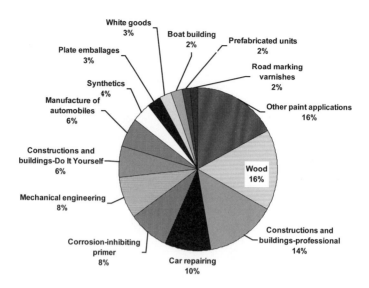

Fig. 2.24. NMVOC emissions from paint application (in total 222 kt)

Approximately 225 kilotonnes were emitted from the use of thinners without exactly defined application fields. Thinner use is mainly required by paint application, degreasing processes, maintenance and equipment cleaning and printing processes. It is necessary to distinguish between clearly defined application of thinners, for example for the adjustment of paint systems, these thinners are considered at paint application and other thinners with not clearly defined application. As indicated in Fig. 2.25, the domestic consumption of thinners has been increasing continuously in recent years in Germany. In fact, it has doubled since 1994, where the caus is yet unclear.

34 kilotonnes of NMVOC were emitted from degreasing processes or dry cleaning. The largest proportion of these emissions (47 %) is caused by metal degreasing (approx. 15 kilotonnes) here as Fig. 2.26 shows.

In the sector of processing and manufacturing of synthetic materials and other products approx. 40 kilotonnes of NMVOC were emitted in 2000 (see Fig. 2.27). Approximately 45% (approx. 18 kilotonnes) originated from the processing of synthetic materials.

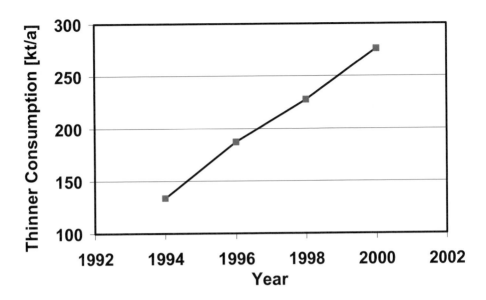

Fig. 2.25. The domestic consumption of thinners has been increasing continuously in recent years

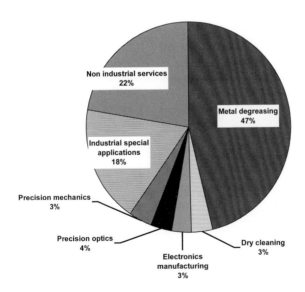

Fig. 2.26. NMVOC emissions from degreasing and dry cleaning (in total: 34 kt)

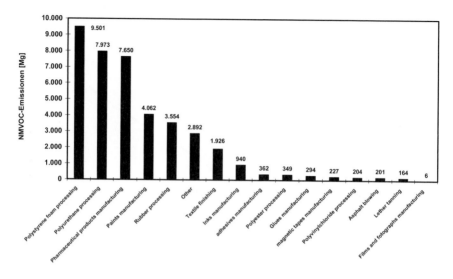

Fig. 2.27. NMVOC emissions from manufacturing and production of solvent borne products and synthetics (in total: 40,5 kt)

Approximately 37% (ca. 340 kilotons) of NMVOC emissions from solvent caused by other applications, for example:

- Domestic solvent use
- Printing industry
- Concrete additives
- Preservation of Wood

About 13% of all NMVOC emissions (approx. 105 kilotonnes) come from the use of consumer goods which are used mainly in private households. Personal hygiene products and detergents as well as motor vehicle antifreeze and cooling media have to be mentioned in particular (see Fig. 2.28 and Fig. 2.29).

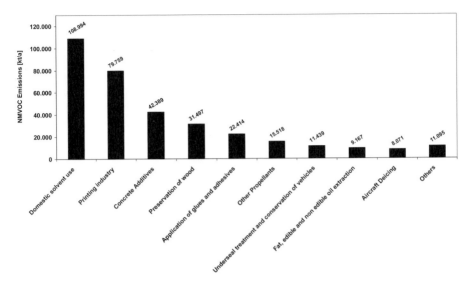

Fig. 2.28. NMVOC emissions from other use of solvents and related activities (in total: 386 kt)

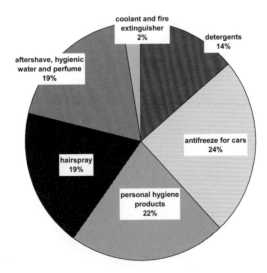

Fig. 2.29. NMVOC emissions from domestic solvent use (other than paint application) (in total: 105 kt)

From the production of printing products approx. 80 kilotonnes NMVOC were emitted in Germany in 2000, with about 40% (31 kilotonnes) emitted from offset printing processes. A relevant share of emissions stemmed from the use of moist media (Isopropanol) and cleaning agents. Approximately 12 kilotons of the NMVOC emissions of the printing industry were caused by rotogravure printing processes (illustration printing products). The emissions originated primarily from the use of toluene as a printing ink thinner. Relief printing methods (letterpress and flexography) and screen printing processes were emitted approximately 3 kilotons of NMVOC. Almost 16 kilotonnes were caused by package printing processes and 17.5 kilotonnes were emitted from heatset offset printing (see Fig. 2.30).

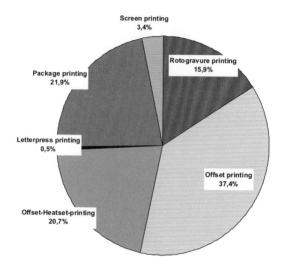

Fig. 2.30. NMVOC emissions from printing processes (in total: 80 kt)

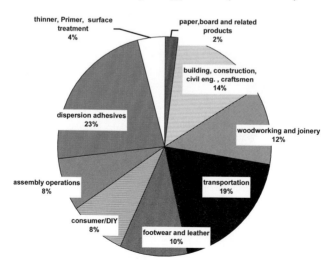

Fig. 2.31. NMVOC emissions from glues and adhesives (in total: 22.4 kt)

Approximately 22.4 kt of VOC originated from the application of glues and adhesives in the year 2000 in Germany. 23% originated from the application of low solvent content dispersion adhesives. Additional 26% were emitted from construction, building and woodworking activities. Nearly 30% caused by shoe production and repair sector and the transport sector (e.g. automobile manufacturing). Further source groups in this sector are assembly operations, DIY applications, the production of paper, cardboard and related products and last but not least the use of thinners and primers for surface treatment (see Fig. 2.31).

2.2.9.5 Speciation to substance classes

To be able to assess the effect of NMVOC emissions, the composition must be known for the speciation of the emissions to substance classes or individual substances. At first a substance speciation was carried out for every sector with the information available and with expert interviews. By breaking down emissions to a detailed split and summarizing emissions by sectors afterwards, consisting checks with regard to solvent consumption could be conducted. The substance class-related speciation thus developed is prone to high uncertainties due to a fragmentary data basis. After this the NMVOC emissions from solvent use consisted in 2000 to 21% of Aliphatics, to 26% of Alcohols and to 22% of Aromatics. Furthermore Esters (10%), Glycol derivates (9%), Ketones (6%), Halogenated Hydrocarbons (2%), Ethers (2%), Terpenes (1%), as well as small amounts of Organic Acids, Aldehydes, Amines und Amides have to be mentioned. Altogether 2% of the NMVOC emissions from solvent use could not be attributed to any substance class.

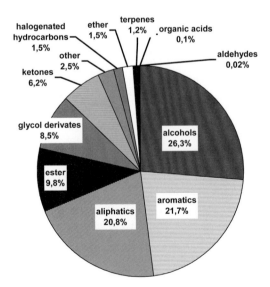

Fig. 2.32. Speciation of the NMVOC emissions from solvent use to substance classes in Germany for the year 2000

2.2.9.6 Plausibility check

The amount of solvent consumption for 1994 was speciated by IER to substance classes and partly to individual substances was compared with the solvent consumption according to production statistics of the solvent industry (minus export plus import).

Table 2.15 shows that the calculated amounts of solvent consumption are plausible with regard to the total sum and also with regard to the speciation to substance classes. However, Table 2.15 shows as well for the example of alcohols that further speciation by individual substances bears considerable uncertainties. In the row "sum" only the amounts are summarized which could be compared in Table 2.15. In Table 2.15 the deviations for substance classes varies between approx. 1.3 percent and 14.4 percent. Comparing the total sum between both approaches, a deviation of about 2 % occurs. But considering the annual consumption of single species, e.g. n-Propanol, the difference is much larger, amounting to several orders of magnitudes.

Table 2.15. Comparison of the substance speciation of the assessed amounts of solvent consumption with domestic consumption amounts of the European solvent industry (for the year 1994) (Theloke et al. 2001).

Substance classes	Domestic consumption [t] (solvent industry)	Domestic consumption [t] (own estimate)	Deviation [%]
Aliphatics	**300 000**	**280 000**	- 6.7
Aromatics	**250 000**	**286 000**	+14.4
Terpenes		10 000	
Halogen. hydrocarbons	**35 000**	**35 000**	
Alcohols	**360 000**	**316 000**	-12.2
-Ethanol	50 000	85 000	
-Isopropanol	250 000	144 000	
-n-Propanol	40 000	200	
-n/i-Butanol	10 000	56 000	
Glykolderivates	**75 000**	**78 000**	+4.0
Esters	**75 000**	**76 000**	+1.3
Ketones	**60 000**	**61 000**	+1.7
Ethers	**30 000**	**29 000**	-3.3
Aldehydes		200	
Organic acids		800	
Phthalates		225 000	
Other VOC's		65 000	
Total *(sum of bold faced classes)*	**1 185 000**	**1 161 000**	**-2.0**

2.2.9.7 Trend scenario

A trend scenario of solvent use emission changes from 1990 to 2010 was calculated under consideration of the national implementation of the EU VOC Directive

(Directive 1999/13/EC 1999) and the pending product related directive for decorative paints and varnishes (Proposal 2002). Emissions had been decreasing from 1990 to 1994, while from 1994 to 2000 a significant increase was calculated. For the year 2010, NMVOC emissions from solvent use are projected to decrease in Germany by approx. 30% in comparison to 2000. Furthermore, it has to be expected that the emission ceiling value of the NEC Directive (Directive 2001/81/EC 2001) is not going to be achieved by Germany. This is mainly caused by the fact that emissions in 2010, taking into account all anthropogenic source groups (e.g. road transport, combustion) and solvent use, will be about 1 400 kt NMVOC emissions.

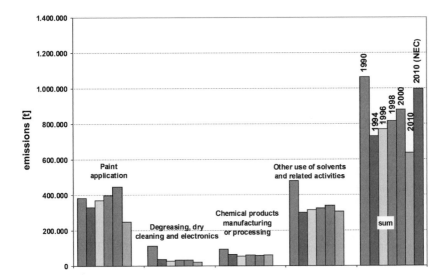

Fig. 2.33. Trend scenario for NMVOC emissions from solvent use from 1990 to 2010

Thus, to be able to reach the emission ceiling of 995 kt of NMVOC, as it has been set for Germany in the National Emission Ceilings (NEC) Directive, additional measures to further reduce emissions of anthropogenic NMVOC emissions have to be undertaken.

2.2.9.8 Uncertainties

The uncertainties of calculated emissions from solvent use are quite difficult to quantify. Annual emissions from solvent use typically contain an uncertainty of ± 30% according to own assessments (Kühlwein et al. 1999). However, there are substantially greater uncertainties in the analysis of individual source sectors. The reasons are the heterogeneous structure of the sources and the quality of available input data. A large variety of sources and emission activities exist with significant uncertainties in the assignment of production, consumption, application and resulting emissions. A chance for the assessment of plausibility is the analysis of surveys, emission declarations, expert surveys (for example from industry associations) and the solvent balance (see Table 2.15). The stated annual amounts of solvent use in different industries varies significantly even within the same source group. Another example of input data uncertainties is the wide distribution of the solvent content of different paint systems (Kühlwein et al. 1999).

2.2.9.9 Conclusions

The product and solvent based method described in this contribution for the calculation of NMVOC emissions has proven to be well designed and was successfully applied for the generation of detailed emission inventories for NMVOCs with high spatial and temporal resolution. By using this method, a detailed assessment of NMVOC emissions from solvent use can be conducted for other OECD countries as well, with acceptable input requirements, as the structure of economic statistics has been standardised. Furthermore, a substance related split of NMVOC emissions can be conducted based on this methodology. The method was applied to calculate the NMVOC emissions from solvent use for Germany in 2000. The total amount of emissions from solvent use was calculated to approximately 863 kilotons in Germany in the year 2000. About 25% (223 kilotonnes) caused by paint application, about 13% originated from antifreeze for cars, hairsprays, washing-up liquids, detergents, aftershave, fragrance, perfumes, cooling mediums and fire extinguishers. 80 kilotonnes were emitted from printing processes and 18 kilotons from the processing of synthetic materials. 40 kilotonnes originated from degreasing processes and dry cleaning. Almost 42 kilotonnes emitted from the application of concrete additives and 31 kilotonnes from wood preservation. The remaining emissions originate from a variety of different source groups, e.g. the application of glues and adhesives (22 kilotonnes), application of propellants (16 kilotonnes), from solvent use in chemical industry (22 kilotonnes). In addition to that, 225 kilotonnes were emitted through the application of thinners without a detailed attribution to sources (mostly for equipment cleaning). The NMVOC emissions from solvent use were speciated to 21% Aliphatics, 26% Alcohols, 21% Aromatics, 10% Esters, 9 % Glycol derivates, 6% Ketones, 2% Halogenated Hydrocarbons, 2% Ethers and 1% Terpenes as well as small amounts of organic acids, Aldehydes, Amines und Amides. A plausibility check resulted in good agreement between the solvent consumptions for 1994 estimated with the product based approach and the same using the solvent based approach (with data from the European solvent industry as a basis). Annual emissions from solvent use have an uncertainty bandwidth of ± 30% according to our assessment with a semi quantitative approach. The increase in emissions from solvent use observed from 1994 to 2000 currently prevails. Thus, it is not clear if Germany will be able to comply with the emissions target of the recently adopted EC National Emissions Ceilings Directive for the year 2010. The implementation of additional measures to further reduce emissions of anthropogenic NMVOC emissions in Germany has to be taken into consideration.

2.3 Emissions of Particulate Matter

2.3.1 Summary

W. Winiwarter

2.3.1.1 Introduction

Historically, emissions of particulate matter were among the first signs of air pollution being noticed. Depending on particle size, pollution can be visible, and at least in form of deposition (on clothes as much as on plant tissues or building facades) easily noted. Health effects, as in working environments (e.g. inside mines) also became obvious. Also, a very direct relationship between source and effect (i.e. air pollution) could be established.

Consequently, particulate matter not only played a large role in atmospheric research, but also abatement technologies were developed quite early. Most major sources at least in Europe are now equipped with abatement devices. PM ceased to be a major topic in air pollution since then. PM was originally not dealt with in long range considerations of air pollution, thus PM was not included in the CORINAIR European emission inventories.

Only with the discovery of health effects and increased mortality due to PM exposure (Pope et al. 1995; Dockery et al. 1993) at quite low ambient concentrations, both science and environmental policy had to react. In Europe, EU-directive 1999/30/EG triggered environmental regulations in the EU member countries. Also emissions of PM became an issue again, and have recently been included in the countries' obligation to report their emissions within the framework of the Convention on Long-range Transboundary Air Pollution. The draft reporting guidelines cover different fractions of PM emission to be reported separately, namely TSP (total suspended particles), PM_{10} (particles with a diameter of 10μm or less), and $PM_{2.5}$ (particles with a diameter of 2.5μm or less).

While most of the health assessment studies have been made in accordance to PM_{10} (determined as total mass) and also the threshold values have been established as PM_{10} mass concentrations, it is not at all clear whether PM_{10} mass is indeed the health related parameter of PM. Other parameters in discussion are the chemical composition, or the particle number concentration which gives a distinctively different weight to the very small particles.

The concept of different modes in PM was originally developed by Whitby (1978), but it is still very relevant. The coarse mode of PM (larger than about 1μm) is created dominantly by breakup of larger solid material. Sedimentation is an important removal process, at least on the large end of the spectrum. The chemical composition is strongly influenced by soil material. The fine aerosol mode is again divided into two modes: the nucleation mode and the accumulation mode. Nucleation mode particles are formed from the gas phase, they are in the size range of nm up to about 0.1µm. Homogeneous gas-to-particle conversion usually involves high concentrations and temperature gradients as are typical for exhaust plumes directly after emissions. While their mass is quite small, they show up strongly in the total particle number. A major sink of nucleation mode particles is their agglomeration into the so-called accumulation mode. The latter is the most stable mode. This is material that has mostly gone through gas phase, it may consist of carbonaceous organic material and soot, but also includes sulphate or nitrate and ammonium ions. Due to their relatively large surface area as well as residence times, accumulation mode particles are also able to adsorb material as PAH's or heavy metals, which are then found together with these particles in the size range of 0.1 - 1µm.

2.3.1.2 PM emissions from combustion processes

As stated before, PM emissions from combustion processes have been dealt with quite early. Abatement equipment is widely in place, also routine emission monitoring may be available. Still not fully resolved is the question of the size fractionation. But as filters and electrostatic precipitators are extremely efficient with larger particles, and somewhat less efficient with fine particles below 1µm, the major fraction of point source emissions is emitted in the fine mode anyway.

Less information is available for small combustion sources. Domestic heating with solid fuels is known to be a considerable source. Consumer behaviour strongly differs with country, with e.g. Sweden having wood combustion strongly in larger installations, and Austria in smaller stoves. As the smaller installations are more relevant in terms of emissions, it is important to have reasonable emission figures available for a specific Austrian study (Spitzer et al. 1998).

Within GENEMIS, single particle analysis of the emissions from a wood fired power plant has been performed using electron probe microanalysis (Török et al., see Sect. 2.3.4). In the power plant stack, after removal of all large particles in a cyclone, typical particle sizes were 1 - 3µm. Five classes of particles were identified, four of them dominated by potassium, the fifth still high in potassium content. Potassium was present mostly in the form of carbonate, but also as chloride. It is characteristic for particles from wood fire.

2.3.1.3 PM emissions from car exhaust

As much as PM from car exhaust is a known issue, data are also available. Exhaust emissions, especially of soot, from diesel engines are known and are now also being regulated. The side effect of such a regulation may be that, while emitted particle mass is reduced, particle number considerably increases.

Such developments have to be closely observed. Thus the contribution by Sturm and Hausberger (Sect. 2.3.3) on the size distribution of exhaust PM emissions presents an excellent report of the status-quo and will become even more relevant in the future, should particle number become considered more relevant than particle mass.

2.3.1.4 PM emissions from non-exhaust emissions of traffic

The issue of non-exhaust emissions from traffic is a hotly debated question, at least as long as course mode aerosol has to be considered health relevant. At this time it is not fully clear which compounds, what magnitude and under which conditions material is moved into the atmosphere by moving cars. Some information, albeit old, exists on tyre and brake wear. But for the road abrasion / suspension emissions only one major set of data is available, put forward by the U.S. EPA (1995). The parameterisation of these measurements has been strongly put in question recently (Venkatram 2000), but it is basically kept up in the most recent version of the EPA guidelines dated October 2002 (see Chap. 13.2.1 at http://www.epa.gov/ttn/chief/ap42/index.html).

Thus measurement data are urgently required, and GENEMIS is able to provide results. Sjödin (Sect. 2.3.2) presents such measurement data and derives composite emission factors (in mg/km travelled). He concludes that the highest emission factors available derive from spring measurements, when road abrasion is high due to the strong use of studded tyres in Sweden. In

Table 2.16 the results given by Sjödin are being compared with the figures Winiwarter (Sect. 2.3.5) applies for Austria (directly derived from the EPA algorithm, using input data from Austria). The latter figures are in all cases consistently higher, even if in Austria typically only little use of studded tyres is made, and this is also not part of the EPA algorithm. The quite variable emission factors from Austria derive first of all from the different "silt load" of Austrian roads, but also of the fleet weight which changed over the years. It needs to be added that the author himself considers this estimation to be clearly too high, but he insists that the value has to be kept for a profound uncertainty assessment.

Pregger and Friedrich (Sect. 2.3.6) take a somewhat more moderate approach: The numbers given in the EPA algorithm were alleviated by excluding the results from extremely high road pollution (silt load). This has first been suggested by Venkatram (2000), even if the intention of this author was rather to put the results in question than to create an improved approach. Consequently the emission factors for Germany are in the same area as the high values given for Swedish tunnel measurements (also given in

Table 2.16).

It has been questioned how far tunnel measurements can reflect the free-flow situation. The entrainment of material onto a tunnel surface certainly is limited to the road abrasion as such. While the abrased material will stay with the road surface even if present in much larger particles, where it may be grinded to airborne matter subsequently, only little other material will become available for suspension. Therefore measurement along open roads have also been performed, even if this requires more elaborate measurement setup and the air flow is not as controlled as in tunnel experiments. A first compilation of such experiments indicates an overestimation of the real PM_{10} emissions by the EPA approach by about a factor of 2 (Gamez et al. 2001). In terms of composite emission factors, the same authors report results between 100 and 250 (and one case of 700) mg PM_{10}/km travelled (Düring et al. 2002) and agree reasonably well with the emission factors used for the German inventory (Pregger and Friedrich, Sect. 2.3.6).

Table 2.16. Comparison of composite emission factors derived in GENEMIS (Sjödin, Sect. 2.3.2) vs. those used for GENEMIS country inventories – all in mg/km. Ranges are given for deviating sets of data and do not express uncertainty, which is considerably larger.

	TSP	PM_{10}	$PM_{2.5}$
Tunnel measurements – Fall	96	44 – 54	12
Tunnel measurements – Spring	520	290	79
Austria – Hwy/rural*	1200 – 2500	230 – 470	60 – 110
Austria – Urban*	5500 – 7700	1100 – 1500	250 – 350
Germany – Hwy/class 1 road**	340 – 360	140 – 150	70
Germany – urban**	550	230	80 – 90
Germany – other roads**	400 – 430	170 – 180	110

* Derived by EPA equation. Road classes differ by their silt load. The ranges express the shift in average fleet weight over the years and between hwy and rural, respectively.
** German composite emission factors have been made available by T. Pregger.

2.3.1.5 PM from other fugitive sources

The process of fugitive emission requires a motion that elevates particles into the air. For non-exhaust road emissions this is the motion of the vehicles, for bulk material and handling it is the air motion caused by the handling operation itself. In these cases the flow is relatively well defined and can be assessed, even if – looking at the industry sector – information on the procedures as such is sparse within industry. For construction (including demolition) of buildings, waste dumps, and especially for wind blown dust (e.g. from uncovered agricultural soil) the flow is much less defined and primarily wind dependent. For these source types it is very difficult to obtain reliable emission factors, even more so to estimate emissions. At this time results refer to an order of magnitude rather than to a specific number. It will be important to confirm the current assumptions, that these sources do not contribute a major fraction to total PM emissions.

2.3.1.6 Country emission inventories

Within GENEMIS two country inventories were established, for Germany and for Austria. A comparison of these two inventories to the national total given in international studies (excluding the larger part of the non-exhaust road emissions) resulted in both cases in very good agreement for total TSP, PM_{10} and $PM_{2.5}$. The situation was quite different however when detailed source groups are looked at. The largest difference between Austria and Germany obviously (see above) was the different treatment of non-exhaust traffic emissions. But approaches differ somewhat for a number of sectors, with the Austrian approach concentrating on measurements and self-reporting of industry, and for Germany using a structural approach on a process basis. A detailed assessment of the methodological differences would need to apply both to one country and therefore a more thorough comparison is beyond the scope of this paper.

2.3.1.7 Outlook

In a project like GENEMIS, results are directed towards questions asked previously. The outcome successfully shows that the involved scientists have been very fast to answer the upcoming issue of particulate matter. Still there are some important open questions, which only start to be filled. The most important of these seems to be the non-exhaust road emissions, where useful results just start to come in. Here a coordinated program on assessing real-world emissions under a variety of environmental conditions would be most desirable.

At this time, the questions are posed towards emitted mass or number concentrations. In future, issues such as chemical composition, specifically albedo, volatility, or even toxicity and mutagenicity of compounds may be raised. As the efforts already taken on single particle analysis as well as on particle number counts prove, it is likewise that answers can be given as soon as the health relevant parameters of PM have been adequately identified.

GENEMIS has helped to establish a firm basis for our understanding of PM emissions. While it was able to contribute to quantify emissions like non-exhaust car emissions, much still has to be done in order to understand the emitting processes. Only if the pathway leading to an emission is understood, efficient reduction measures can be established.

2.3.2 PM emission factors for road vehicles from tunnel measurements

Å. Sjödin

2.3.2.1 Introduction

Road traffic is a major source for particulate matter (PM), especially in urban areas. The source receptor relationship is very complex, since PM from road vehicles is not only attributed to exhaust emissions, but also to road and tyre wear and wear of brake linings as well as resuspension of road dust. Therefore it is very difficult to assess and analyse this problem, based only on conventional emission measurements in chassis dynamometer facilities. An alternative approach is offered by road tunnel measurements which will yield the sum of all the various emission sources related to road traffic. By measuring the difference Δc_i between the tunnel outlet and inlet air concentrations of compound i, the traffic flow T and the tunnel ventilation flux V, and by knowing the distance L between the tunnel inlet and outlet, composite emission factors e_i in mass units per unit length travelled can be derived from a mass balance equation:

$$e_i = m_i / (T * L) = (\Delta c_i * V) / (T * L) \qquad (2.6)$$

Vehicle specific emission factors (for light-duty vehicles LDV and heavy-duty vehicles HDV) can be derived from a linear regression analysis if the fraction of heavy-duty vehicles a_{HDV} is known:

$$e_i = e_{i,LDV} + a_{HDV} * e_{i,HDV} \qquad (2.7)$$

As part of the GENEMIS project real-world PM emission factors were derived from measurements in three urban tunnels, reflecting different traffic conditions and grades, in 1999 and 2000 in Gothenburg, Sweden. The main characteristics of each of the tunnels are given by Table 2.17. Each tunnel consists of two bores, each bore carrying traffic in one direction only. All tunnels are self-ventilated by the movement of traffic (piston effect).

Table 2.17. Characteristics of the two tunnels included in this study. Data on speed and %HDV are from actual observations during each study.

Name of tunnel	Length (m)	Lanes per bore	Average grade inlet-outlet (geometry)	Average speed (km/h)	Average %HDV (range)
Gnistäng tunnel	700	2	0.8% (flat or up)	69±4	11(2-17)
Tingstad tunnel	450	3	-0.2% (down 3%, up 3%)	62±11	12(8-18)
Lundby tunnel	2060	2	0.6% (down 3%, flat, up 3.5%)	78±2	15(7-31)

2.3.2.2 Composite PM emission factors from tunnel measurements

Composite emission factors for TSP, PM_{10}, $PM_{2.5}$ and PM_1 derived from the measurements in the three tunnels are presented in
Table 2.18. As seen there are large differences between the tunnels, especially for the Lundby tunnel compared with the two other tunnels, the former exhibiting much higher values. Analysis of the content of various metals in the particulate samples from the tunnels showed that the observed differences can mainly be explained by a larger resuspension of road dust in the Lundby tunnel compared to the two other tunnels (Sternbeck et al. 2002). The reason for the much higher degree of resuspension in the Lundby tunnel may be explained by the fact that these measurements were carried out in the early spring, when the amount of dust on the road surface reaches its maximum, at the same time when a large share of the cars are still using studded tyres, promoting resuspension even further. On the contrary the measurements in the two other tunnels were carried out during fall when studded tyres are not used and the road surface is much cleaner with regard to dust.

Table 2.18. Composite emission factors for various fractions of particulate matter in mg/km derived from the tunnel measurements.

Name of tunnel	TSP	PM_{10}	$PM_{2.5}$	PM_1
Gnistäng tunnel	-	54±14	-	-
Tingstad tunnel	96±58	44±33	12±11	-
Lundby tunnel	518±324	285±115	79±77	91±35

The values in Table 2.18. can be compared with results from other tunnel studies in North America and Europe during the 90's, ranging from 30 to 60 mg/km for PM_{10} and 10 to 50 mg/km for $PM_{2.5}$ (Gertler et al. 1997; Rogak 1996; Weingartner et al. 1997).

2.3.2.3 Vehicle specific emission factors

Linear regression analysis was carried out for the measurements in the Gnistäng and Lundby tunnel, in order to derive vehicle specific emission factors for the various PM fractions. The relationships between the measured composite emission factors and the fraction of heavy duty vehicles in the tunnels are displayed in Fig. 2.34.

One of the measurements in the Lundby tunnel was carried out during a heavy rainfall ("Lundby wet"), resulting in a wet road surface and likely less resuspension. It can be seen from the fact that in particular the PM_{10} emission factor decreases drastically compared to the emission factors derived during the dry road surface conditions ("Lundby dry"), whereas the PM_1 emission factors are not affected by the wet road surface, since it relates mainly to the exhaust particles.

The inserted regression lines in Fig. 2.34 represent the measurements "PM_{10} Lundby dry", "PM_1 Lundby dry" and "PM_{10} Gnistäng". Assuming that the PM_1 fraction in the Lundby tunnel and the PM_{10} fraction in the Gnistäng tunnel represent merely exhaust particles, then the linear relationships according to Fig. 2.34 yield a PM_1 emission factor of 20 mg/km and a PM_{10} emission factor of 26 mg/km for light-duty vehicles, and a PM_1 emission factor of 320 mg/km and a PM_{10} emission factor of 690 mg/km for heavy duty vehicles. The large difference between the emission factors for heavy duty vehicles in the two tunnels may be explained by the differences in both traffic speed and road grade between the two tunnels.

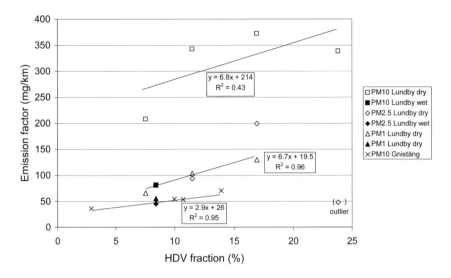

Fig. 2.34. Measured composite emission factors PM$_{10}$, PM$_{2.5}$ and PM$_1$ as a function of the fraction of heavy duty vehicles in two of the tunnels.

2.3.3 PM Size distribution measurements in engine exhaust in context to road tunnel and ambient air measurements

P.J. Sturm, S. Hausberger

2.3.3.1 Background

An increasing number of studies indicate that particulate matter air pollution can have a severe effect on human health. These findings were followed by a discussion on introducing standards for size specific particulate mass emissions. Especially particulate matter smaller than 10 μm (PM$_{10}$ – coarse particles) or less than 2.5 μm (PM$_{2.5}$ – fine particles) has to be considered. There are strong indications that the number of particles may be more relevant for human health effects than the particle mass, as the human immune system is not well adapted to particles with diameters in the nanometre range. In ambient air the number of particles increases with lower diameter and one peak occurs in the range of 10 to 100 nm. However, these particles contribute very little to the total suspended mass and thus are not taken into account by emission standards related to the particle mass only, even though they may have severe health effects.

Air quality standards limit either TSP, PM$_{10}$ or PM$_{2.5}$ (mass fraction of PM sizes below 2.5 μm). Recent epidemiological studies have indicated that the adverse health effects may not be only dependent on total PM mass. Fine and ultrafine PM can cause more adverse effects than large particles. In recent years it has been recognized that a large portion of PM emitted from road traffic lie in the ultrafine region which is defined with a cut-off of 100 nm or 0,1 μm ($PM_{0.1}$).

The understanding of the conversion process of particles in the atmosphere is a fundamental pre-requisite for defining proper methods for measurements on test beds and to set corresponding emission limits to minimize adverse health effects.

The following section will concentrate on PM emissions from road vehicles. Although the technology of measuring ultra fine particles was improved within the last few years, there is very little data available on the production rate, airborne concentrations and the time development of the particle size distribution. On top of that there is up to now no standardized method for PM size distribution measurements for exhaust emissions. It is well known that the sampling conditions influence strongly the shape of the size distribution and the existence of nucleation mode PM. Hence, it is always necessary to know the conditions under which measurements were performed, before using such data in a general way.

2.3.3.2 PM size distribution of vehicle emissions

Vehicle emissions may contribute significantly to PM number concentrations in urban areas. In 1998, the European Council of Environment Ministers reacted a political agreement which resulted in the final EURO III standard and adopted the so called EURO IV and EURO V standard for the years 2005/2008. These standards foresee strong emission reductions for gaseous pollutants as well as for PM.

Despite the fact, that the emission per vehicle will decrease, the number of vehicles on the roads still increases in total. In addition diesel fuelled PC became very popular. As a follow up of such facts the emission situation may not improve as strong as it should. Of course improvements will be visible on a per mass bases. If improvements can be made on a "number" bases regarding ultrafine particles is at least questionable.

But the main problem is not whether there is a reduction or not. The main problem is still a more fundamental one, it is the particle formation and structure problem. Diesel exhaust PM are usually agglomerated. Their core is from solid carbonaceous material or ash covered by a larger of volatile organic and sulphur compounds (Steiner and Burtscher 1993).

PM undergoes various phases when being emitted and released into the atmosphere (i.e. from the engine into the atmosphere).
- Formation of soot and VOC during combustion processes
- Exhaust dilution processes as the gas exits into the atmosphere.

During the formation, PM undergoes nucleation, surface growth and agglomeration. The size range in diluted exhaust is between 50 and 300 nm with a peak around 80 to 100 nm. Dilution takes place in atmosphere as soon as the tailpipe emission is mixed with ambient air. During this process a formation of new particles has been observed. In nucleation, number count increases rapidly and the number of nucleated particles is usually much bigger than that of aggregated ones. In addition, the new particles have a much smaller size (10 –

50 nm) than the aggregated ones. The onset of nucleation is very sensitive to dilution ratios (Abdul-Khalek and Kittelson 1998; Kittelson and Abdul-Khalek 1999). During standardized emission tests (CVS-tests) the nucleation mode often does not exist.

2.3.3.3 Measurements of vehicle and engine emissions

There are several research projects ongoing aiming at unifying measurement methods for PM size distribution measurements of diesel vehicles. One of them is the PARTICULATES project (http://vergina.eng.auth.gr/mech/lat/particulates/). During this project a harmonized method was developed and applied to chassis dynamometer measurements for diesel passenger cars (PC) and heavy duty vehicles (HDV). In addition, HDV engines have been tested on test beds. The test procedure requires a constant dilution ratio in the range of 1:15 and measurements of "wet" and "dry" (using a thermo-denuder) PM.

If one takes some examples from HDV engines measured, size distributions similar to those shown in Fig. 2.35 can be expected. As Fig. 2.35 shows, the size distribution of a modern HDV EURO III engine strongly depends on the engine load. The different curves represent different engine loads. During idling (0% load) and low load (10%) conditions nucleation in the range of 5 to 8 nm occurs. But as soon as the engine load rises the soot mode (80 nm range) becomes dominant.

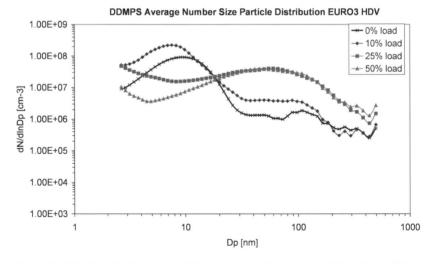

Fig. 2.35. PM size distribution at different engine loads done with a dual differential mobility particle sizer (DDMPS) under stationary conditions.

As already mentioned above, the measured size distribution is very sensitive to the dilution ratio as the primary dilution may already freeze the conditions for the formation of very small PM and only coagulation and agglomeration can contribute to further changes in the distribution shape. The possibilities of varying the dilution ratio on test facilities is limited and never reaches situations which occur in free atmosphere.

2.3.3.4. Road tunnel measurements

The PM size distribution in road tunnels shall of course reflect the emission situation of the vehicles, mainly diesel vehicles and there especially the HDV's. As already mentioned above, particle formation is a function of engine load, engine technology, fuel quality and unfortunately also of sampling conditions. In order to do investigations including bigger dilution ratios and real world vehicle operation conditions, measurements in road tunnels were performed. Road tunnels are more or less big laboratories, as the boundary conditions such as emission rate, air flow (and hence the dilution rate) can be described with sufficient accuracy. Nevertheless the environment is still not the same as in open space. Such measurements were already performed in different street tunnels (e.g. Sturm 2002; Gertler et al. 2001; John et al. 1999). One thing which changes frequently in street tunnels is the traffic volume, and hence the emission quantity. In general the fresh air volume in the tunnel varies with traffic flow, but only indirectly as the decisive parameters are either CO concentration inside the tunnel or visibility. Visibility corresponds of course to PM emissions in general, but different sizes have different light extinction characteristics.

Fig. 2.36 shows the PM size distribution as function of traffic volume (here expressed as daytime). The passenger car traffic has a pronounced morning and evening peak, whereas HDV traffic starts in the morning and remains relatively constant throughout the day (Sturm 2002). This is reflected in the size distribution. There are almost two different characteristics visible. During daytime the fraction in the 60 to 100 nm region dominates, whereas during nighttime the nucleation part (10-20 nm) is most pronounced. This can be explained with the high HDV share during the day. In addition, the ventilation rate (fresh air inflow) is at its maximum during daytime, but due to high emission quantities the dilution ratio is lower than during night time. Hence more aerosol surface is available which promotes coagulation and agglomeration.

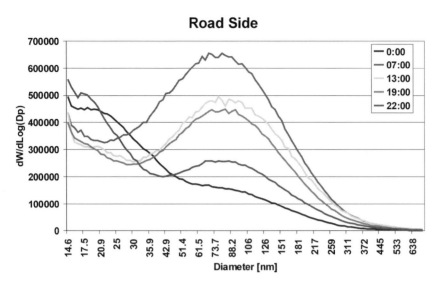

Fig. 2.36. PM size distribution in a street tunnel at curbside as a function of daytime; diameter in [nm], concentrations in [#/cm³] (Sturm 2002).

2.3.3.5 Measurements in urban atmosphere

If measurements are undertaken in (urban) ambient air many influences occur on PM size distributions. Traffic is a major source for PM_{10} emissions, nevertheless it is not the only PM source. In addition background concentrations with "aged PM" contribute to the measured values. Otherwise they reflect real world conditions as emissions from various sources are emitted under high dilution ratios. When traffic is concerned, the size distribution shows a strong dependency on the distance from the emitter, i.e. the road. Fig. 2.37 shows curbside measurements at a crossroad with some 45,000 vehicles passing per day. The size distribution contains nucleation PM as well as very small PM peaking at 30 nm. Again this behaviour is time dependent. During the morning peak hours (6:00 and 7:00) the highest concentrations appear, while during the remaining time the distribution is almost uniform. But as soon as the distance and thus the travel time of PM grow, the nucleation PM disappear and only the 30 nm peak prevails (Fig. 2.38). The soot mode particles (80 nm range) do not appear at all, neither in the curbside measurements nor in the urban background.

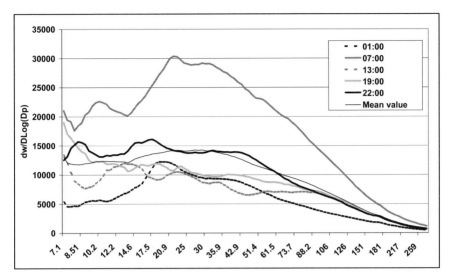

Fig. 2.37. PM size distribution at an intersection, curbside measurement; diameter in [nm], concentrations in [#/cm³].

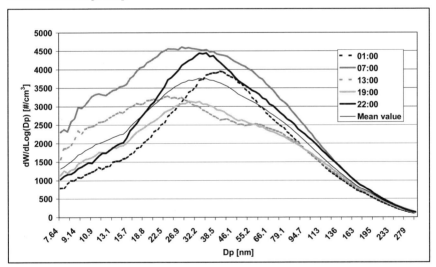

Fig. 2.38. PM size distribution at urban background; diameter in [nm], concentrations in [#/cm³].

2.3.3.6 Conclusions

The size distribution of road traffic related PM depends on a lot of parameters. The investigations done so far can be summarized as follows:

• The engine measurements show clear dependencies of the PM size distribution on engine load and sampling conditions (dilution ratio).

- The curbside measurements in the tunnel show a distribution containing nucleation mode particles as well as soot mode particles. This distribution is similar to the engine measurements.
- The curbside measurements at an intersection show also a bimodal distribution having nucleation mode PM as well as a second fraction peaking at 30 to 40 nm. The pure soot mode (80 nm range) is not visible anymore. Measurements in the urban background do not show nucleation at all. The peak of the size distribution is in the 30 nm region.

It seems that the nucleation mode PM can be observed at roadside locations. Differences in the size distribution at the engine and in the tunnel compared to the distribution in urban atmosphere suggest completely different reaction processes, and make it difficult to utilize PM size distribution data from engine test beds or chassis dynamometers for considerations in ambient air.

2.3.4 Atmospheric emission from wood combustion

S. Török, J. Osán, G. Falkenberg

2.3.4.1 Introduction

Wood combustion particles have significant impact on human health worldwide. Their effect is different in industrialised countries and third-world areas. In developing countries, wood is extensively used in households for cooking or space heating (Dufournaud et al. 1994). The open fire typically used for these activities exposes around 2 billion people to indoor air pollution. The amount of the suspended particulate matter can be $10 - 20$ times larger than in outdoor air (Kaarakka et al. 1989). In developing countries, the most important problem related to wood combustion is the indoor air quality. In Europe, however, wood and biomass burning is regarded as a renewable source of energy (Bhattacharya 1998). The contribution of renewable energy sources to the total energy budget is currently rather low (around 5%), and the European Commission plans to double this contribution by 2010. Biomass combustion in recently developed boilers is continuously increasing, especially in those countries that have huge forests or agricultural areas but no coal resources (e.g. Sweden) (Palola 1991).

The most important environmental problem connected with wood or biomass burning is the high probability of incomplete combustion, producing particulate matter and hazardous organic compounds (Chagger et al. 1998). The majority of the operating wood and biomass combustion boilers installed in the past decade(s) are not equipped with any flue gas filtering system, releasing all of the produced particles into the air. The public living in the vicinity of such plants is very much concerned about the visible particulate emission. At other utilities, which have a simple fly-ash collection device like a cyclone, the collected toxic material is often handled improperly. The resuspension of the deposited particles can also very significantly affect air quality. The environmental impact of wood combustion is most important in small settlements where central heating is scarce and natural gas is not commonly used (Osán et al. 2002). The most severe air quality problems occur during winter periods, when inversion is typical or the boundary layer is at low elevation. It is of great importance to investigate the physical and chemical properties of combustion residues in order to improve combustion technology and to assess the effects that such residues may have on the environment. Due to the increasing use of biomass combustion for energy generation in the past decade, several researchers have attempted to characterise particulate emissions, such as inorganic and organic components (Someshwar 1996). Most of these studies have aimed to obtain information on the average chemical characteristics of wood combustion particles, and used bulk analytical techniques (Sheffield et al. 1994). Very few techniques are capable of relating the chemical composition to the aerodynamic behaviour (size) of single particles.

Electron probe microanalysis (EPMA) method is capable of providing detailed morphological information and semi-quantitative composition data for each measured individual particle in a non-destructive way. The technique is even capable of detecting inhomogeneities within individual particles. Another microchemical technique is the non-destructive microscopic X-ray absorption near edge structure (μ-XANES) spectrometry that allows even the speciation of selected elements from a microscopic volume of a single fly ash particle (Osán et al. 1997; Huggins et al. 2000). In other words it means that the chemical state of the element can be identified that can be directly related to the leachability and toxicity. This chapter aims to show that these techniques deliver unique and relevant information on the emission of wood combustion particles by determining the major chemical components and speciation of selected metals in individual particles.

2.3.4.2 Materials and methods

Samples

The particulate and gaseous emissions of a 400 kW heating boiler for an industrial facility processing wood, at Sződliget (a small town at 30 km North of Budapest), were measured during standard winter operation. The utility burned the chippings residue of the on-site wood processing. Below the gas flow duct, a cyclone collected mostly large particles. The sampling of flue gas particles was isokinetic in the 1 m diameter vertical stack at a flow rate of 7 m s^{-1}. The particles were collected at 5 m height; the stack height of the heating plant was 10 m. For EPMA, one part of the Nuclepore filter containing stack gas or atmospheric aerosol particles was glued to a Cu-Zn sample holder and coated by a 25 nm carbon layer for electron microprobe analyses. Around 300 particles were measured in each sample. No special sample treatment was necessary for the μ-XANES measurements.

EPMA measurements

The analyses of particles collected on Nuclepore filter were performed using a PHILIPS 505 scanning electron microscope (SEM) equipped with a Link EDX detector. The detector has a 7.62 µm thick beryllium window, and its energy resolution is 150 eV at 5.9 keV. The EPMA measurements were carried out at an accelerating voltage of 20 kV, and a beam current of 1 nA. The single particles were measured automatically, controlled by a home-made computer software, only scanning the electron beam over the whole projected area of the particles. Around 300 particles were measured in each sample. Morphological parameters such as diameter and shape factor were calculated using the image processing routine of the measuring program. The method commonly used for classification of the particles is hierarchical cluster analysis (HCA). The particles to be classified can be regarded as points in a multi-dimensional space defined by the chemical composition. At each step of the procedure, a new cluster is formed from the two most similar particles or clusters. The measure of similarity of the points is in most of the cases their Euclidean distance. The strategy of the clustering is based on the calculation of the distance of two clusters. The particles were classified using the normalised X-ray intensities, as it was shown that the groups obtained are the same whether the HCA is based on the intensity or the concentration data obtained by conventional EPMA (Bernard et al. 1986). The average intensities and particle diameters were calculated for each particle group. The average intensities of each group were converted to concentration values using the MULTI software (Trincavelli and Castellano 1999) employing a standardless peak-to-background method (Trincavelli and Van Grieken 1994).

Micro-XANES measurements

The experiments were performed at the micro-fluorescence beamline L at the HASYLAB synchrotron facility (Hamburg, Germany). The white beam of a bending magnet was monochromatised by a Si(111) double monochromator. A focussed X-ray beam of 15 µm diameter was used for the measurements. The absorption spectra were recorded in fluorescent mode, tuning the excitation energy near the K absorption edge of Cr, Cu and Zn (5989, 8980 and 9659 eV) by stepping the Si(111) monochromator. The fluorescence yield was detected at an angle of 90° to the incoming beam using an energy-dispersive Si(Li) detector. Individual wood fly ash particles of 20 – 40 µm diameter were selected for XANES measurements. The evaluation of the Zn XANES spectra was performed by linear combination of standard spectra. Compounds generally present in environmental samples were selected as standards.

2.3.4.3 Results

Particulate and exhaust gas emissions were monitored. Analysis of CO, O_2, CO_2, NO_X and total hydrocarbon (THC) was carried out by a mobile laboratory. Since the generated heat is used for space heating the plant is regulated by a thermostat.

The temporal resolution of the emitted gaseous compounds is presented in Fig. 2.39. The measured CO concentration was over 600 mg/Nm^3 in the peaks.

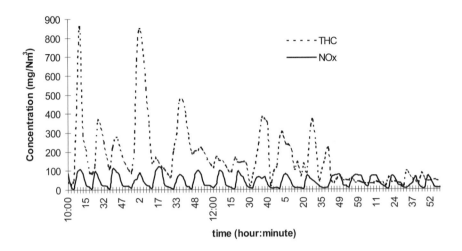

Fig. 2.39. Temporal resolution of THC and NO$_x$ concentration in exhaust gas of wood combustion plant during a regular winter day.

Though only conventional EPMA was possible for stack gas particles and hence no low-Z elements could be determined, the Bremsstrahlung background of the X-ray spectra of most particles indicated an organic matrix, as a result of incomplete burning. The majority of the particles emitted from the stack were in the respirable size range. Particle classes obtained for the stack-gas sample are presented in Table 2.19. The concentration values were calculated only for the inorganic composition, C and O concentrations were obtained from stoichiometry. Potassium compounds are the dominant constituents in all classes. The two most abundant classes (29 and 24%) are composed mostly of potassium carbonate and potassium sulphate. Classes 3 and 5 consist of particles internally mixed of calcium and potassium compounds, mostly carbonates and sulphates. Class 4 obviously contains KCl-rich particles with 1.2 μm average diameter. Fine-mode ($d < 2.5$ μm) KCl particles were found to be characteristic for biomass burning (Liu et al. 2000) and wood combustion (Valmari et al. 1999). These particles were most probably formed by condensation of alkali chloride vapours present in the stack gas during quenching.

In order to estimate the environmental hazard of the heavy metals contained in the particles, speciation analysis of Cr, Cu and Zn was performed on characteristic single wood fly ash particles using μ-XANES. The concentration of Cr and Cu was in the 100 μg/g range. Fig. 2.40 shows typical spectra collected at the K absorption edge of the three elements of interest. The zero of the energy scale is set to the absorption edge energy for each element. The absence of the pre-edge peak characteristic for the VI oxidation state in the Cr XANES spectrum indicates that chromium is present mostly (> 95%) in the less toxic Cr(III) form. Cu is present mostly as Cu(II), but the presence of Cu(I) is not negligible due to the shoulder appearing at 0 eV. The evaluation of the Zn K-edge XANES spectrum resulted in that more than 60% of Zn is present in the particle as Zn silicate, around 20% as ZnS and 20% as the most mobile (Fe, Zn) oxide form.

Table 2.19. Average composition and diameter for the major particle classes obtained for the stack gas fly ash collected at Sződliget.

Abundance (%)	29	24	21	15	12
Diameter (µm)	0.8	2.9	2.7	1.2	4.1
Concentration (wt%)					
Al	1.2	0.2	0.8	0.3	0.0
Si	6.7	1.4	3.2	2.3	2.6
P	0.5	0.1	0.5	0.0	1.5
S	3.9	11.4	8.3	3.8	1.8
Cl	0.8	0.3	1.0	15.5	0.9
K	31.9	34.3	20.5	34.3	10.1
Ca	2.1	1.7	11.0	1.7	25.1
Zn	7.4	1.9	2.8	3.3	2.7

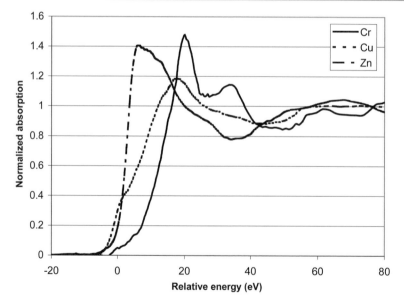

Fig. 2.40. Micro-XANES spectra of an individual wood fly ash particle collected at the K absorption edges of Cr, Cu and Zn. The zero of the energy scale is set to the edge energy for each element.

2.3.4.4 Conclusions

Since space heating small boilers are regulated by thermostat and are not operated continuously, the pulsed emission of plume is of great concern for the public in dwelling areas. The results showed the typical particle classes in the wood combustion flue gas, namely potassium, calcium and zinc rich particles. The ash also contained a considerable amount of unburned carbon.

The chemical state analysis, μ-XANES results indicate that less than 20% of the heavy metal content of the wood combustion particles is related to mobile compounds. It means that after deposition to the ground their metal content will be leached in a short period of time. Since the majority of the particles emitted from the wood burning plant was found to be in the respirable size range, they can have significant effect to the human health. Therefore the extension of the studies by trace element and organic analysis combined with modelling the deposition in the human lung provides useful information to toxicological research.

2.3.4.4 Acknowledgements

The present research was partially supported by the National Science Foundation (OTKA) through Contract No. T034195 and by the IHP-Contract HPRI-CT-1999-00040 of the European Commission. The János Bolyai Research Fellowship for J. Osán is also appreciated.

2.3.5 A national emission inventory of Particulate Matter (TSP, PM_{10}, $PM_{2.5}$)

W. Winiwarter

2.3.5.1 Summary

Emissions of total suspended particles (TSP) have been assessed from all anthropogenic sources for Austria. National statistics have been combined with emission factors, if possible from national origin, to yield annual emissions for three years (1990, 1995, 1999). Both combustion emissions and fugitive emissions were assessed. Emissions of the smaller fractions of particulate matter (PM_{10}, $PM_{2.5}$) were derived from TSP emissions and size distributions available in the literature for specific source groups or processes. In some cases an appropriate size distribution had to be estimated. In general, reasonable knowledge exists on emissions from combustion sources. Especially industry equipped with efficient PM abatement devices is able and willing to share information on their emissions. But also for domestic heating and car exhaust emission factors exist, even if in the latter case particles are also formed in or immediately after the exhaust pipe, causing problems in defining the appropriate transfer point and in assessing the dependence of emissions on environmental conditions. A totally different situation exists for fugitive emissions. Industry knows very little about their emissions, little information is available for agriculture or construction work. A well documented, but probably unreliable approach is available for suspension (or possibly just road abrasion) of particles from road traffic, published by the U.S. EPA. Using this approach, Austrian emissions amount to 77,000 t PM_{10} in 1995, 32,000 t are attributed to road resuspension alone. This number is probably an overestimation and will have to be revised. At this time it is however still supported by the only large set of measurements available. The overall uncertainty of the PM emission inventory is determined almost exclusively by the uncertainty of resuspension emissions. Additional uncertainty introduced by estimating the fractions PM_{10} or $PM_{2.5}$ within TSP hardly contributes to overall uncertainty.

2.3.5.2 Introduction

Recent studies on health effects of ambient concentrations of particulate matter (PM) caused raise of public awareness as well as found the way into legislation. Consequently, very recently international efforts to assess the emissions of PM started (Berdowski et al. 2001; Lükewille et al. 2001), PM had not been covered by international activities to assess emissions (e.g. CORINAIR), consequently little information is available on a national scale and the international studies are in wide areas the only information available.

In order to balance the situation at least for Austria, a study on the national emissions was initiated. PM emissions were assessed for different size classes, as particle size is a potentially health relevant parameter: Total suspended particles (TSP), particles with an aerodynamic diameter of less than 10 μm (PM_{10}) or 2.5 μm ($PM_{2.5}$) - see U.S. EPA (1995). Emissions were derived for the base years 1990, 1995 and 1999. Natural sources, specifically wind blown dust, were not considered in this study. Full details of the methods and all input data used, of the assumptions and the conclusions were reported by Winiwarter et al. (2001, 2002) and the references therein.

The current paper refers on the results of this study, which will also be compared to the results of the international studies mentioned above as far as relevant for Austria. A qualitative assessment of the uncertainty, followed by a concept to diminish this uncertainty, is presented in addition.

2.3.5.3 Methods to assess emissions

As PM in the very beginning had been a local air quality problem, industrial installations were equipped with abatement installations and on-line control of TSP emissions within the last decades. Emissions from large single sources have not only been reduced considerably, but also stack emissions are frequently measured or even monitored continuously for very large point sources.

TSP emissions from power plants have been assessed using specific emission information by power plant, if available. A large fraction of the power plant emissions have been covered this way. The rest used energy statistics and nationally derived emission factors. Only due to recently increased use of biogenic fuels in small plants, the fraction of power plant emissions which is directly measured, decreased. Energy statistics was also used to assess the exhaust emissions of road traffic.

For large industrial sites (steel mills, refinery, cement factories) specific company information as environmental reports in combination with specific inquiries was used. This information was in part cross-checked with ambient measurements.

Due to a high proportion of solid fuels (wood) in Austrian residential heating, this source required specific consideration. A recent study (Spitzer et al. 1998) established Austrian emission factors, but also emphasized the data uncertainty. At least this is the best background one may think of.

In contrast to the point source emissions, fugitive emissions do not have a well-defined point of entry into the atmosphere. For non-exhaust emissions from road traffic, agricultural and construction operations, calculation schemes were available from the EPA (1995). Bulk material handling and landfill operations were calculated according to the German VDI (1999). All these procedures were tested previously for applicability in Austria. Still the quality of the emission factors and calculation procedures was quite poor, but the approaches had to be used due to lack of any better information. For activity data, national statistics were applied, specifically those also used previously in the Austrian air pollution inventory, where applicable.

For all sources, both point sources and fugitive sources, emissions of smaller size fractions (PM_{10}, $PM_{2.5}$) were calculated as fractions of the TSP emissions. The respective data per source group were taken primarily from a series of German measurements on industry and power plant stacks. For fugitive sources, the split into size fractions was given as part of the respective equations in the literature cited.

2.3.5.4 Results and discussion

The resulting emissions of TSP, PM_{10} and $PM_{2.5}$ were grouped according to the CORINAIR SNAP source sectors (Fig. 2.41). For TSP, the most important sectors (between 25,000 t/yr and 10,000 t/yr) were industrial processes, agriculture and residential combustion. These emission sources are clearly exceeded however by road transport, which alone contributed 150,000 t/yr and more. For road transport as well as most of the other major source sectors, PM_{10} and $PM_{2.5}$ emissions are distinctively smaller. This indicates towards the importance of fugitive emissions (which tend to be in the coarser size spectrum) over the combustion and point source emissions. The only notable exception is residential combustion, where TSP emissions seem to consist mostly of $PM_{2.5}$. Consequently, domestic combustion becomes a more important contributor in this fraction of smaller particles. Still road traffic emissions (a large part of which derives from non-exhaust emissions) make up almost half of $PM_{2.5}$ emissions and clearly are the largest individual contribution to the total.

Within the period studied, the sum of TSP emissions as well as PM_{10} emissions show a clearly increasing trend. An increase of $PM_{2.5}$ is hardly noticeable. A detailed look reveals that most of this increase needs to be attributed to road traffic. With the high importance of traffic emissions and clearly increasing traffic (as well as vehicle weight, which is another important input parameter) this increase can be immediately explained.

There is however a serious drawback to be considered when comparing the figures. While fugitive emissions in general need to be considered rather unreliable, it is especially the non-exhaust emissions from road traffic, which have become matter of scientific dispute. In an approach to assess road dust resuspension, the U.S. EPA undertook a series of detailed measurements (see EPA 1995 and references therein). Recent literature strongly questions the approach (Venkatram 2000), not only in terms of the statistics used, but also for the measurement setup. TSP emissions from resuspension were calculated for Austria according to the EPA approach at 170,000 t, PM_{10} at 32,000 t and $PM_{2.5}$ at 8,000 t (all numbers for 1995). Thus resuspension emissions alone dominate TSP emissions, and still are at a similar magnitude as all other emissions for PM_{10}. The figures have to be put in question, however.

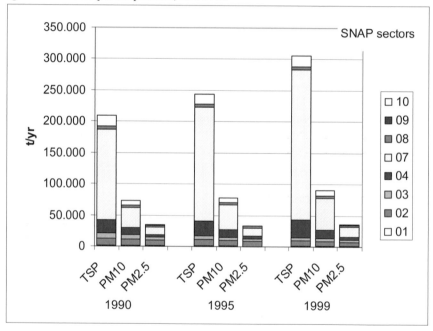

Fig. 2.41. Emissions of TSP, PM_{10} and $PM_{2.5}$ by SNAP source sectors. Presented are the results for 1990, 1995 and 1999 for the total emissions (including the full account for non-exhaust emissions from road traffic). The SNAP source sectors are: *1 - Power plants; 2 - residential combustion; 3 - industry combustion; 4 - industrial processes; 5 - fossil fuel extraction and treatment; 6 - solvents; 7 - road traffic; 8 - other traffic; 9 - waste treatment; 10 - agriculture; 11 - other sources and sinks* (not used here)

Due to the lack of knowledge, the above mentioned international studies preferred to not include the full consequence of the EPA studies. Instead, they present relatively small contributions from road abrasion, potentially kept as a wildcard for future, better data. For a comparison with the results derived in this study, the emissions from resuspension have been neglected (Fig. 2.42), but fugitive emissions from tyre and brake wear are still included.

This representation also allows a more detailed look at some of the source sectors derived within this study. Still combustion emissions only become important in the $PM_{2.5}$ fraction. Both domestic heating (most of its emissions is $PM_{2.5}$, there is only a small proportion of larger particles) and the exhaust fraction of road traffic (revealed by the same level of PM_{10} and $PM_{2.5}$ emissions, which indicates that abrasive emissions have lost their importance at this point) need to be considered. The importance of domestic heating may be attributed to a high proportion of small wood-fired heating systems.

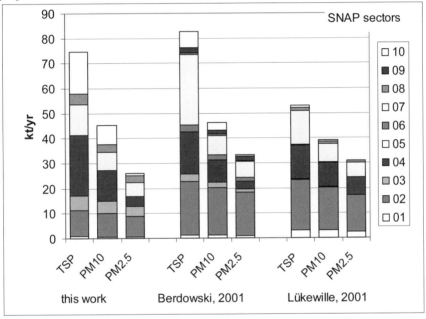

Fig. 2.42. Emissions of PM in Austria by source sector (without considering resuspension of road dust) - comparison with the respective results for Austria from all-European assessments - all for 1995.

Also in Fig. 2.42, the results from all-European assessments are presented in the identical manner. It is most interesting to note that the emission totals for all three size classes seem to agree reasonably well. Discrepancies occur for TSP between Lükewille et al. (2001) and the two other studies. A more detailed assessment reveals however that there is striking difference in the source sector detail. Especially for bulk material handling, approaches seem to be very different. While Lükewille et al. (2001) neglect most of this type of sources, it was included by Berdowski et al. (2001) as well as in this work. However, it is relatively difficult to reproduce the assumptions taken by Berdowski, as the documentation is not complete.

2.3.5.5 Uncertainty and considerations on non-exhaust road traffic emissions

The standard implicit procedure for dealing with uncertain estimates is to look into the details of a process. By separating individual contributions, it may be possible to establish also statistically independent information on the uncertainty of each of the contributions. Using any procedure to trace the uncertainty propagation will show that overall uncertainty tends to decrease when uncertain information is put together. This procedure has in the past been successfully applied also to emission information (Winiwarter and Rypdal 2001; Kühlwein and Friedrich 2000).

In this situation, however, we encounter the larger part of emissions which cannot be further separated, associated with the largest uncertainty. The non-exhaust emissions from traffic have to be considered somewhere between the mere tyre and brake wear, and the full extent according to the EPA equation. Consequently, the separation technique fails, as the total uncertainty is determined almost exclusively by this single source.

At the same time, the laws of error propagation allow conclusions on another issue. The reported range for the fraction of TSP which is PM_{10} (or $PM_{2.5}$) is relatively large for point sources. Consequently, the uncertainty is quite low. Similarly, while the uncertainty of the respective contribution for fugitive sources is much larger, also the TSP emissions from fugitive sources are very uncertain. Consequently the total uncertainty of $PM_{2.5}$ emissions is dominated by the uncertainty of TSP emissions, with the uncertainty of the respective fraction contributing to the minor part. This qualitative derivation allows concluding that the uncertainty to $PM_{2.5}$ or PM_{10} emissions is not much larger than the uncertainty of TSP emissions.

In order to reduce the uncertainty, the knowledge on the processes that lead to suspension of particles due to road traffic needs to be improved. Roadside monitoring provides evidence for the importance of this source (12 out of 15 Austrian monitors where a violation of the PM_{10} threshold was observed in 2001 can be considered roadside stations). Possible materials to be crushed by passing cars and subsequently elevated into the air are: road abrasion (both concrete and asphalt surfaces will predominantly exhibit a rocky surface, as the bitumen content in asphalt is no more than 3-7%), dust, sand, gridding, or organic litter such as leaves. It needs to be noted that most of the materials are Aluminosilicates, making it quite difficult to apply standard chemical analysis for a source differentiation. A recent German study (Düring and Lohmeyer 2001) indicates that road abrasion alone would clearly be sufficient to explain the amount of material being emitted as PM_{10}. But other potential input pathways also need to be considered.

It has been discussed that under all circumstances double-counting of PM emissions needs to be avoided, i.e. resuspension also causes a problem as the same particle may be counted several times. This is however just a matter of the interface to an atmospheric model. If this model adequately treats the deposition of this particle, the atmospheric residence time will be covered correctly even if the same particle is being emitted more than once. Most probably direct resuspension of PM plays a minor role anyway, as the amount of PM entrained from the atmosphere is expected to be fairly small.

2.3.5.6 Conclusions and recommendations

Further work is required to assess the most critical of the uncertain components of PM emissions, the non-exhaust emissions from road traffic. While the approach currently available has become the focus of scientific discussion and its results seem unrealistically high, also compared to ambient air data, the practice of ignoring this source is also far from acceptable. A complete assessment of the actual emissions from non-exhaust road traffic including suggestions for emission reductions requires a better theoretical understanding of the process leading to the emissions. This understanding consequently needs to be combined with an empirical evaluation of the emissions.

A number of European studies are under way (see e.g. Sjödin, Sect. 2.3.2) which eventually can balance the weight of the EPA data set. At this time no specific effort in coordination of the individual studies is visible yet. Knowledge on magnitude of emissions at different places, under varied circumstances is required however in order to replace the existing highly uncertain estimation procedure.

2.3.5.7 Acknowledgements

The underlying study has been performed under contract for the Austrian Federal Environment Agency (UBA), who also provided a considerable portion of the statistical data that is also used in the Austrian inventory system OLI. The contributions of Christian Trenker and Wilhelm Höflinger in quantifying fugitive emissions within the original study are gratefully acknowledged.

2.3.6 Anthropogenic Particulate Matter emissions in Germany

T. Pregger, R. Friedrich

2.3.6.1 Introduction

Identification of major emission sources is essential for clean air policy. A detailed emission inventory of primary TSP, PM_{10} and $PM_{2.5}$ has been developed to identify significant anthropogenic sources in Germany. A first compilation of available basic data and first results were published in 1999 (Dreiseidler et al. 1999). Since then, several new emission measurement projects have been undertaken in Germany and other countries so that the inventory database could be enhanced and improved continuously (see also Friedrich et al. 1999; Pregger and Friedrich 2001). The structure of the inventory and calculated emission data for 1998 are presented in this section. The results can be a basis for the evaluation of further abatement options for a future air pollution control strategy. It can also provide essential input data for atmospheric dispersion modelling and emission calculations of toxic substances by combining TSP emission data with information on the chemical composition of emitted particles.

2.3.6.2 Methodology

Sectoral emission inventories for the years 1996 and 1998 have been developed by multiplying emission factors, particle size factors and activity data (e.g. production volume, fuel input, daily traffic volume). These process specific parameters have been combined on different aggregation levels, depending on data availability. Extensive data investigations have been undertaken to reduce uncertainties and to get a sound database of typical process specific emission factors. Nevertheless, available and used emission factors may not be representative of processes in Germany. This is the main cause for uncertainties and has to be taken into consideration when interpreting the resulting emission data.

Due to incomplete information not all anthropogenic emission sources could be covered. Emissions from several fugitive sources especially in industry cannot yet be quantified with a sound methodology. However, main processes emitting fine particles could be taken into account. For road dust emissions, only very uncertain estimations can be done at present. An empiric equation was used for this purpose, developed by the U.S. EPA (EPA 1997). This equation has been modified for normal road conditions without dirt on the road using results from Venkatram (2000). The equation has been adjusted to German conditions by determined values for road surface silt loading and average weight of vehicle fleet for different road categories. A complete documentation of the methodology and data sources for the inventory will be published soon (Pregger 2003). Fig. 2.43 shows all sectors taken into account and the structure and aggregation level of the basic data used.

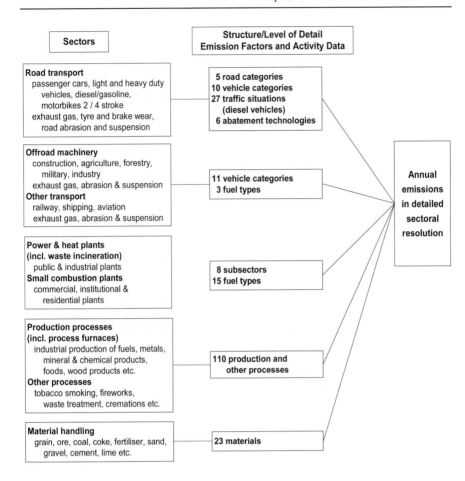

Fig. 2.43. Sectors and data structure for the emission calculation.

2.3.6.3 Results

Fig. 2.44 presents total sums and aggregated results sorted by $PM_{2.5}$ emissions. Predominant contributors to $PM_{2.5}$ are transport, production processes and power and heat plants. Road abrasion and suspension by road transport, offroad-machinery and other transport have not been included in this Figure.

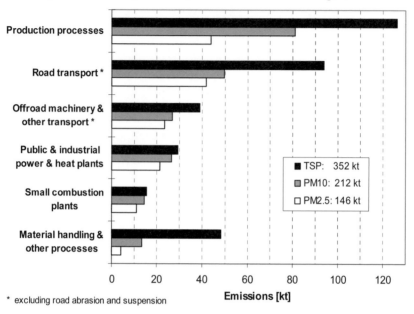

* excluding road abrasion and suspension

Fig. 2.44. PM emissions in Germany 1998 - all sectors.

Fig. 2.45 shows more detailed emission data for transport processes. PM_{10} and $PM_{2.5}$ emissions are mainly attributed to diesel vehicles on the road. About 70% of these emissions originate from heavy-duty vehicles. Brake and particularly tyre wear cause as well a large contribution to TSP emissions. Since tyre wear generates mostly coarse particle emissions, this process does not result in high emissions of PM_{10} and especially $PM_{2.5}$. Fig. 2.45 shows also estimated emissions from road abrasion and suspension. These results are quite uncertain due to insufficient measurement data. However, the results show that road dust may contribute significantly to PM_{10} and also $PM_{2.5}$ emissions in Germany. This has to be further investigated by experimental studies. Non-exhaust emissions from offroad-machinery and other transport (nonroad) are dominated by abrasion processes from railway traffic, whereas suspension causes few emissions only. Table 2.20 shows a comparison between calculated road dust PM_{10} emissions and calculations with emission factors from other studies. The results show the same order of emissions from road dust suspension.

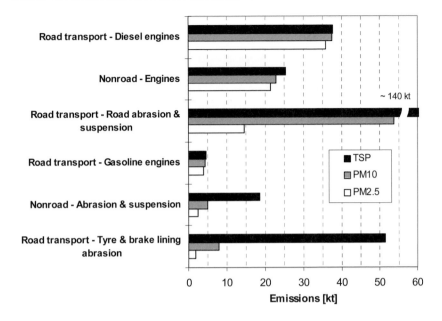

Fig. 2.45. PM emissions in Germany 1998 - transport processes.

Table 2.20. PM_{10} emissions from road dust - comparison with other studies.

Germany 1998	IER, modified EPA-equation	BUWAL 2001, emission factors	Hüglin et al. 2000, emission factors
PM_{10} [kt]	52	46	45

Fig. 2.46 shows results for combustion processes sorted by fuel type. These include power and heat plants as well as small combustion plants. Emissions are mainly attributed to solid fuel combustion. Above all, the use of lignite in power and heat plants causes significant fine particle emissions. Wood and coal combustion causes most emissions from small combustion plants. Wood combustion occurs mainly in households.

In Fig. 2.47 emissions from production processes, other processes and material handling are given. Highest emissions are attributed to production processes in metal industry like sinter, pig iron and steel production. Production of building materials like cement, glass, crushed stones, bricks and tiles causes a high emission contribution as well. Considered industrial processes often include fuel combustion in process furnaces. Fireworks and tobacco smoking are main contributors to fine particle emissions from other considered anthropogenic processes. Material handling is a major source of TSP but makes only a small contribution to fine particle emissions.

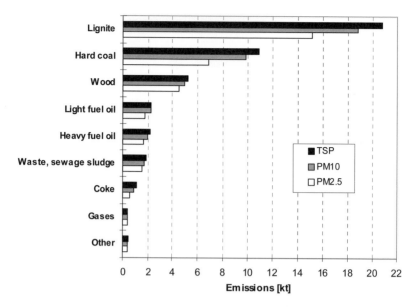

Fig. 2.46. PM emissions in Germany 1998 - combustion processes.

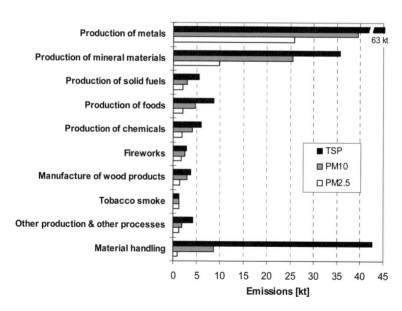

Fig. 2.47. PM emissions in Germany 1998 - production & other processes, material handling.

In Fig. 2.48 total results for the year 1996 are compared with results from other studies. Emission data for Germany were calculated within the European emission inventories from TNO and IIASA (TNO 2001; IIASA 2002) as well as by the German Federal Environmental Agency (UBA 2000). The European inventories have a simpler sectoral structure. In every study a different process inventory was taken into account and different process specific emission factors were often used. IER results on the right side include emissions due to road abrasion and suspension. Relative small differences occur for the total sums of PM_{10} and $PM_{2.5}$, especially between IER and IIASA.

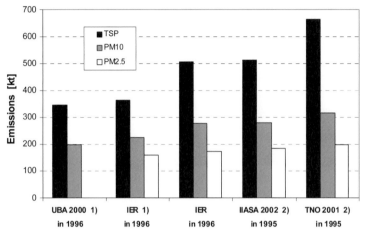

1) excluding road abrasion and suspension
2) including road abrasion, excluding suspension

Fig. 2.48. Total results for Germany in 1996 in comparison with other studies.

Fig. 2.48 shows as an example sums of PM_{10} emissions for different sectors, including road dust. This comparison shows relatively small differences (< 30%) only for tyre and brake wear and production processes. However results for road dust differs greatly because IER inventory takes suspension of road dust into account. Emission data for the other sectors have significant disagreements as well especially emissions from material handling. TNO results for combustion processes are in general much higher than calculated by IER and IIASA. A more detailed comparison of the inventories would show far higher differences for various processes.

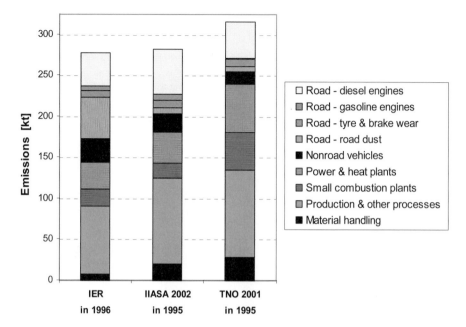

Fig. 2.49. Sectoral results for PM_{10} emissions in Germany 1996 (including road dust) compared with other studies.

2.3.6.4 Conclusions and outlook

This work presents the current state of knowledge about emissions of fine particulate matter in Germany. Major sources identified are diesel engines, solid fuel combustion and several production processes in primary industry. Diesel engines in road transport cause approx. 18% of the total PM_{10} and 25% of the total $PM_{2.5}$ emissions (excluding road dust). 16% of PM_{10} and 19% of $PM_{2.5}$ are attributed to solid fuel use in combustion plants (without process furnaces). Emissions from all production processes represent 38% of total PM_{10} and 30% of total $PM_{2.5}$. In addition, road dust may contribute significantly to PM_{10} and also $PM_{2.5}$ emissions. First estimations show PM_{10} emissions of about 25% and $PM_{2.5}$ emissions of about 10% of the total emissions from all other anthropogenic sources in Germany. The uncertainties of these estimations are very high. Further experimental studies are necessary to better assess the significance of road dust for fine particle emissions.

A comparison with emission data from other studies shows only small differences for country totals, although a detailed comparison indicates large differences for some sectors and certain processes. The reasons are different sectoral structures and the use of various emission factors that can be found in literature for many processes with a wide range of values. Further research is needed in this area to reduce uncertainties, especially concerning particle size distributions.

This sectoral emission inventory can give detailed information about sources of fine particulate matter emissions in Germany. Emission data in high spatial and temporal resolution can be generated from these results as essential input data for atmospheric dispersion modelling. Sectoral resolved data can be a basis for the evaluation of emission abatement potentials. Furthermore, combining TSP emission data with information on the chemical composition of emitted particles can provide detailed emission estimations of toxic substances.

2.4 Emissions of Ammonia

W. Asman, N. Hutchings, S. Sommer, J. Andersen, B. Münier, S. Génermont, P. Cellier

2.4.1 Introduction

Ammonia (NH_3) is mainly released into the atmosphere by man-made agricultural sources. On a global scale, about 80% originates from agricultural sources and 20% from natural sources (Bouwman et al. 1997). NH_3 is a gas and it is the only base that is released into the atmosphere in large quantities. In the atmosphere NH_3 neutralizes to a large extent the acid formed by oxidation of sulphur dioxide (SO_2) and nitrogen oxides ($NO_x = NO + NO_2$) to form particulate ammonium (NH_4^+). NH_4^+ is therefore an important constituent in atmospheric particles. Due to uptake of NH_3 and NH_4^+ containing particles by cloud- and raindrops it is also an important component in precipitation. The pH of cloud water is changed by the uptake of NH_3, which has some influence on other reactions in cloud water, such as the oxidation of dissolved SO_2 by O_3. When NH_4^+ is present in sulphate (SO_4^{2-}) containing particles it reduces water uptake and contributes in that way to global warming (Jacobson 2001). NH_3 and NH_4^+ are collectively termed NH_x.

In the atmosphere NH_3 neutralizes acids, but when the reaction product NH_4^+ is deposited and enters the soil, nitrification can occur. The atmospheric acid that was neutralized by NH_3 in the atmosphere is then released and additional acid is formed (Van Breemen et al. 1982):

$$NH_4^+ + 2\,O_2 \rightarrow 2\,H^+ + NO_3^- + H_2O \qquad (2.8)$$

As a result, NH_4^+ deposition and subsequent nitrification leads directly to soil acidification. In addition, the NO_3^- created can be leached from the soil, taking with it an anion, thereby further reducing the soil pH. Dry deposition of NH_3 to acidic soils also leads to the formation of NH_4^+ in the soil, which can be converted to NO_3^- by the above reaction, leading to soil acidification.

Excessive deposition of nitrogen compounds to ecosystems not only leads to soil acidification, but also to eutrophication. This leads to a change in plant species from N-intolerant and often rare species to those that are more N-tolerant, common on agricultural soils. Selective uptake of NH_4^+ by plant species leads to a reduced uptake of other cations, which can lead to growth disturbances. Moreover, the susceptibility of plant species to secondary stress factors (pathogens, frost and drought) may be affected by nitrogen deposition (Fangmeier et al. 1994; Bobbink et al. 1998). In many sea areas nitrogen is the limiting factor for plant and algal growth. Increase of the nitrogen loading of these sea areas can lead to excessive plant and algal growth thereby increasing the amount of organic material. When the plants and algae die they decompose in which process oxygen from the water is used. Fertilization of sea areas can therefore lead to such low oxygen concentrations in the water that fish and other organisms die.

Almost no NH_4^+ is released into the atmosphere. This means that all atmospheric NH_4^+ has been formed from NH_3. NH_3 and its reaction product NH_4^+ are not the only nitrogen compounds that cause acidification or act as fertilizers. Also deposition of nitrogen oxides (NO_x) and their reaction products (HNO_3, NO_3^- etc.) can lead to acidification and fertilization. Expressed as emissions of nitrogen, the emission of NH_3 and the emission of NO_x are about equally important in Europe (Asman et al. 1998). The same holds for deposition, because most of these compounds emitted in Europe are also deposited within Europe, but a larger fraction of the NH_3 is deposited locally than is the case for NO_x.

The European countries have decided by ratifying the Gothenburg Protocol that not only the NO_x emission in Europe should be reduced, but also the NH_3 emission. This will require a reduction of about 15% in the European Community emissions by the year 2010, as compared to the situation in 1990. However, the reduction varies between countries. For example, Denmark and the Netherlands have to reduce their NH_3 emission by 43%, Germany by 28% and the U.K. by 11%. Some countries have already reduced their emission since 1990, so that the reduction to reach in the period 2002-2010 is substantially less than for the period 1990-2010. Atmospheric transport models play an important role in developing strategies to reduce the effects of nitrogen deposition (Amann et al. 1999). These models need as an input the geographical distribution of the NH_3 emission as a function of time, usually on a regular grid.

In the next sections we will discuss NH_3 emission processes, and the progress made in their description as well as in developing such emission inventories.

Table 2.21. Global NH_3 emissions for 1990 (10^6 tons N yr^{-1}).

Source	Emission
Dairy cattle	4.3
Non-dairy cattle	8.6
Buffaloes	1.2
Camels	0.2
Horses, mules and asses	0.5
Sheep and goats	1.5
Pigs	3.4
Poultry	1.9
Subtotal domestic animals	21.6
Human and pet excrements	2.6
Synthetic fertilizers	9.0
Agricultural crops	3.6
Biomass burning in agriculture and biofuel use	2.7
Deforestation	1.4
Fossil fuel combustion	0.1
Industry	0.2
Subtotal anthropogenic emissions	41.2
Wild animals	0.1
Soils under natural vegetation	2.4
Biomass burning in natural ecosystems	1.8
Seas	8.2
Subtotal natural emissions	12.5
Total	53.7

2.4.2 Overview of ammonia emissions

Table 2.21 shows an overview of the global ammonia emissions. It can be seen that anthropogenic sources are more important than natural sources. In Europe, the most important sources are animal husbandry, use of synthetic fertilizers and agricultural crops. Natural sources are less important in Europe than at a global scale. For 1990, it was estimated that 2/3 of the emission from animal husbandry for all European countries was caused by cattle (Asman 1992). The emission from the application of fertilizers was estimated to be only 20% of the emission from animal husbandry. The emission from agricultural crops is much less.

2.4.3 Emission from animal husbandry

An overview of the emission of NH_3 from animal husbandry is given by Asman (1992) and Sommer and Hutchings (1997). Animals use only a fraction of the nitrogen given to them in feed. The remaining part is excreted. Animals excrete faeces and urine separately, with the exception of poultry, which excrete nitrogen as uric acid. The major source of NH_3 emission is NH_4^+ derived from the hydrolysis of urea in urine but some NH_3 is also released during the decomposition of organic nitrogen in solid manures.

The excretion of nitrogen by animals depends on:

- The animal feed intake (depends on the weight and whether the animal is used for milk, egg and meat production or for labour).
- The quantity and quality of the nitrogen in the feed.
- The conversion factor between N in the feed and N in the meat and in the milk (this determines the amount of N in the waste available). This depends on the weight/age and function of the animal.

Animal housing systems vary greatly. The floor can be fully slatted, partially slatted or solid. In housings with slatted floors NH_3 is lost from the floor and from the storage pit below. In these housings about equal amounts of NH_3 are lost per surface area from the floor and from the surface of the manure stored underneath (Voorburg and Kroodsma 1992). The animals can be tied or allowed to move freely, leading to variations in the area covered by urine and faeces. The ventilation can be natural (by the wind) or mechanic. Naturally ventilated housings can be open to various degrees. There are different types of manure: slurry, liquid manure and solid manure and deep litter.

The NH_3 emission from housings depends on:

- Housing system. In general the emission increases with the fraction of the surface covered by manure that is in contact with the atmosphere and the ventilation rate. The ventilation rate of naturally ventilated housings depends on the wind speed outside the buildings. The ventilation rate of mechanically ventilated housings depends on the temperature within the housing. The ventilation rate is increased with temperature until a maximum ventilation rate is obtained.
- Animal behaviour.
- Housing management e.g. frequency of cleaning of the floor.

- The pH of the manure. The pH varies the first days after production varies due to changes in the acid and base components. The emission rate increases with pH.
- Temperature of the manure. This depends partly on the temperature in the housing, which is a function of meteorological conditions and the ventilation rate.
- Losses of other nitrogen compounds (N_2, N_2O etc.), as the fraction that is lost in that way cannot be lost as NH_3.

The NH_3 emission rate from housings is higher during summertime due to the higher indoor temperature. Moreover, higher temperatures will lead to an increase in the ventilation rate of mechanically ventilated houses, which also leads to a higher emission rate.

Manure is after removal from animals houses stored either in storage tanks or in piles. In the USA a system exists where the slurry temporary stored underneath the housing is flushed into a lagoon (storage basin) with water from that lagoon. Storage tanks can be open, but in some countries they have to be covered by plastic, textile or material that forms a crust such as straw, peat or expanded clay particles to reduce the NH_3 emission. Covering with a crust not only reduces the transport of NH_3 into the air, but it may also reduce the natural mixing of the slurry in the tank, which also reduces the emission (Olesen and Sommer 1993).

The NH_3 emission during storage depends on:

- Type of storage (tanks, lagoons, piles) and the way they are covered.
- Meteorological conditions (wind speed, temperature). The emission increases with wind speed and temperature (Olesen and Sommer 1993).

After storage manure is applied to land (bare soil, soil covered with crops including grass). The emission during the application itself is of minor importance and can be neglected. This means that almost all NH_3 is emitted after application. The emission after application depends on:

- Meteorological conditions: solar radiation, temperature, turbulence (wind speed), air humidity and precipitation. The emission rate generally increases with temperature and turbulence. It decreases with air humidity, because the evaporation of water from the manure is then slowed, which leads to a lower NH_3 concentration in the manure. The emission rate and during and after precipitation periods is generally lower, because part of the nitrogen is washed into the soil.
- Irrigation. If a field is irrigated the manure is diluted and enters the soil at a higher rate. This leads to a lower emission rate.
- Properties of the soil (pH, calcium content, cation exchange capacity, water content, buffer capacity, porosity etc.) The emission rate increases with pH, calcium content and porosity but is decreases with buffer capacity and the water content.
- Properties of the manure (pH, viscosity, content of dry matter). The emission rate increases with pH, viscosity and content of dry matter. A high viscosity and dry matter content prevents the manure from entering the soil and leads therefore to a higher emission rate.

- Amount applied pr ha. The fraction of N that volatilises as NH_3 decreases with increasing amount per m^2 applied.
- The way the manure is applied (broadspread, trailing hoses, injection).
- The time between application and incorporation into the soil. The emission is generally largest during the first hours after application. Incorporation into the soil shortly after application reduces the emission considerably.

If animals are grazing, their excrement is deposited directly onto the grass. The urine and faeces are deposited more or less separately and not mixed, as in slurry. During urination a large amount is deposited onto a small area and for that reason much of the urine has entered the soil before it is hydrolysed and NH_3 is generated. For that reason is the emission during grazing lower than when the same amount of nitrogen is used in the form of slurry. Experiments have shown that the emission from urine is much more important than the emission from dung pats (Petersen et al. 1998). The emission during grazing is influenced by the same meteorological conditions, soil properties as the emission after application of manure. The fraction of the nitrogen that volatilises increases with the nitrogen input per m^2. This is possibly because the soil has a limited ability to bind NH_4^+ and if excess NH_4^+ is added then the remaining NH_4^+ will be at risk of volatilisation.

In some countries animals are always housed, in other countries they are always outside. There are, however, also situations where the animals are only outside during part of the year or part of the day. This complicates estimates of the emission and its diurnal and seasonal variation.

Many European countries have already taken steps to reduce the NH_3 emission by restricting manure application to some periods, by requiring covering of storage tanks and by making it obligatory to apply manure with special equipment such as trailing hoses and injectors.

The emission from a whole farm system is usually not measured. What is measured is the emission from one type of activity: housing, storage, application or grazing. This means usually that these emissions are measured independently and that they are not measured during the same time and often even not on the same farm. If the emissions from each step are added to form the emission from the whole farm one might in exceptional cases calculate losses over 100%. This is of course not realistic and is the result of not taking into account the relation between the *subsequent* steps in the emission chain.

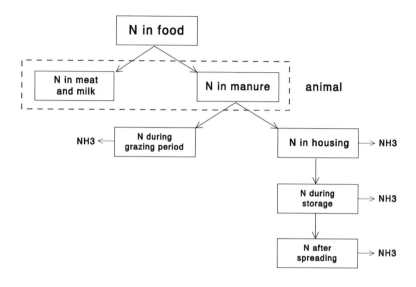

Fig. 2.50. Schematic nitrogen flow with subsequent step: from nitrogen in animal feed to emission of NH$_3$ to the atmosphere. Losses of N other than as NH$_3$ to the atmosphere are not shown.

The reality is that if less NH$_3$ is emitted during one stage (e.g. storage), there will be more nitrogen available for volatilisation in later stages (e.g. after application). Measures taken to reduce the emission during storage may in that case even lead to an increase in the total emission from the whole system. To avoid this type of problems a mass balance approach is used in the calculations: the amount of nitrogen that is left after each step should be equal to the sum of the amount of nitrogen that was input to this step and the emission during the step (Fig. 2.50). The emission in this system is always expressed as a fraction of the nitrogen that is input to the system.

Within a country there are many different combinations of housing, storage and application systems and restrictions are possible. This leads to quite different emissions per animal present for the various possible combinations. For that reason, information on the frequency of occurrence of the different possible combinations of systems is needed for accurate calculation of the emission for a country or for whole Europe. Moreover, differences in climatology and soil properties play a role. Fig. 2.51 shows the NH$_3$ emission for different combination of animals, housing and manure handling system for Denmark. The emission is expressed per 100 kg N excreted, so that emissions from different animals and different combinations of housing, storage and application systems can be compared easily.

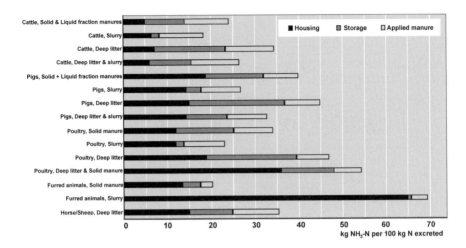

Fig. 2.51. NH$_3$ emission for different combinations of animals and housing and manure application (kg NH$_3$-N per 100 kg N excreted) (Hutchings et al. 2001). Reprinted from Atmospheric Environment 35:1959-1968, with permission from Elsevier Science.

Fig. 2.51 shows that the emission from the same animal category can be a factor two different for different systems (Hutchings et al. 2001).

2.4.4 Emission from fertilizers

The volatilisation of NH$_3$ from the application of fertilizers depends on:

- Properties of the soil (pH, calcium content, cation exchange capacity, water content, buffer capacity, porosity etc.) The emission rate increases with pH, calcium content and porosity but is decreases with buffer capacity and the water content.
- Chemical properties of the fertilizer (e.g. losses from urea are generally much higher than from other types of fertiliser).
- The time between application and ploughing. Emission is not always highest just after application. Urea for example has first to be hydrolysed by the enzyme urease, which is present in the soil.
- Meteorological conditions: temperature, turbulence (wind speed), air humidity and precipitation. The emission rate generally increases with temperature and turbulence. The emission rate and during and after precipitation periods is lower, because part of the nitrogen is then leached into the soil.
- Irrigation. If a field is irrigated the fertilizer is diluted and enters the soil at a higher rate. This leads to a lower emission rate.

Table 2.22 gives an overview of some average emission factors used for fertilizers (Asman 1992; Bouwman et al. 1997).

Table 2.22. Emission factors for N fertilizers under European conditions (Asman 1992; Bouwman et al. 1997).

Fertilizer	Loss of N (%)
Urea	15
Ammonium sulphate	8
Ammonium phosphate	4
Ammonium nitrate	2
Calcium ammonium nitrate	2
Gaseous ammonia, direct application	1

In many countries, different types waste products or compost such as sewage sludge or landfill products are more and more applied to agricultural fields. The NH_4^+ content in these waste products can be very variable. For that reason there is a need to assess the environmental impact of this application including the volatilisation of NH_3.

2.4.5 Crops

Vegetation can act as a source of NH_3. Whether the vegetation acts as a source or as a sink depends on the difference between the concentration of NH_3 in the air and the concentration of NH_3 in the gas phase in the leaves. The ambient NH_3 concentration at which there is no net exchange of NH_3 between the air and the leaves is called compensation point (e.g. expressed in $\mu g\ m^{-3}$). The compensation point increases with temperature. It varies with plant species, growth stage, the soil and plant nitrogen status as well as the meteorological conditions. For that reason it will also show diurnal and seasonal variations. Net emission occurs when the compensation point of the plant is higher than the concentration in the air (Schjoerring et al. 1991; Holtan-Hartwig and Bøckman 1994). The emission has only been determined for a limited number of species and climates and is for that reason rather uncertain. Emissions could vary between 0 and 15 kg N $ha^{-1}\ yr^{-1}$. Even natural vegetation can emit NH_3 under exceptional circumstances such as drought.

2.4.6 Spatial resolution required

NH_3 is emitted from numerous sources with low source heights that are scattered both in time and space. The emission from these sources is usually not monitored but is estimated from a limited number of experimental studies. The local deposition of NH_3 is substantial (20-30% of the emission) compared to that of NO_x. This difference is the result of a much higher dry deposition velocity and the lower source heights for NH3, compared to those of NO_x. NH_3 sources are located in agricultural areas, often close to semi-natural ecosystems. A single farm usually makes a substantial contribution to the total nitrogen deposition up to a few hundred metres distance from its sources. NH_x is, however, not only a local problem:

European countries typically export 40-60% of the emitted NH_3 (for NO_x this number is typically 80-95%).

Some surfaces, such as agricultural crops can act as sources as well as sinks for NH_3. Part of the emitted NH_3 is deposited very locally (within 10-100 m). For these reasons it is useful to have models that can describe the transport and surface exchange in both directions on a local scale. Such models have been developed within the EUROTRAC-GENEMIS framework (Asman 1998; Loubet et al. 2001).

The spatial resolution required of an emission inventory depends e.g. on the resolution of the atmospheric transport model, which will typically depend on the purpose of the calculations and the computational resources available. However, it will also depend on the spatial distribution and emission density of the sources as well as the atmospheric chemistry of the compound.

In areas with a medium to high NH_3 emission density (e.g. The Netherlands, Belgium, Denmark, parts of Germany and France) the N deposition will to a large extent be influenced by the dry deposition of NH_3 from local sources and show a high spatial variation. Due to the high spatial variation an accurate estimate of the nitrogen deposition would require an extremely large number of monitoring stations, which is technically and financially not feasible. The only way to model this situation is to use atmospheric transport models with a relatively high spatial resolution, also in the emissions (a 1×1 km^2 resolution or even higher).

In areas with a relative low NH_3 emission density such as in large parts of Sweden the N deposition will to a large extent be determined by foreign emissions and will show relatively small variations in the spatial distribution. If the deposition in such an area has to be calculated a high spatial resolution in the emission is not needed (a 50×50 km^2 resolution would be sufficient).

2.4.7 Temporal resolution required

Emission factors give the annually averaged emission e.g. per animal, kg fertilizer applied, per crop etc. The NH_3 emission rate is, however, not constant but shows considerable diurnal and seasonal variations. Normally the same emission factors are applied for subsequent years. In reality the emission factors are not the same for subsequent years because the emission is also a function of the meteorological conditions that will vary from year to year. This is important if the emission is used in atmospheric transport models and the results of these models are compared with actual measurements for these years. For policy studies, however, it is more useful to use emissions for average meteorological conditions, as the effect of measures to reduce the NH_3 emission on the NH_x deposition is studied and not the influence of meteorological variations on the deposition.

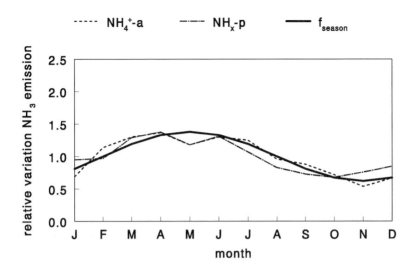

Fig. 2.52. Relative seasonal variation in NH_3 emission rate for The Netherlands for around 1990, derived from measured NH_4^+ aerosol concentrations (NH_4^+-a) or measured NH_x deposition (NH_x-p). The average values are 1. f_{seas} is a function that gives a reasonable description of the variation $f_{seas} = 1 + 0.38\sin(2\pi(d-45)/365)$, where d is the Julian day number (Asman 1992).

Meteorological parameters such as wind speed and temperature show often diurnal variations, which will not only affect the NH_3 emission rate, but also the dry deposition velocity of compounds, the atmospheric dispersion and reaction rate. For that reason atmospheric transport and deposition models need information on the diurnal variations in the NH_3 emission rates. In almost all atmospheric transport models for NH_x this co-dependence of the emission rate, the dry deposition velocity and dispersion rate on the meteorological conditions is not taken into account. Taking this co-dependence into account leads to higher emission rates at higher wind speeds, and a higher wind speeds also long-range transport is favoured. For that reason it is useful to develop subroutines by which (part of) the NH_3 emission rate can be generated using the same meteorology as the atmospheric transport model. Such subroutines could be embedded in atmospheric transport models or be generated by a pre-processor previous to the model run.

The NH_3 emission rate shows also seasonal variations caused by seasonal variations in the application of manure and fertilizer, grazing, crop cover and temperature. Other compounds that are involved in the atmospheric chemistry of NH_3 such as SO_2 and NO_x also show seasonal variations. Moreover, the effects of nitrogen deposition on ecosystems show also seasonal variations. Nitrogen deposited during spring can easily be taken up by vegetation and algae, whereas nitrogen deposited during winter cannot. For those reasons it is important also to know the seasonal variation in the NH_3 emission. Until now the seasonal variation in the emission rate has been estimated from the results of atmospheric transport models, where the emission rate was kept the constant during the whole year. The seasonal variation in the emission rate was then found from the ratio measured vs. calculated NH_4^+ concentrations in air and wet deposition of NH_4^+. Fig. 2.52 shows an example of the seasonal variation for the Netherlands derived in this way (Asman 1992). Present seasonal variations will certainly be different due to the measures that have been taken since 1990 to reduce the emission. The diurnal and seasonal variations in the emission rate will also be different in different parts of Europe due to differences in agricultural practise, soil properties and climatology.

2.4.8 Possibilities for reducing emissions

The simplest way of reducing NH_3 emissions is usually not discussed, because it seems to be politically difficult to accept: reducing the numbers of animals. The conversion of protein from plant products to body protein and products (milk, eggs) in animals is not very efficient ($18\pm10\%$; Isermann 1993). Humans in Western Europe and the USA eat already more (animal) protein than they need. This consumption also leads to surplus of nitrogen in the environment, which can have detrimental effects to ecosystems (due to e.g. NH_3 emission, nitrate leaching, N_2O and CH_4 emission etc.).

There are technical possibilities to reduce NH_3 emissions (Menzi and Achermann 2000). Some of the measures can be taken rather easily and relatively fast, whereas others are may take much time and are expensive. We will now shortly discuss the possibilities of emission reduction in the various steps.

Feed: It is possible to give the animals a more optimal diet with regard to their need of amino acids and thereby reducing the excretion of proteins that they do not digest and the related NH_3 emission. It is also possible to optimise the diet of the animals with regard to their age and to select animals during breeding that are more efficient at using proteins in fodder.

Animal housing: It is possible to reduce the emission from housings by careful design of the floors and the underlying pits, by having a ventilation system that is constructed in such a way that the temperature and ventilation rate at floor height and in the pit are relatively low. Emission can also be reduced by capturing NH_3 in the exhaust ducting, although this is currently an expensive option. It should be noted that efforts to reduce the emission by changing the design of the building take time because animal housing is normally replaced only every 20-30 years. Emission can also be reduced by regularly flushing the floor (Monteny and Erisman 1998). For poultry there are special possibilities of reducing the emission due to the fact that poultry do not excrete urea, but uric acid. Drying of the poultry manure leads to a reduction of the NH_3 emission, because the microbial growth needed to form NH_3 from uric acid stops at low moisture contents (Groot Koerkamp 1994).

Emission during storage in storage tanks can be reduced by covering the tanks with a tent or a layer on top of the slurry by adding straw etc. (Sommer and Hutchings 1995).

Emission after application can be reduced relatively easily by incorporating the manure into the soil as soon as possible after application, or better by applying manure by trailing hoses/shoes or injection in stead of by broad spreading (Sommer and Hutchings 2001).

Emissions caused by grazing animals cannot be reduced, but they are relatively low.

Emission from fertilizers can be reduced by applying fertilizers with a low volatilisation potential. For that reason, the use of urea should be avoided, especially in areas with high temperatures.

Agricultural activities can lead to other environmental and related problems than NH_3 emission. It will often not be possible to find methods that optimise agricultural activities with regard all involved aspects such as the emission of NH_3, N_2O (nitrous oxide), CH_4 (methane), odour, energy, animal welfare and economics (see e.g. Brink et al. 2001a, 2001b). This has to be kept in mind when reducing the NH_3 emission, so that e.g. care is taken that the reduction of the NH_3 emission does not lead to higher nitrate leaching rates, due to an increase in nitrogen of the manure that is incorporated in the soil.

2.4.9 A framework for the calculation of NH_3 emissions in Europe

Different types of emission information can be distinguished, e.g.:
- Information on the annually averaged emission, on a country basis.
- Information on the actual emission rate as a function of time on a regular grid for use in atmospheric transport models. In this case subroutines should be provided so that the emission rate can be calculated with the same values of the meteorological parameters that the atmospheric transport model uses.

Table 2.23 and Table 2.24 give an overview of the knowledge needed to calculate the emission rate for NH_3 as a function of time on a regular grid. In principle the same information has to be known to calculate the annually averaged emission on a country basis, but in practice the results of the accumulated emission calculated with process based models can be used instead of the emission rate as a function of time. There are large differences between different countries with regard to the availability of information that is needed to generate reliable emission estimates.

Table 2.23. Availability of information on processes needed to calculate the NH$_3$ emission rate for use in atmospheric models.

Type of process	Availability of information
Animal excretion	good
Emission from housings	poor
Emission from manure storage:	
– slurry tanks	good
– manure heaps	poor
Emission during spreading	not important
Emission after spreading of:	
– liquid manure	good
– solid manure	poor
– fertilizer	good
Emission during grazing	good
Emission from crops	poor

Table 2.24. Availability of information on the geographical distribution of factors that are important for the calculation of the NH$_3$ emission rate in atmospheric transport models.

Geographical distribution	Availability of information
Number of animals	for country level OK; for most countries also data for the distribution within the country (county of municipality level)
Fodder	poor
Housing types	poor
Type of manure/manure handling system	poor
Type of fertilizer	for country OK, within countries poor
Fraction of time animals are outside	poor
Soil types	good
Crops	from EU-support data; satellite
Meteorology	good
Regulations	possible to collect, but not yet done

So far, insufficient information is available on processes that describe the emission rate as a function of time, but some progress has been made (Monteny et al. 1998; Ni 1999). Information on the geographical distribution of parameters necessary to calculate the emission is also often lacking. Some progress has, however, been made in the past decade and in the next sections some examples of new results are given.

2.4.10 ALFAM: an EU database on the NH₃ emission from manure applied to fields

In different European countries experiments have been made to determine the NH_3 emission after application of manure to fields. NH_3 volatilisation experiments are expensive and are typically designed to investigate one or two factors that influence the NH_3 emission rate. For policy decision purposes information is needed on many more factors that influence the NH_3 emission rate, addressing the range in agricultural practices, soil and weather conditions that can occur within Europe or even within a single country. A single institute has usually not the data available to do this. For that reason the EU concerted action ALFAM was started, in which a database was created with experimental data from Denmark, Italy, The Netherlands, Norway, Sweden, Switzerland and the United Kingdom (Søgaard et al. 2002). Moreover, the data were used to develop a model that predicts the NH_3 emission after cattle and pig slurry application as a function of time.

Table 2.25. Factors affecting NH_3 emission from animal slurry applied to fields (Søgaard et al. 2002).

Experimental factor	Effect on NH₃ emission
Soil moisture	Wet soil 10% higher than dry soil
Air temperature	+2% per °C
Wind speed	+4% per m s⁻¹
Slurry type	Pig slurry 14% less than cattle slurry
Dry matter content	+11% per % dry matter
Total ammoniacal nitrogen content	-17% per g N kg⁻¹ slurry
Application method:	
– Band spreader/trailing hose	42% less than broadcast spreader
– Open slot injection	72% less than broadcast spreader
Manure incorporation	No incorporation 11 times higher than shallow cultivation

Reprinted from Atmospheric Environment 36: 3309-3319. Copyright 2002, with permission from Elsevier Science.

Table 2.25. shows how the emission depends on the most important factors. The report of the ALFAM project gives also an overview of the methods to measure fluxes (see www.alfam.dk). The ALFAM project was not done within the EUROTRAC-GENEMIS framework.

2.4.11 Detailed emission map of Vejle county, Denmark

In this section an example is given of an emission inventory with a high resolution that is used to calculate the nitrogen deposition with a high resolution in areas with high NH_3 emission densities. The area studied was the county of Vejle in Denmark. In the county of Vejle and in Denmark in general a significant part of the biodiversity in terrestrial ecosystems can be found in minor, extensively used patches, consisting of forests, fens, wet meadows, dry grasslands, heathlands and salt marshes. These patches are usually small (< 200 ha) and embedded in intensively managed agricultural areas with such high NH_3 emission densities that the critical load for nitrogen deposition often is exceeded. In Denmark counties have to assess the environmental impact of larger farms (> 250 livestock units; 1 livestock unit is defined in Denmark as 100 kg yr^{-1} after storage) on local nature areas. For these reasons a decision support tool is needed that is able to estimate the impact of local sources on the local nitrogen deposition, but at the same time is able to take more distant and foreign nitrogen sources into account. This type of tool was developed for the county of Vejle, Denmark, which has a size of 2997 km^2. The contribution from NH_3 emissions in the county of Vejle to the nitrogen deposition in the county of Vejle was calculated with a spatial resolution of 100 m, using the results of the atmospheric transport model DEPO1 at a 100×100 m^2 grid scale (Asman 1998). To speed up the calculations a matrix was calculated with DEPO1 that gives the contribution of the NH_3 emission of one grid element of 100×100 m^2 to the nitrogen deposition in all other grid elements, as well as to the grid element itself. The calculations have been implemented into a geographical information system (GIS), using a grid convolution extension built into a desktop-GIS (ArcView GIS with Spatial Analyst from ESRI). The contribution from other Danish and foreign NH_3 sources to the nitrogen deposition as well as all Danish and foreign NO_x sources was calculated on a 5×5 km^2 resolution for the county of Vejle and was added as a background nitrogen deposition.

The NH_3 emission density was calculated on a 100×100 m^2 grid for Vejle County using emission factors for housing, storage and application of manure that are specific for Denmark, taking into account the actual fodder, housing systems and climate in Denmark (Hutchings et al. 2001). A geographical distribution of the emission was obtained by using the following information for each individual farm and its fields:

- Number and kind of animals from the Danish Central Husbandry Register (CHR).
- Information on the crops grown within a "field block" (groups of 1-8 adjacent fields, mapped as one spatial unit) from the General Agricultural Register (GAR).
- Farm type, derived from the combination of CHR and GAR data at farm level (pig, cattle, mixed, and cereal farming).
- Geographical position of animal housings and storage facilities.

NH_3 emissions from housings and storage facilities were calculated as point sources, assuming that all sources are located at the housing's location. Field emissions were calculated as emission densities, based upon the sum of agricultural activities within each field block, depending on farm type for every field allocated within one field block. Additional information on crop type and soil type, as provided by GAR, did not match census data used for calculating emission coefficients and had to be excluded from the approach.

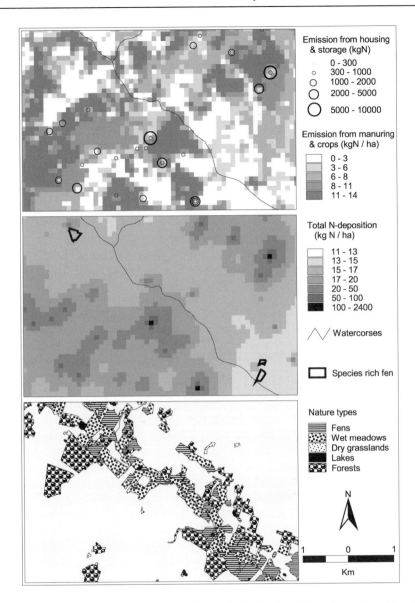

Fig. 2.53. NH₃ emission, total N deposition (originating from dry and wet deposition from Danish and foreign NH₃ and NOₓ emissions) and nature areas in one part of the county of Vejle, Denmark.

Furthermore, emission coefficients used were calculated anticipating an average Danish situation regarding type of housing and manure handling at farm level, as no other information was available. For the deposition calculation the emissions were redistributed onto a regular 100×100 m² grid.

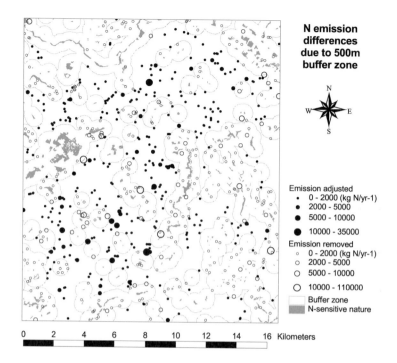

Fig. 2.54. Differences in NH₃ emission due to removal of livestock within a 500 m zone from bogs, fens, dry grasslands and heathlands (kg N yr⁻¹).

The NH₃ emission shows a high spatial variation and this is reflected in the total N deposition (Fig. 2.54). From the figure it can be seen that the spatial resolution of the calculated emission and deposition is now so high that one grid element $(100 \times 100 \text{ m}^2)$ has a size that is less or comparable to the size of the nature areas. Information on a $10 \times 10 \text{ m}^2$ scale on nature elements in the landscape was used to calculate the average total N deposition to different categories of elements.

An example is presented below to illustrate the potential of the developed decision tool. In this example it was assumed that:

- All livestock was removed from a zone of 500 m around bogs, fens, dry grasslands and heathlands and were relocated elsewhere in the area. Livestock in a zone of 500 m from forests, wet meadows, lakes and ponds were *not* relocated, because there would be insufficient area remaining in which they could be relocated.
- Emissions due to application of manure onto fields were not relocated.
- Livestock and their associated emissions were relocated to the remaining farms in the area, in proportion to their original livestock and emissions.

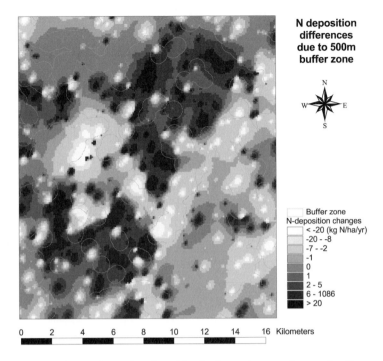

Fig. 2.55. Difference in total N deposition in the study area after moving livestock outside the 500 m zone from bogs, fens, dry grasslands and heathlands (kg N ha^{-1} yr^{-1}).

Fig. 2.54 and Fig. 2.55 show the change in total N deposition in the study area due to this relocation. This decision tool and emission inventory presented in this section was developed within the EUROTRAC-GENEMIS framework.

2.4.12 Modelling of the NH$_3$ emission after spreading of manure

Génermont and Cellier (1997) developed a mechanistic model for calculating the NH$_3$ emission after spreading of manure to bare soil as a function of time. This model takes into account all processes in the soil and in the atmosphere that lead to NH$_3$ volatilisation:

- Physical and chemical equilibria in the soil.
- Aqueous and gaseous transfer of NH$_x$ in the soil.
  Transfer of gaseous NH$_3$ from the soil to the atmosphere.

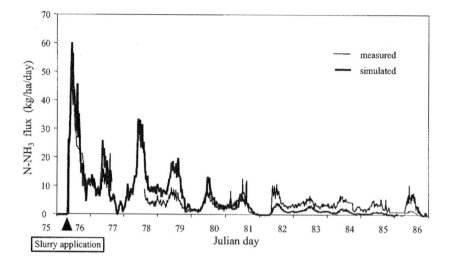

Fig. 2.56. Measured (thin line) and calculated (thick line) NH₃ emission after slurry application in Britanny, France in 1994.

As these processes are a function of the soil temperature, soil moisture content and the meteorological conditions the following processes have also to be modelled:

- Water transport in the soil.
- Heat transport in the soil.
- Water and heat exchange between the soil and the atmosphere.

Fig. 2.56 shows that this model able to reproduce measured fluxes at the INRA experimental station of Le Rheu (near Rennes, France) reasonably well. The problem with this type of mechanistic models is that they need information on many more input parameters than are usually available. For that reason there is a need for simple mechanistic models that can mimic the results of the more complicated mechanistic models, but which need information on input parameters that is normally available.

This model was developed within the EUROTRAC-GENEMIS framework and is presently being extended to treat the volatilisation of other nitrogen sources such as mineral fertilizers and urban compost.

2.4.13 Need to check reported national NH₃ emissions

The European countries that have ratified the Gothenburg Protocol have promised to reduce their NH₃ emissions by 2010 and are obliged to report estimates of their annual emission. Ideally, there should be an independent assessment of the emissions for each separate country. In principle it is possible to verify the emissions with measured atmospheric concentrations and depositions (Sutton et al. 2002).

However, this is not straightforward. One reason is that there are often variations of the order of 30% in concentrations and depositions between years that are
only caused by variations in meteorological conditions. Moreover, the NH_3 emissions themselves are also influenced by meteorological conditions. For these reasons measurements over many years are needed before a statistically significant
conclusion can be drawn on a trend in atmospheric concentrations and depositions
and hence on a trend in NH_3 emissions.

The other problem is that in areas with high NH_3 emission densities NH_3 concentrations are to a large extent influenced by very local sources. For that reason
an extremely large number of measuring stations would be needed (of the order of
1 per 3 km^2; Duyzer et al. 2001), even if the position of the stations with regard to
surrounding sources were chosen carefully. This is practically and economically
not feasible. This means that it is only possible to detect trends in measured NH_3
concentrations in a very limited number of areas in this way.

The concentrations of NH_4^+ in air and wet deposition of NH_4^+ are usually determined by the NH_3 emissions by more than one country. This means that it is
usually impossible to determine the trends in the emission from only one country
from these measurements.

Then there is another problem. NH_3 reacts with H_2SO_4 (sulphuric acid) formed
by oxidation of SO_2 or with HNO_3 (nitric acid) formed by oxidation of nitrogen
oxides (NO_x). As a result changes in emissions of SO_2 and NO_x may lead to
changes in the ratio NH_3/NH_4^+ and as a result to changes in dry deposition of NH_3
vs. wet deposition of NH_x.

For these reasons it is unlikely that a trend in NH_3 emissions of one country
over the period 2002-2010 can be estimated from measured concentrations and
depositions. It is, however, likely that trends in emissions larger parts in Europe
can be detected from measured NH_4^+ concentrations and wet deposition of NH_x,
provided emissions of SO_2 and NO_x only show slow changes. But even in this
case the measurement period up to 2010 is not likely to be long enough to detect
changes of 10-20% in emission.

2.4.14 Conclusions

In recent years, progress has been made to obtain better NH_3 emission estimates.
Much information is now available that can help to estimate emissions for northwest Europe. There is, however, still more information needed. The following information is needed to estimate annually averaged emissions (see also Sommer
and Hutchings 1997):

Animal feed:
• The typical quantity and quality of feed given to animals of different types.

Emission from manure in animal housings:
• Effect of animal house design
• Effect of ventilation and temperature on the emission rate
• Changes in manure composition in animal houses (pH, NH_x)
• Emission from deep litter

Emission from storage facilities:
- Emission from solid manure and composting manure
- Crust formation processes

Emission after application of manure:
- Emission from solid manure and composted manure
- Change in pH of the slurry surface
- Influence of slurry viscosity and soil characteristics on infiltration

Emission during grazing:
- Grazing period during the year.
- Proportion of day spent at pasture (as opposed to animal housing or milking parlour) during the grazing period.

Emission from fertilizers:
- Measurement of volatilisation in the field

Emission from crops:
- Dependence of the NH_3 compensation point on plant species, meteorological conditions and harvesting practise (emission occurs also after harvesting)

As much less is known for south and west Europe, additional research should be done in these regions.

The simple models currently available are unable to respond to short-term changes in meteorology. Conversely, the mechanistic models currently available, e.g. for losses from animal housing or after field application, demand input data that are not available at regional or national scales. There is a need for the development of models with an intermediate complexity i.e. models that are capable of simulating the major mechanisms but have a more limited demand for data inputs.

In the past spatially detailed emission inventories were made for countries where standard emission factors were applied, that did not take into account the geographical distribution of various types of fodder, housing and manure application systems. Often the geographical distribution of the emissions was found using the number of animals for each municipality in the country and standard emission factors for each kind of animal (e.g. emission of 1 pig of average size = 5.4 kg NH_3 animal^{-1} yr^{-1}) (Buijsman et al. 1987; Asman et al. 1998).

The consumption of fertilizers was usually not known on this scale, but often on a county scale. As the emission from the same kind of animal can be up to a factor of two different depending on which combination of housing and manure handling system is used, it is not only sufficient to know the geographical distribution of the number of animals, but also the geographical distribution of housing and application systems as well as the type of farm (cattle, pigs, mixed, cereal) and the crops. If measures are taken to reduce the emission this type of information becomes even more important, because the emission in some systems can be reduced rather easily, whereas it is more difficult to reduce emissions in other systems. For that reason the geographical distribution of the systems should also be known. At present this information is not readily available. It would therefore be worthwhile to investigate and the possibilities to find the geographical distribution of housing, manure handling systems, type of farm and crops for the different European countries. Once such a system is set up it will not take much time to provide updates, provided the input parameters and the structure of the information do not change.

In order to calculate the NH_3 emission rate as a function of time and space the results of mechanistic models and information on the geographical distribution of different agricultural systems should be integrated. This is not a simple task. Moreover, additional measurements are needed to test such models on a larger scale (e.g. 50×50 km^2, the size of a grid element of the EMEP model).

2.5 References

4.BImSchV zur Durchführung des Bundes- Immissionschutzgesetzes (Verordnung über genehmigungsbedürftige Anlagen – 4. BImschV) vom 24.July 1985 (BGBl. I S. 1586), last Change 1999, Bonn 1999

20. BImSchV (1998) Zwanzigste Verordnung zur Durchführung des Bundes-Immissionsschutzgesetzes (Verordnung zur Begrenzung der Emission flüchtiger organischer Verbindungen beim Umfüllen und Lagern von Ottokraftstoffen - 20. BImSchV), Änderungsverordnung vom 27. Mai 1998. BGBl. I, Germany, 1174.

20. BimSchV (2001) Zwanzigste Verordnung zur Durchführung des Bundes-Immissionsschutzgesetzes (Verordnung zur Begrenzung der Emission flüchtiger organischer Verbindungen beim Umfüllen und Lagern von Ottokraftstoffen - 20. BImSchV), Änderungsverordnung vom 24. August 2001. BGBl. I, Germany, 2213.

21. BImSchV (1992) Einundzwanzigste Verordnung zur Durchführung des Bundes-Immissionsschutzgesetzes (Verordnung zur Begrenzung der Kohlenwasserstoffemissionen bei der Betankung von Kraftfahrzeugen - 21. BImSchV) vom 7. Oktober 1992. BGBl. I, 1730.

Abdul-Khalek IS, Kittelson DB, Graskow BR, Wei Q, Brear F (1998) Diesel Exhaust Particle Size : Measurement Issued and Trends, SAE paper 980525.

Amann M, Bertok I, Cofala J, Gyarfas F, Heyes C, Klimont Z, Schöpp W (1999) Integrated assessment modelling for the protocol to abate acidification, eutrofication and ground-level ozone in Europe. Report prepared by IIASA. Report Lucht & Energie 132. Ministry of Housing, Spatial Planning and the Environment, The Hague, The Netherlands. Available at http://www.iasa.ac.at/~rains

Asman WAH (1992) Ammonia emission in Europe: updated emission and emission variations. Report 228471008. National Institute of Public Health and Environmental Protection (RIVM), Bilthoven, The Netherlands. Available at: http://www.rivm.nl/bibliotheek/rapporten/228471008.html

Asman WAH (1998) Factors influencing local dry deposition of gases with special reference to ammonia. Atmospheric Environment 32:415-421.

Asman WAH, Sutton MA, Schjoerring JK (1998) Ammonia: emission, atmospheric transport and deposition. New Phytol. 139:27-48

Atkinson R (2000) Atmospheric chemistry of VOCs and NO_x. Atmos. Environ. 34, 2063–2101.

Bailey JC, Schmidl B, Williams ML (1990) Speciated hydrocarbon emissions from vehicles operated over the normal speed range on the road. Atmos Environ 24:43–52

Bernard PC, Van Grieken RE, Eisma D (1986) Classification of estuarine particles using automated electron microprobe analysis and multivariate techniques. Environ Sci Technol 20:467–472

Berdowski JJM, Visschedijk AJH and Pulles T (2001) Presentation at the UNECE TFEIP and EIONET Workshop, Geneva

Bhattacharya SC (1998) State of the Art of Biomass Combustion. Energy Sources 20:113–130

Bishop GA, Stackey JR, Ihlenfeldt A, Williams WJ, Stedman DH (1989) IR long-path photometry: A remote sensing tool for automobile emissions, Anal. Chem., 61, 671A.

Bobbink, R, Hornung M, Roelofs JGM (1998) The effects of air-borne nitrogen pollutants on species diversity in natural and semi-natural European vegetation. J. Ecol. 86:717-738

Bouwman AF, Lee DS, Asman WAH, Dentener FJ, van der Hoek KW, Olivier JGJ (1997) A global high-resolution emission inventory for ammonia. Global Biogeochem. Cycles 11:561-587

Bräutigam M, Kruse DK Ermittlung der Emissionen organischer Lösemittel in der Bundesrepublik Deutschland. Köln: Consulting Services, 1992-Forschungsbericht 104 04 116/01 im Auftrag des Umweltbundesamtes

Brandt A, Adelt F, Schulz T, Motz GB Ozonrelevante VOC-Emissionen aus Kleingewerbe und Privathaushalten in Nordrhein-Westfalen, Gefahrstoffe-Reinhaltung der Luft 60 (2000)), Nr.11/12-November/Dezember

Brasseur GP, Orlando JJ, Tyndall GS (1999) Atmospheric Chemistry and Global Change. Oxford University Press, New York.

Brink C, Kroeze C, Klimont Z (2001a) Ammonia abatement and its impact on emission of nitrous oxide and methane in Europe – Part 1: method. Atmospheric Environment 35:6299-6312

Brink C, Kroeze C, Klimont Z (2001b) Ammonia abatement and its impact on emission of nitrous oxide and methane in Europe – Part 2: application for Europe. Atmospheric Environment 35:6313-6325

Brüggemann N, Schnitzler JP (2002) Comparison of isoprene emission, intercellular isoprene concentration and photosynthetic performance in water-limited oak (Quercus pubescens Willd. and Quercus robur L.) saplings. Plant Biology 4, 456-463.

Buijsman E, Maas JFM, Asman WAH (1987) Anthropogenic NH_3 emissions in Europe. Atmospheric Environment 21:1009-1022

BUWAL (1995) Emissionsfaktoren von Leichten Motorwagen in der Schweiz. Schlußbericht Bearbeitet durch: TÜV Rheinland Sicherheit und Umweltschutz GmbH, Abteilung Verkehr und Umwelt. Herausgegeben vom Bundesamt für Umwelt, Wald und Landschaft (BUWAL), Bern

Carter WPL (1994) Development of ozone reactivity scales for volatile organic compounds. J Air Waste Manage Assoc 44:881–899

Carter RE, Lane DD, Marotz GA, Chaffin CT, Marshall TL, Tucker M, Witkowski MR, Hammaker RM, Fataley WG, Thomas MJ, Hudson JL (1993) A method of prediction point and path-averaged ambient air VOC concentrations using meteorological data. Journal of the Air Waste Management Association 43, 480-488.

Chagger HK, Kendall A, Williams A (1998) Formation of dioxins and other semi-volatile organic compounds in biomass combustion. Appl Energy 60:101–114

Decker G, Beyersdorf J, Schulze J, Wegener R, Weidmann K (1996) Das Ozonbildungspotential unterschiedlicher Fahrzeug- und Kraftstoffkonzepte. ATZ 98:280–289

Derwent R, Jenkin ME, Saunders SM, Pilling J (1998) Photochemical ozone creation potentials for organic compounds in Northwest Europe calculated with a master chemical
mechanism. Atmos. Environ. 32, 2429-2441.

De Vlieger I (1996) On-board emission and fuel consumption measurement campaign on
petrol-driven passenger cars. Atmos Environ 31:3753–3761

Directive 1999/13/EC of 11 March 1999 on the limitation of emissions of volatile organic
compounds due to the use of organic solvents in certain activities and installations,
Brüssel, 1999

Directive 2001/81/EC of the European Parliament and the Council of 23.October 2001 on
National Emission Ceilings for certain atmospheric pollutants, Office Journal of the
European Communities, L309/22, 27.11.2001

Dockery DW, Pope CA, Xu X (1993) An association between air pollution and mortality in
six US cities. New England Journal for Medicine 329:1753-1759

Dreiseidler A, Baumbach G, Pregger T, Obermeier A (1999) Studie zur
Korngrößenverteilung ($<PM_{10}$ und $<PM_{2,5}$) von Staubemissionen. UBA-FB 297 44
853, Report by IVD and IER of University of Stuttgart, Umweltbundesamt Berlin

Düring I and Lohmeyer A (2001) Validierung von PM_{10} Immissionsberechnungen im
Nahbereich von Straßen und Quantifizierung der Feinstaubbildung von Straßen. Im
Auftrag der Senatsverwaltung für Stadtentwicklung, Berlin und des Sächsischen
Landesamtes für Umwelt und Geologie, Ing. Büro Lohmeyer, Radebeul, Germany

Düring I, Jacob J, Lohmeyer A, Lutz M, Reichenbächer W (2002) Estimation of the "Non
Exhaust Pipe" PM10 Emissions of Streets for Practical Traffic Air Pollution Modelling. In: eds? Proceedings of the 11th Int. Transport And Air Pollution, Graz, pp 309-
316

Duffy BL, Nelson PF, Ye Y, Weeks IA (1999) Speciated hydrocarbon profiles and calculated reactivities of exhaust and evaporative emissions from 82 in-use light-duty
Australian vehicles. Atmos Environ 33:291–307

Dufournaud CM, Quinn JT, Harrington JJ (1994) A partial equilibrium analysis of the impact of introducing more-efficient wood-burning stoves into households in the Sahe-
lian region. Environ Planning A26:407–414

Duyzer JH, Nijenhuis B, Weststrate H (2001) Monitoring and modelling of ammonia concentrations and deposition in agricultural areas of The Netherlands. Wat. Air Soil. Pollut.: Focus 1:131-144

ECE (Economic Commission for Europe) (1990) Emissions of Volatile Organic Compounds (VOC) from Stationary Sources and Possibilities of their Control", Final Report, Economic Commission for Europe - VOC Task Force, Karlsruhe.

Ehrlich C and Kalkoff W VDI-Aktuelle Aufgaben der Meßtechnik in der Luftreinhaltung.
VDI-Berichte Nr. 1059. VDI-Verlag. Düsseldorf. 1993. 183-208

Eichhorn J (1989) Entwicklung und Anwendung eines dreidimensionalen mikroskaligen
Stadtklima-Modells. Dissertation am FB Physik der Johannes Gutenberg-Universität
Mainz.

EPA (1995) AP-42, Volume I, 5th edn. Office of Air Quality Planning and Standards,
United States Environmental Protection Agency, Washington.

EPA (1997) Compilation of Air Pollutant Emission Factors AP-42. 5th Edition, Volume I:
Stationary Point and Area Sources, Section 13.2.1. Paved Roads, Office of Air Quality
Planning and Standards, United States Environmental Protection Agency, Washington

Fangmeier A, Hadwiger-Fangmeier A, van der Eerden L, Jäger H-J (1994) Effects of atmospheric ammonia on vegetation – a review. Envir. Pollut. 86:43-82

Forkel R, Steinbrecher R, Stockwell WR (2001) Modelling of the Effect of In-Canopy Chemical Reactions on the Emission Rates of Biogenic VOC, In: Proceedings from the EUROTRAC Symposium 2000, P.M. Midgley, M. Reuther, M. Williams (Eds.), Springer Verlag Berlin, Heidelberg, GENEMIS, 725-728.

Friedrich R, Pregger T, Droste-Franke B (1999) Sources of Particulate Matter. Proceedings of the HART Conference "Particulate Matter, Science, Sources & Solutions", 20.-21.10.1999, Brussels

Gámez AJ, Berkowicz R, Ketzel M, Lohmeyer A, Reichenbächer W (2001) Determination of the non exhaust pipe PM10 emissions of streets for practical traffic air pollution modelling. In: Proceedings of the 7th Int. Conference on Harmonisation within Atmospheric Dispersion Modelling for Regulatory Purposes, Belgirate, pp 271 - 275

Gemeinsamer Abschlußbericht zum Dialog des BMU und des VCI zu Umweltzielen am Beispiel VOC, erarbeitet von Vertretern des Bundesumweltministeriums, des Umweltbundesamtes, des Verbandes der Lackindustrie e.V., des Verbandes der Druckfarbenindustrie im Verband der Mineralfarbenindustrie e. V., des Bundesverbandes Druck, des Industrieverbandes Klebstoffe e.V. und des Verbandes der Chemischen Industrie e.V., Frankfurt, 1997

Génermont S, Cellier P (1997) A mechanistic model for estimating ammonia volatilization from slurry applied to bare soil. Agric. For. Meteorol. 88:145-167

German Association of Chemical Industry (VCI), „Chemiewirtschaft in Zahlen 2000", Franfurt, Germany, 2001

German Technical Instructions for air pollution control (German TA Luft), Last Modification 2002, Berlin

Gertler AW, Sagebiel J C, Wittorff DN, Pierson WR, Dippel WA, Freeman D, Sheetz L (1997) Vehicle emissions in five urban tunnels. CRC Final Report No. E-5, Desert Research Institute, P.O. Box 60220, Reno, Nevada 89506, March 1997

Gertler AW, Abu-Allaban M, Coulombe W, Gillies JA, Pierson WP, Rogers FC, Sagebiel JC, Tarnay, L, Cahill TA (2001) Measurements of mobile source particulate emissions in a highway tunnel. IJVD 27:86 – 93

Grell GA, Emeis S, Stockwell WR, Schoenemeyer T, Forkel R, Michalakes J, Knoche R, Seidl W Application of a multiscale, coupled MM5/Chemistry Model to the complex terrain of the VOTALP Valley Campaign. Atmospheric Environment *34, (2000), 1435 – 1453*

Groot Koerkamp PWG (1994) Review on emissions of ammonia from housing systems for laying hens in relation to sources, processes, building design and manure handling. J. Agric. Engng. Res. 59:73-87

Guenther AB, Zimmerman PR, Harley PC, Monson RK, Fall R Isoprene and Monoterpene Emission Rate Variability: Model Evaluations and Sensitivity Analysis J. Geophys. Res. 98D, (1993), 12609-12617.

Guenther A, Hewitt NC, Erickson D, Fall R, Geron C, Graedel T, Harley P, Klinger L, Lerdau M, McKay WA, Pierce T, Scholes B, Steinbrecher R, Tallamraju R, Taylor J, Zimmerman PA (1995) Global model of natural volatile organic compound emissions. J. Geophys. Res. 100, 8873-8892.

Hassel D, Jost P, Weber FJ, Dursbeck K, Sonnborn S, Plettau D (1994) Abgasemissionsfaktoren von Pkw in der Bundesrepublik Deutschland. Umweltbundesamt (Ed) Erich Schmidt Verlag Berlin

Hassel D, Jost P, Weber FJ, Dursbeck F, Sonnborn KS, Plettau D (1993) Abgas-Emissionsfaktoren von Pkw in der Bundesrepublik Deutschland - Abgasemissionen von Fahrzeugen der Baujahre 1986 bis 1990. Untersuchung im Auftrag des Umwelt-bundesamtes, Berlin Köln

Hassel D (1997) Eignungsprüfung und Überwachung von Gasrückführsystemen. TÜ 38, 26-31.

Hempel D, Hirsch W, Kutzer HJ (2000) Explosionsgefährdung Binnentankschiffe. Ab-schlussbericht zum Forschungsvorhaben 96519/97 des Bundesministeriums für Ver-kehr, Bau- und Wohnungswesen, Physikalisch-Technische Bundesanstalt, Braun-schweig.

Helmig D, Klinger LF, Guenhter A, Vierling L, Geron CH, Zimmerman P (1998) Biogenic volatile organic compound emissions (BVOCs) I. Identifications from three conti-nental sites in the U.S. Chemosphere 38, 2163-2187.

Hewitt NC (1999) Reactive hydrocarbons in the atmosphere. Academic Press, San Diego.

Hoekman SK (1992) Speciated measurements and calculated reactivities of vehicle exhaust emissions from conventional and reformulated gasolines, *Environ. Sci. Technol.* 26, 1206.

Holtan-Hartwig L, Bøckman OC (1994) Ammonia exchange between crops and air. Norw. Agric. Sci., Suppl. 14.

Hüglin C, Gehrig R, Hofer P, Monn C, Baltensperger U (2000) Partikelemissionen (PM_{10} und $PM_{2,5}$) des Straßenverkehrs. Chemische Zusammensetzung des Feinstaubes und Quellenzuordnung mit einem Rezeptormodell. EMPA, Abteilung Luftfremdstoffe/Umwelttechnik, Bericht C4 des NFP41 "Verkehr und Umwelt", CH-Dübendorf

Huggins FE, Shah N, Huffman GP, Robertson JD (2000) XAFS spectroscopic characteriza-tion of elements in combustion ash and fine particulate matter. Fuel Proc Technol 65–66:203–218

Hutchings, NJ, Sommer SG, Andersen JM, Asman WAH (2001) A detailed ammonia emis-sion inventory for Denmark. Atmospheric Environment 35:1959-1968

IIASA (2002) RAINS PM Modul. International Institute for Applied Systems Analysis (IIASA) Laxenburg (12/19/2002) (http://www.iiasa.ac.at/~rains/PM/pm-home.html)

Isermann K (1993) Ammoniakemissionen der Landwirtschaft als Bestandteil ihrer Stick-stoffbilanz und Lösungsansätze zur hinreichenden Minderung. In: Ammoniak in der Umwelt. KTBL, Darmstadt-Kranichstein, Germany, 1-1-1-76.

Jacobson MZ (2001) Global direct radiative forcing due to multicomponent anthropogenic and natural aerosols. J. Geophys. Res. 106:1551-1568

Jenkin ME, Clemitshaw KC (2000) Ozone and other secondary photochemical pollutants: chemical processes governing their formation in the planetary boundary layer. Atmos. Environ. 34, 2499 – 2527.

Jemma CA, Shore PR, Widdicombe KA (1995) Analysis of C_1 –C_{16} hydrocarbons using dual-column capillary GC: application to exhaust emissions from passenger car and motor cycle engines. J Chrom Sci 33:34–38

John C, Friedrich R, Staehelin J, Schläpfer K, Stahel WA (1999) Comparison of emission factors for road traffic from a tunnel study (Gubrist tunnel, Switzerland). Atmospheric Environment, Vol. 33, pp 3367 – 3376

Kaarakka P, Kanarek MS, Lawrence JR (1989) Assessment and control of indoor air pollu-tion resulting from wood burning appliance use. Environ International 15:635–642

Kittelson DB, Abdul-Khalek I (1999) Formation of nanoparticles during exhaust dilution. EFI members conference. In: "Fuels, Lubricants, Engines & Emissions". University of Minnesota, Department of mechanical engineering

Knobloch TH, Engewald WJ High Resol. Chromatogr. 18 (1995) 635-642

Kühlwein J, Friedrich R, Obermeier A, Theloke J (1999) Estimation and Assessment of uncertainties for high resolution NO_x and NMVOC emission data, PEF - Projekt Europäisches Forschungszentrum für Maßnahmen zur Luftreinhaltung, research report PEF 2 96 002, Karlsruhe

Kühlwein J, Friedrich, R (2000) Uncertainties of modelling emissions from road transport. Atmospheric Environment 34:4603–4610

Lal R (1999) World soils and the greenhouse effect. IGBP - Newsletter 37, 4-5.

Lehning M, Chang DPY, Shonnard DR, Bell RL (1994) An Inversion Algorithm for Determining Area-Source Emissions from Downwind Concentration Measurements. Journal of the Air Waste Management Association 44, 1204-1213.

Lenaers G (1996) On-board real life emission measurements on a 3 way catalyst gasoline car in motor way- rural- and city traffic and on two Euro-1 diesel city buses. Sci Tot Environ 189/190:139–147

Liu XD, Van Espen P, Adams F, Cafmeyer J, Maenhaut W (2000) Biomass Burning in Southern Africa: Individual Particle Characterization of Atmospheric Aerosols and Savanna Fire Samples. J Atmos Chem 36:135–156

Loubet, B, Milford C, Sutton MA, Cellier P (2001) Investigation of the interaction between sources and sinks of atmospheric ammonia in an upland landscape using a simplified dispersion-exchange model. J. Geophys. Res. 106:24183-24195

Lükewille A, Bertok I, Amann M, Cofala J, Gyarfas F, Heyes C, Karvosenoja N, Klimont Z, Schöpp W (2001) A framework to estimate the potential and costs for the control of fine particulate emissions in Europe. Interim Report IR-01-023, IIASA, Laxenburg.

McKee DJ (1994) Tropospheric Ozone, Human Health and Agricultural Impacts. CRC Press, Boca Raton, Florida.

Menzi H, Achermann B (eds) (2000) Proceedings Meeting UN/ECE Ammonia Expert Group. Berne, 18-20 September 2000. Swiss Agency for the Environment, Forests and Landscape (SAEFL), Bern, Switzerland

Milne PJ, Prados AI, Dickerson RR, Doddridge BG, Riemer DD, Zika RG, Merrill JT, Moody JL (1999) Nonmethane hydrocarbon mixing ratios in continental outflow air from eastern North America: Export of ozone precursors to Bermuda. J. Geophys. Res. 105, 9981.

Monteny GJ, Erisman JW (1998) Ammonia emission from airy cow buildings: a review of measurement techniques, influencing factors, and possibilities for reduction. Neth. J. Agr. Sci. 46:225-247

Monteny GJ, Schulte DD, Elzing A, Lamaker EJJ (1998) A conceptual mechanistic model for the ammonia emissions from free stall cubicle dairy cow houses. Transactions of the ASAE 41:193-201

Mueller F (1992) Geographical distribution and seasonal variation of surface emissions an deposition velocities of atmospheric trace gases. J. Geophys. Res. 97, 3787-3804.

Ni J (1999) Mechanistic models of ammonia release from liquid manure: a review. J. Agric. Engng. Res. 72:1-17

Obermeier A (1995) Ermittlung und Analyse von Emissionen flüchtiger organischer Verbindungen in Baden-Württemberg. Institut für Energiewirtschaft und Rationelle Energieanwendung der Universität Stuttgart, Forschungsberichte IER (ISSN 0938-1228), Band 19, Stuttgart.

Olesen JE, Sommer SG (1993) Modelling effects of wind-speed and surface cover on ammonia volatilization from stored pig slurry. Atmospheric Environment 27:2567-2574

Opsis (1997) DOAS User Guide. Opsis AB, Furulund, Sweden.

Osán J, Török B, Török S, Jones KW (1997) Study of the chemical state of toxic metals during the life cycle of fly ash using x-ray absorption near edge structure. X-Ray Spectrom 26:37–44

Osán J, Alföldy B, Török S, Van Grieken R (2002) Characterisation of wood combustion particles using electron probe microanalysis. Atmos Environ 36:2207–2214

Own calculation from IER, biogenic data from R. Steinbrecher, IFU, Garmisch-Partenkirchen, Germany

Palola E (1991) Wood Opportunities. Summary Burning wood under the new Clean Air Act. Independent Energy 21(6):49–53

Petersen SO, Sommer SG, Aaes O (1998) Ammonia losses from urine and dung of grazing cattle: effect of N intake. Atmospheric Environment 32:295-300

Piccot SD, Masemore SS, Herget WF (1994) Validation of a method for estimating pollution emission rates from area sources using open-path FTIR spectroscopy and dispersion modeling techniques. Journal of the Air Waste Management Association 44, 271-279.

Piccot SD, Masemore SS, Lewis-Bevan W, Ringler ES, and Harris DB (1996) Field Assessment of a new Method for Estimating Emission Rates from Volume Sources Using Open-Path FTIR Spectroscopy. Journal of the Air Waste Management Association 46, 159-171.

Pierson WR, Gertler AW, Robinson NF, Sagebiel JC, Zielinska B, Bishop GA, Stedman DH, Zweidinger RB, Ray WD (1996) Real-world automotive emissions – Summary of studies in the Fort McHenry and Tuscarora mountain tunnels. Atmos Environ 30:2233–2256

Poisson N, Kanakidou M, Crutzen PJ (2000) Impact of non-methane tropospheric chemistry and the oxidizing power of the global troposphere: 3-dimensional modelling results. Atmos. Chem. 36, 157-230.

Pope CA, Thun MU, Nambodim MM (1995) Particulate air pollution as a predictor of mortality in a prospective study of US adults. American Journal of Respiratory and Critical Care Medicine 151:669-674

Pregger T, Friedrich R (2001) Anthropogene Quellen und Quellstärken von Primärpartikeln. Proceedings of the Workshop from DECHEMA e.V. "Herausforderung Aerosole vor dem Hintergrund der europäischen Umweltgesetzgebung" Frankfurt a.M., 30.-31.05.2001

Pregger T (2003) Ermittlung und Analyse von Partikelemissionen in Deutschland und Maßnahmen zu deren Minderung. Dissertation, Universität Stuttgart, Institut für Energiewirtschaft und Rationelle Energieanwendung (in preparation)

Preußer C, Dech S, Tungalagsaikhan P, Steinbrecher R (1999) Development of leaf area index (LAI) distributions for Germany from NOAA/AVHRR NDVI satellite data, in Borrell M, Borrell P (eds.), *Proc. EUROTRAC Symposium 1998*, Vol 2, WITpress, Southamton, pp. 50-54.

Priemé A, Knudsen TB, Glasius M, Christensen S (2000) Herbivory by the weevil, *Strophosoma melanogrammum*, causes severalfold increase in emission of monoterpenes from young Norway spruce (*Picea abies*). Atmos. Environ. 34, 711-718.

Proposal for a Directive of the European Parliament and the Council „On the limitation of emissions of volatile organic compounds due to the use of organic solvents in decorative paints and varnishes and vehicle refinishing products and amending Directive 1999/13/EC, COM (2002) 750 final, 2002/0301 (COD), 23.12.2002, Bruessel

Richter K, Knoche R, Schoenemeyer T, Smiatek G, Steinbrecher R (1998) Abschätzung der biogenen Kohlenwasserstoffemissionen in den neuen Bundesländern Zeitschrift für Umweltchemie und Ökotoxikologie 10, 319-324.

Rijkeboer RC, Hendriksen P (1993) Regulated and unregulated exhaust gas components from LD vehicles on petrol diesel LPG and CNG TNO-report 93ORVM: 029/1/PHE/PR TNO Road Vehicles Research Institute Delft, The Netherlands

Rogak SN (1996) Particulate and gaseous pollutant emission factors measured in a traffic tunnel in Vancouver BC. In: 6th CRC On-Road Vehicle Emissions Workshop. San Diego, 3–21

Sandermann H, Wellburn AR, Heath RL (1997) Forest Decline an Ozone. A Comparison of Controlled Chamber and Field Experiments. Ecological Studies 127, Springer, Berlin.

Schaab G, Steinbrecher R, Lacaze B (2003) Influence of seasonality, canopy light extinction and terrain on potential isoprenoid emission from a Mediterranean-type ecosystem in France. JGR, in press

Schäfer K, Steinecke I, Emeis S, Stockhause M, Sussmann R, Trickl T, Reitebuch O, Hoechstetter K, Sedlmaier A, Depta G, Gronauer A, Seedorf J, and Hartung J (1998a) Inverse Modelling on the Basis of Remote Sensing to Determine Emission Rates. Meteorologische Zeitschrift, Neue Folge 7, 7-10.

Schjoerring JK (1991) Ammonia emissions from the foliage of growing plants. In: Sharkey TD, Holland EA, Mooney HA (eds) Trace gas emissions by plants. Academic Press, San Diego, USA, pp 267-292

Schmitz T, Hassel D, Weber FJ (1999) Zusammensetzung der Kohlenwasserstoffe im Abgas unterschiedlicher Fahrzeugkonzepte. Berichte des Forschungszentrum Jülich JÜL-3646 ISSN 0944-2952

Schmitz T, Hassel D, Weber F-J (2000) Determination of VOC-components in the exhaust of gasoline and diesel passenger cars. Atmos Environ34:4639–4647

Schmitz T, Klemp D, Kley D (1997) Messung der Immissionskonzentrationen verschiedener Ozonvorläufersubstanzen in Ballungsgebieten und an Autobahnen. Berichte des Forschungszentrums Jülich JÜL-3457 ISSN 0944-2952

Schnitzler JP, Bauknecht N, Brüggemann N, Einig W, Forkel R, Hampp R, Heiden AC, Heizmann U, Hoffmann T, Holzke C, Jaeger L, Klauer M, Komenda M, Koppmann K, Kreuzwieser, J, Mayer H, Rennenberg H, Smiatek G, Steinbrecher R, Wildt J, Zimmer W "Emission of biogenic volatile organic compounds: An overview of field, laboratory and modelling studies performed during the `Tropospheric Research Program' (TFS) 1997–2000, J. Atmos. Chem. 42(1): 159-177, 2002.

Série E and Joumard R (1997) Modelling of cold start emissions for road vehicles. Institut national de recherche sur les transport et leur sécurité, INRETS report LEN 9706, Februar 1997

Sheffield AE, Gordon GE, Currie LA, Riederer GE (1994) Organic, elemental, and isotopic tracers of air pollution sources in Albuquerque, NM. Atmos Environ 28:1371–1384

Siegl WO, Hammerle RH, Herrmann HM, Wenclawiak BW, Luers-Jongen B (1999) Organic emissions profile for a light-duty diesel vehicle. Atmos Environ 33:797–805

Simpson D, Guenther A, Hewitt CN, Steinbrecher R Biogenic emissions in Europe. 1. Estimates and uncertainties. J. Geophys. Res. 100D, (1995), 22875-22890

Simpson D, Winiwarter W, Börjesson G, Cinderby S, Ferreiro A, Guenther A, Hewitt N, Janson R, Khalil MAK, Owen S, Pierce T, Puxbaum H, Shearer M, Skiba U, Steinbrecher R, Tarrason L, Öquist MG (1999) Inventorying emissions from nature in Europe. J. Geophys. Res. 104, 8113-8152.

Sjödin Å, Sjöberg K, Svanberg PA, Backström H (1996) Verification of Expected Trends in Urban Traffic NOx Emissions from Long-Term Measurements of Ambient NO2 Concentrations in Urban Air. Sc. Total Environ. 189/190, 213.

Smiatek G, Stockwell WR (1999) Application of a geographical information system (GIS) to land use mapping for biogenic emission modelling in Germany and Europe, Proc. 8th A&WMA Conference 8-10 December 1998: *The Emission Inventory: Living in a Global Environment*, New Orleans, 626-636

Smiatek G (2001) GIS and RDBMS class system in support of BVOC Emission Inventories. Proceedings of the 5th GLOREAM Workshop, Wengen 24-26.2001,(2001)

Søgaard HT, Sommer SG, Hutchings NJ, Huijsmans JFM, Bussink DW, Nicholson F (2002) Ammonia volatilization from field-applied animal slurry – the ALFAM model. Atmospheric Environment 36:3309-3319

Someshwar AV (1996) Wood and Combination Wood-Fired Boiler Ash Characterization. J Environ Quality 25:962–972

Sommer SG, Hutchings NJ (1997) Components of ammonia volatilization from cattle and sheep production. In: Jarvis SC, Pain BF (eds) Gaseous nitrogen emissions from grasslands. CAB International, Wallingford, Oxon, UK, pp 79-93

Sommer SG, Hutchings NJ (2001) Review. Ammonia emission from field applied manure and its reduction – invited paper. Eur. J. Agron. 15:1-15

Sommer, S.G., Hutchings, N.J.. Techniques and strategies for the reduction of ammonia emission from agriculture (1995) Wat. Air Soil Pollut. 85:237-248

Spitzer J, Enzinger P, Fankhauser G, Fritz W, Golja F, Stiglbrunner R (1998) Emissionsfaktoren für feste Brennstoffe. Bericht Nr.: IEF-B-07/98, Joanneum Research Graz

Staehelin J, Keller C, Stahel W, Schläpfer K, Wunderli S (1998) Emission factors from road traffic from a tunnel study (Gubrist tunnel, Switzerland). Part III: Results of organic compounds, SO_2 and speciation of organic exhaust emission. Atmos Environ 32:999–1009.

Staehelin J, Locher R, Mönkeberg S and Stahel WA (2001) Contribution of road traffic emissions to ambient air concentrations of hydrocarbons: the interpretation of monitoring measurements of Switzerland by Principal Component Analysis and road tunnel measurements, Int. J. Veh. Design, V27, N1-4, 161-172.

Steinbrecher R (1998) Emission biogener und anthropogener flüchtiger organischer Verbindungen (VOC) und ihre Bedeutung für die Chemie der Atmosphäre. In: Herbsttatung des DGMK-Fachbereichs Verarbeitung und Anwendung, DGMK, Tagungsbericht 9804, Hamburg, ISBN 3-931850-47-1, p. 19-30.

Steinbrecher R, Hauff K, Hakola H, Rössler A revised parametrisation for emission modelling of isoprenoids for boreal plants, in Biogenic VOC Emission and Photochemistry in the Boreal Regionas of Europe , Th. Laurila, V. Lindfors (eds) European Communities 1999 EUR 18910 EN, ISBN 92-828-6990-3, 29-43.

Steinbrecher R, Schaab G, Smiatek G, Zimmer W (2001) Biogenic emission modelling on a regional scale: Some recent improvements, In: Proceedings from the EUROTRAC Symposium 2000, P.M. Midgley, M. Reuther, M. Williams (Eds.), Springer Verlag Berlin, Heidelberg, 203-209.

Steinbrecher (1994) Emission of VOC from selected European ecosystems, in Proc. EUROTRAC Symposium 1994, Transport and Transformation of Pollutants in the Troposphere, eds. Borell PM et al., SPB Academic Publishing the Hague, pp. 448-454.

Steiner D, Burtscher H (1993) Studies on dynamics of adsorption and desorption from combustion particles, by temperature dependent measurement of size, mass and photo-electric yield. Water, Air and Soil Pollution 68:159 – 176

Sternbeck J, Sjödin Å, Andréasson K (2002) Metal emissions from road traffic and the influence of resuspension: results from two tunnel studies. Atmospheric Environment, 36:4735

Stewart EH, Hewitt CN, Bunce RGH, Steinbrecher R, Smiatek G, Schoenemeyer T (2003) A highly spatially and temporally resolved inventory for biogenic isoprene and monoterpene emissions - model description and application to Great Britain. JGR, in press.

Struschka M, Straub D, Baumbach G Emissionsverhalten von Holzbriketts, Braunkohlebri-ketts und Buchenscheitholz. Forschungsbericht Nr. 34-1995. Universität Stuttgart

Sturm PJ (2002) Road size measurements of PM size distribution, Deliverable 7 of the FP5 project PARTICULATES (http://vergina.eng.auth.gr/mech/lat/particulates/)

Sutton MA, Asman WAH, Ellermann T, van Jaarsveld JA, Acker K, Aneja V, Duyzer J, Horvath L, Paramonov S, Mitosinkova M, Sim Tang Y, Achermann B, Gauger T, Bartnicki J, Neftel A, Erisman JW (2002) Establising the link between ammonia emis-sion control and measurements of reduced nitrogen concentrations and deposition. Ac-cepted for publication in Environ. Monit. and Assessment.

TNO (2001) CEPMEIP Database. TNO Institute of Environmental Sciences, Energy Research and Process Innovations, Apeldoorn (NL) (http://www.air.sk/tno/cepmeip/)

Theloke J and Friedrich R (2003) Development of an improved product based approach for the calculation of NMVOC emissions from solvent use in Germany and uncertainty analysis, Contribution to the 12th International Emission Inventory Conference "Emission Inventories – Applying New Technologies", San Diego, 28th april – 1st May 2003, EPA

Theloke J, Obermeier A, Friedrich R Ermittlung der Lösemittelemissionen 1994 in Deutschland und Methoden zur Fortschreibung, Forschungsbericht 295 42 628 im Auftrag des Umweltbundesamtes, Juni 2000

Theloke J, Obermeier A, Friedrich R Assessment of emissions from solvent use in Germa-ny, Gefahrstoffe- Reinhaltung der Luft 61 (2001) Nr.3-März

Trincavelli J, Castellano G (1999) MULTI: an interactive program for quantitation in EPMA. X-Ray Spectrom 28:194–197

Trincavelli J, Van Grieken R (1994) Peak-to-background method for standardless electron microprobe analysis of particles. X-Ray Spectrom 23:254–260

UBA (2000) German Federal Environmental Agency (UBA), personal communication of Mr. Remus 31.03.2000, UBA II 2.2

UNECE "Protocol to the 1979 Convention on Long-Range Transboundary Air Pollution to abate acidification, eutrophication and ground-level ozone", Gothenburg, 30. Novem-ber 1999

UNFCCC „The Protocol to the United Nations Framework Convention on Climatic Change", FCCC/CP/1997/L.7/Add.1, 10 December 1997

Valmari T, Lind TM, Kauppinen EI, Sfiris G, Nilsson K, Maenhaut W (1999) Volatiliza-tion of the Heavy Metals during Circulating Fluidized Bed Combustion of Forest Resi-due. Energy Fuels 13:390–395

Van Breemen N, Burrough PA, Velthorst EJ, van Dobben HF, de Wit T, Ridder TB, Rei-jnders HFR (1982) Soil acidification from atmospheric ammonium sulphate in forest canopy throughfall. Nature 299:548-550

VDI 3945/3, (1999) Umweltmeteorologie - Atmosphärische Ausbreitungsmodelle – Partikelmodell, VDI-Handbuch Reinhaltung der Luft, Band 1. Deutsches Institut für Normung e.V. (DIN), Verein Deutscher Ingenieure (VDI), Düsseldorf.

VDI (1999) Umweltmeteorologie, Emissionen von Gasen, Gerüchen und Stäuben aus diffusen Quellen, Lagerung, Umschlag und Transport von Schüttgütern. VDI 3790, Blatt 3, Düsseldorf

Veldt C in: Baars HP et al. (eds) Proc. TNO/EURASAP workshop on the realiability of VOC emission data bases, IMW-TNO Publication P93/040, Delft 1993, p.5

Venkatram A (2000) A critique of empirical emission factor models: a case study of the AP-42 model for estimating PM-10 emissions from paved roads. Atmospheric Environment 34:1-11

Voorburg JH, Kroodsma W (1992) Volatile emissions of housing systems for cattle. Livestock Production Science 31:57-70

Wang KY, Shallcross DE (2000) Modelling terrestrial biogenic isoprene fluxes and their potential impact on global chemical species using a coupled LSM-CTM model. Atmos. Environ. 34, 2909–2925.

Weingartner E, Keller C, Stahel WA, Burtscher H, Baltensperger U (1997) Aerosol emission in a road tunnel, Atmospheric Environment 31:451

Whitby KT (1978) The physical characteristics of sulphur aerosols. Atmospheric Environment 12:135-159

Winiwarter W, Trenker C, Höflinger W (2001) Österreichische Emissionsinventur für Staub. ARC Seibersdorf research Report, ARC—S-0151, Seibersdorf

Winiwarter W, Trenker C, Höflinger W (2002) Emission Inventory of Particulate Matter (TSP, PM10, PM2.5) for Austria. In: P.M. Midgley, M. Reuther (Eds.), Proceedings from the EUROTRAC-2 Symposium 2002, paper GEN12 (CD-ROM supplement). Margraf-Verlag, Weikersheim

Winiwarter W, Rypdal K (2001) Assessing the uncertainty associated with national greenhouse gas emission inventories: a case study for Austria. Atmospheric Environment. 35:5425–5440

Xiaoshan Z, Yujing M, Wenzhi S, Yahui Z (2000) Seasonal variations of isoprene emissions from deciduous trees. Atmos. Environ. 34, 3027–3032.

Zimmer W, Steinbrecher R, Körner C, Schnitzler JP (2003) The recently developed process-based model SIM-BIM realistically predicts isoprene emissions from adult *Quercus petraea* forest trees. Atmos. Environ, in press

3 Uncertainties, Validation and Verification

3.1 Introduction

J. Kühlwein

As the previous chapter has indicated, significant improvements in the quantification of emissions could be made through the improvement of emission factors in various areas. The next focus of the GENEMIS work was on evaluation of emission data, in particular to assess uncertainties of data and methods and to validate and verify emissions. In the following sections, these aspects of the GENEMIS work will be elaborated on in detail.

Sect. 3.2 deals with the analysis of uncertainties, first establishing methodologies and describing state-of-the-art approaches. In 3.2.1, for instance, a harmonised method for the compilation of urban emission inventories is described, both for bottom up and for top down approaches. The following Sect. 3.2.2. then elaborates on statistical uncertainty analysis, describing detailed methods for road transport and solvent use.

In Sect. 3.3., results of work conducted on the comparison of emissions in general is presented with a spotlight on the comparison of emission inventories. Sect. 3.3.2 is of particular importance as results of comparisons between calculated emissions and measurements are described, with road transport as a specific focus. Several research groups have applied tunnel and road side measurements and remote sensing in intensive measurement campaigns to investigate this area of interest. A central piece of work within GENEMIS is illustrated in Sect. 3.3.3, a large scale evaluation experiment covering a whole city in Germany, the Augsburg experiment (EVA). Emission rates in high spatial, temporal and substance resolution have been calculated for all relevant sources within the city (stationary and mobile), while extensive measurement campaigns took place using tethered ballons, aircraft and ground based measurements. Quality assurance and quality control (QA/QC) form a vital part of the work related to the Augsburg experiment, as a large number of measurement methods, different equipment and the design and implementation of measurements had to be harmonised and the quality of the results needed to be assured.

In Sect. 3.4 finally, an approach for sensitivity analysis is introduced. On the example of tropospheric ozone, model calculations are performed using the EURAD chemical transport model (CTM) with the aim to determine the impact of variations of individual input data on the results of the model calculations, in this case on ambient ozone concentrations.

3.2 Uncertainty Analysis

3.2.1 Harmonized method for the compilation of urban emission inventories for urban air modelling

P. Sturm, W. Winiwarter

3.2.1.1 Introduction

Emission Inventories are important tools to describe the emission situation and eventually to manage air quality. They provide comprehensive information on emission sources and emission fluxes in the area under consideration. With the use of air pollution modelling it is possible to relate air emissions to air quality.

State of the art are so called "top-down" emission inventories describing the emission situation at a regional or national scale. On a Europe-wide basis, CORINAIR/EMEP provide annual emission information. More detailed emission inventories are available for certain regions and/or cities. When looking at pollutant dispersion and air quality modelling it is necessary to describe the dynamic behaviour of the emission sources. During the EUROTRAC I-Subproject GENEMIS the main emphasis was on the temporal resolution of emissions in order to provide data for macro-scale dispersion modelling. Within the SATURN project, dispersion modelling should be undertaken for urban areas.

A prerequisite for urban air models is an emission inventory which is able to deliver high quality emission information with a high resolution in time and space.

Within the Saturn working group 4, validations of air quality models are planned. A prerequisite for this activity is the availability of an appropriate emission inventory. To ensure consistency among the emission inventories used for the validation attempts, guidelines with minimum requirements for emission inventories will be defined by the working group "Urban Emission Inventories" of GEMEMIS and the working group "Val. 3 of Saturn".

The aim of this section is to
- present a harmonized method for the compilation of urban emission data,
- provide sources for suitable emission factors,
- provide tools for producing emission inventories.

3.2.1.2 Methodologies

Different methodologies can be applied to establish an emission inventory. While an inquiry for single sources – based on in site information on emissions and activity data – is defined as bottom-up methodology, a use of statistic data results in a top-down estimation. Bottom-up approaches are connected to big effort in data collection. Therefore this approach is restricted to a certain amount of emitters. On the other hand delivers top-down approaches information on a statistical basis, which is often not detailed enough for urban emission inventories. The big effort for bottom-up methodologies and the inaccuracy of top-down approaches are limitations for both methods. Therefore a combination of both methods is normally used for urban emission estimates. The combination should be made in a way that ensures that the majority of the emission quantities are estimated with the use of bottom-up approaches while the rest is based on a top-down method.

3.2.1.3 General requirements

Pollutants

Council Directive 96/62/EC of 27 September 1996 on ambient air quality assessment and management provides in Annex I a list with atmospheric pollutants that are to be taken into consideration in the assessment and management of ambient air quality. Pollutants to be studied at an initial stage, including pollutants governed by existing ambient air quality directives are:
1. Sulphur dioxide
2. Nitrogen oxides
3. Fine particulate matter such as soot (including PM_{10}, $PM_{2.5}$)
4. Suspended particulate matter
5. Lead
6. Ozone
Other air pollutants to be considered are
7. Benzene
8. Carbon monoxide
9. Poly-aromatic hydrocarbons
10. Cadmium
11. Arsenic
12. Nickel
13. Mercury

Since "assessment" is described as any method used to measure, calculate, predict or estimate the level of a pollutant in the ambient air, it is important that the urban emission inventory includes at least the pollutants that are referred to in the first part of the list, i.e. SO_2, NO_2, PM_{10}, SPM, Pb and the ozone precursors NO_x and VOCs. With regard to health effects of urban air quality, benzene and CO are relevant as well. This first group of air pollutants has thus the highest priority within urban air quality assessment and urban air quality modelling. In addition, the quality and availability of emission data for these pollutants is much better than for the rest (except CO_2).

Although the other air pollutants mentioned above, PAH, Cd, As, Ni and Hg are relevant with respect to health effects as well, they probably appear less frequently in urban environments. However in specific urban situations an assessment of these pollutants might be important and inclusion of one or more pollutants from this second group might be desirable.

Pollutants with regard to regional or global impact that are not included in the list are: CO_2, CH_4, N_2O, etc. These pollutants might not be considered for urban chemistry, nevertheless information about CO_2 emissions should be to be included as well.

Emission sources

Usually, emission inventories classify emissions into three source types, namely point, area, and line sources. For point sources, emission estimates are provided on an individual plant or emission outlet basis. Small or diffuse sources are normally dealt with within area sources, where the necessary information is provided on an area basis. Line sources are used to estimate emissions along routes, like streets, waterways, etc. Line and area sources are a statistical description of a large number of point sources. When dealing with area sources it is always a question of the spatial resolution and source strength if an accumulation of point sources (or even line sources) is dealt with as area source or point source.

According to the CORINAIR Emission Inventory Handbook point, area and line sources are defined as follows:

Point sources - emission estimates are provided on an individual plant or emission outlet (usually large) usually in conjunction with data on location, capacity or throughput, operating conditions etc.

Area sources - smaller or more diffuse sources of pollution are provided on an area basis.

Line sources - in inventories, vehicle emissions from road transport, railways, inland navigation, shipping or aviation etc are provided for sections along the line of the road, railway-track, sea-lane etc.

In order to cover all activities which are related to emissions, a nomenclature must be used. The most recent emission source nomenclature for anthropogenic and natural activities is the SNAP 94 (Selected Nomenclature for Air Pollution) developed by the EEA. Together with the EMEP/CORINAIR "Atmospheric Emission Inventory Guidebook", a framework for estimating emissions can be defined.

The following contains a source oriented listing as well as requirements for data capturing for each of these source groups. Note that - in contrast to national inventories - industrial activities will be looked at on a plant-by-plant basis rather than using economic statistics. Thus emission figures will be taken from numbers given directly for installations within the plant, rather than applying top-down emission factors. This way the CORINAIR SNAP code cannot be fully applied here, as on the specific installation level measurement data may be more readily available than production figures.

This list will be used throughout the document describing the requirements, methodologies, and data structure for emission estimates.

Stationary sources

- Power, cogeneration and district heating plants (SNAP CODE 01, 02)
  This group covers emissions from power and heating plants. There is a need for a threshold value to separate this group from small combustion processes for space and water heating which will be covered in the group "domestic heating, private households". Due to the fact that such emitters contribute strongly to the overall emissions (if they are present in urban areas) in situ information has to be used to define the source strength.
- Trade and industry (SNAP CODE 03, 04, 05, 06)
  This group covers emissions from industrial combustion and production from the industry and trade emission sector.
- Farming/agriculture (SNAP CODE 10, 0203)
  Emissions from farming and agriculture, greenhouses with heating systems, life stock, use of fertilizer, etc. are covered by this group.
- Domestic heating, private households (small combustion for space and water heating) (SNAP CODE 02, 06)
  Emissions resulting from space and water heating for households, public buildings, and small business as well as solvent use in private households are covered in this group.
- Nature (SNAP CODE 11)
  This group contains biogenic emissions as far as they are not covered in the agriculture group.
- Other sources
  Emissions which cannot directly be assigned to one of the above mentioned groups will be covered here as well as waste water treatment and disposal (SNAP CODE 09 except 0902).

Mobile sources

This section covers the emissions from road transport, off-road vehicles and machinery, railways, waterways and aircrafts.
- Road traffic (SNAP Code 07)
  Road traffic covers emissions from vehicles on the road including parking lots, garages, tunnel ventilation stacks and evaporative emissions. Emissions from gasoline stations belong to stationary sources.
- Off-road vehicles and machinery (SNAP Code 0801, 0806, 0807, 0808, 0810)
  This group covers emissions from agriculture, forestry, industry, military and household / gardening. In winter sport resorts relevant emission sources like diesel engines for lifts, etc., must be taken into account.
- Railway traffic (SNAP Code 0802)
  This group includes the non-electrified activities on railroads (diesel powered trains) and shunting activities caused by non-electrified engines. Manipulation of goods at the stations or container terminals must be treated as area sources according to road or off-road activities.
- Maritime and inland traffic on waterways (SNAP Code 0803 and 0804)
  Here, the emissions on waterways, lakes and harbours are included

(commercial and sporting activities). Motorized ground traffic at harbours must be treated as an area source according to road or off-road activities.

- Air traffic (SNAP Code 0805)
  Traffic at airports and emissions from landing/take-off (LTO-cycle) must be considered. Motorized ground traffic at airports must be treated as an area source according to road or off-road activities.

Resolution (temporal, spatial)

Temporal resolution

The distinction between high and low time resolution of emission data is important. For micro-scale or local scale estimation as well as for the simulation of air pollution episodes, high time resolution emission inventories are needed. To estimate the background concentration of primary pollutants, a lower temporal resolution is sufficient. To this aim, the distinction between continuous and discontinuous sources, taking into account seasonal variations, can be appropriate. In principle the time variation of the emission quantity follows the time dependency of the activity pattern.

The standard for urban air modelling is a time resolution of one hour. This time resolution is necessary for activity parameters like traffic density, as well as for ambient parameters, like temperature. It should be noticed that time dependent activity patterns are a function of local time (day light saving time), while ambient parameters like temperature are often given for standard time (UTC).

Spatial resolution

A characteristic of urban air emission inventories is the high spatial resolution in order to simulate the emission situation in an appropriate way. That means the smaller the region of interest is the finer the spatial resolution of the emission inventory has to be. When focusing on local problems a spatial resolution of 250 m to 500 m in square is recommended, while for regional problems (O_3) a resolution of 1 km by 1 km is sufficient. For micro-scale models a much finer resolution is needed in order to represent the location and emission behaviour as accurate as possible. But it has to be noted that for such applications a bottom up approach has to be applied. A disaggregation of statistical emission data may not be sufficient.

In order to ensure usability of gridded emission information a reference for the origin of the grid and a conversion to the national or an international grid or co-ordinate system has to be given.

Emission factors

Data used to compile emission inventories can originate from a wide range of sources: in the case of large point sources actual continuous measurements are usually available; emission estimates may also be based on discontinuous measurements or on emission factors. For many activities, direct measurements are not available, therefore there is a wide utilization of emission factors in the compilation of emission inventories, especially referring to those activities characterized by emissions from traffic, small combustion, etc.

In general, a reference must be given for all the emission factors used.

VOC-Split

To simulate photochemical processes it is necessary to have information about different reactive components of VOC. There is only little information available about direct measurements of VOC species within emission inventories. Therefore, a disaggregation of VOC is performed according to different relations within different source categories.

For chemical reaction models a split into alkanes, alkenes, aromatics and carbonyls is a necessary minimum. In order to make best use of state-of-the-art chemical mechanisms a further split is required:

Alkanes: methane, ethane, propane and higher alkanes

Alkenes: ethene, terminal alkenes, internal alkenes, dienes and biogenic alkenes

Aromatics: benzene, toluene and xylene

Carbonyls: formaldehyde acetaldehyde and higher aldehydes; ketones

In addition, information about organic nitrogen, organic peroxides and organic acids is desirable.

It has to be noted that many models for urban chemistry need a very detailed VOC specification. On the other hand the knowledge about VOC splits for different source types has to be remarkably improved to deliver VOC split data with the required accuracy.

NO_x Split

In emission inventories, emission quantities from nitrogen oxides (NO_x) are always given as a total amount of NO_2 although they consist mainly of NO. NO reacts in the atmosphere eventually to NO_2 using O_3 or radicals as oxidants. For photochemical models it is necessary to have as detailed as possible the fraction of NO and NO_2 as a starting condition. In the emission inventory, the NO and NO_2 proportion at the exhaust stack, pipe, etc. must be given. Which share is eventually used within the urban air model is a function of the grid size of the model.

3.2.1.4 Generation of urban emission inventories

Bottom-up methodology

The bottom-up methodology is based on a source oriented inquiry of all activity and emission data needed for describing the emission behaviour of a single source. Due to the fact that this is connected to a big effort in data collection, real bottom-up approaches are limited to certain emission sources and types.

Although the following requirements and recommendations are listed under bottom-up methodology, they include also information for top-down approaches.

Data structure

This section gives a short summary of a typical data structure for an urban emission inventory. The necessary data to be collected, such as source categories, geographical data and administrative data, will be described.

Data types

The data can be categorized as 'static lookup data', 'lookup data' and 'dynamic input data'.

'Static lookup data' are data such as units, unit conversion factors, compounds, etc. These data are normally entered into the database before installation.

'Lookup Data' include emission factors, building registers, road link definitions, source complex and company registers, administrative district boundaries, etc. Although the 'lookup data' may change over time, they are considered to change less frequently then the 'dynamic input data'.

'Dynamic input data' are data that needs to be updated regularly as they have only a limited validity period. This group includes all dynamic data that are not considered to be lookup data, such as e.g. area distributed consumption data, population data, point source emission/consumption/production data, time variations, etc.

Time variation

Most of the activities which create air emissions in a city are time dependant. Normally, emission or energy consumption values are averaged over longer period such as month or year. To describe the emissions to air as a function of time of the day, week, month, year; time variation functions must be established.

The structure of Time Variations is hierarchical. This means that one Time Variation may have one or several Sub Time Variations. This structure takes care of the periodicity and logical structure that is often found in such Time Variations.

Point source sata

The data describing point source emissions can be divided into several groups. Some of the data describes the stacks through which the emissions are released to the atmosphere. Other data describes any cleaning devices that may be installed to reduce the emissions, and yet other data describe the activity that generate the emission, in the form of emission, consumption or production data for a given process.

- Stack Data: The information of stack data contains standard parameters such as co-ordinates, identification numbers, name, height, diameter, gas flow rate, building dimensions and emissions.
- Cleaning Device Data: The following data may be given for each cleaning device, and for each component of which the emission is reduced by the cleaning device: identification number, physical links between units, cleaning efficiency factor etc.
- Process Data (Activity): Industrial activities can be divided into different processes. For each process data can be collected on consumption, production

or directly on emissions. The parameters to be collected for the three alternatives are similar, such as identification numbers of the process. For energy consumption and production, fuel type or product type, respectively, must be defined.

In addition to or instead of consumption or production data, one may have measured data for the process emissions. These data may be in the form of measured emission, or as measured flue gas concentration and flow rate.

Area source data

Area Source Data may include emission data and energy consumption data distributed in administrative sub districts or in a grid. Necessary grids or administrative district level must be defined before import of these data.

This includes the geographical boundary of each district.

Traffic source data

For the definition of traffic source data, a number of 'Lookup Data' concerning traffic data and the modelling of traffic induced emissions must be defined.

- Road Classes: Road classes are used to categorize road segments. A number of traffic parameters may be assigned default values according to the road class to which the individual road belongs. Typical road classification can be: major roads / transit roads, city centre streets, residential area streets, industrial area streets and secondary streets.
- Vehicle Classes: Before importing data for traffic volumes, it is necessary to define the relevant Vehicle Classes. Such classes may be: personal cars, buses, two-stroke engine motorcycles, trucks etc. The division is such that it is natural to assign emission factors to each Vehicle Class separately, as well as it is possible to obtain traffic counting of such vehicles on the various roads.
- Road Link Definition: The road network consists of several road segments which are called road links. These must be described with identification number, co-ordinates and name. In addition, physical description such as number of lanes, width in each direction, road class type, must be entered.
- Dynamic Traffic Data: For each road link or classes of road links, detailed information on the traffic flow must be collected. Examples of such data are: annual daily traffic, free flow speed, cold start ratio and vehicle class distribution. Most of the parameters above must be given for each lane and direction.

Population data

Population data may be defined as spatially distributed on administrative districts in the same way as area source data. They may also be imported in exactly the same way. Population data is normally used for distribution of data which is related to human activities.

Description of emission sources

In the following, a description of activity data for the different emissions sources according to the definition in section 0 will be given. The description is structured into actions (what has to be done), sources (where is such data normally available), and data needed (which data is needed to estimate the emission quantity - activity).

Stationary sources
- Power Generation and District Heating Plants (SNAP Code 01, 02, 0902)
  Detailed investigation of single sources concerning relevant emission data including data from emission measurements and production.
- Data needed:
  Co-ordinates of the source (x, y, z), source strength (volume flow, emissions per volume, temperature of the effluent, velocity of the effluent, stack diameter, operation hours), fuel data (fuel consumption, type of fuel used), exhaust gas treatment system.
- Industry and Trade (SNAP Code 03, 04, 0506)
  Detailed investigations for emission relevant companies (distinction see section 0) and use of statistic data for energy consumption and productions in industrial processes
- Data needed: point sources see section "data structure"; material throughput, energy demand, statistical data for small sources.
- Agriculture/Farming (SNAP Code 10)
  Use of statistic data about life stock, amount of fertilizers used, energy consumption figures, etc., and detailed investigations for major single emitters.
- Domestic heating, private households, small business (small combustion for space and water heating) (SNAP CODE 02, 06)
  Data needed:
  Point sources: Co-ordinates of the source, energy demand, type of heating system, type of fuel used.
  Area sources: Co-ordinates of the area source, number of persons and/or places / employments per area source, energy needed per m^2 used area and K, type of fuel used, type of heating, heating-degree-days.
- Nature (SNAP Code 11)
  Biogenic emissions from agricultural crops, forest meadows and soil (compounds: isopren, monoterpens and alfa-pinen).
- Data needed: Co-ordinates of area sources, land use including specification of vegetation.
  For biogenic emissions from agricultural crops crop type, seasonal /permanent cultures, emission factors.
  Biogenic emissions from forest: land cover classes, biomass density for each species type, hourly meteorological data (temperature and radiation).
- Other stationary sources
  Use of statistics on energy consumption, number of work places, persons per

company etc.; detailed investigation for relevant single emitters, material throughput, energy demand.

Mobile sources
- Road Traffic (SNAP Code 07)

  Source (road) definition: Main road network as line sources, secondary roads with little traffic as area sources.

  Traffic volume: Use of traffic counts and/or traffic models for the main road network and estimations for area sources; consideration of the fleet distribution.

  Driving characteristic, street characteristic and emission factors: following the recommendation of the MEET deliverable No.14 "Average hot Emission Factors for Passenger Cars and Light Duty Trucks"
  Data needed:
  Hot emissions: Co-ordinates of the endpoints of the individual sections, street characteristics (length, gradient, altitude), traffic characteristics (number of the vehicles, HDV share, vehicle gross mass), driving characteristics (mean velocity, street and traffic parameters).
  Cold start emissions: Number of cold starts, number and length of trips with cold start, number of warm starts, ambient temperature.
  Evaporative emissions: Parking duration per car, trip length before parking, ambient temperature profile.

- Off-road vehicles and machinery
  Methodology following the recommendations of the CORINAIR methodology.
  Data needed: Co-ordinates of area sources, number and type (category) of off-road machinery, fuel consumption per category, time distribution.

- Railway traffic:
  Source definition: Rail network as line sources, railway stations and shunting areas as area sources.
  Traffic volume: Use of schedules for the traffic within the rail-net, use of fuel consumption for shunting activities. For characteristics and emission factors following the recommendations of the MEET-Deliverable No.17 "Railway traffic".
  Data needed: Co-ordinates of the end points of the individual sections (rail net) or co-ordinates of the area (shunting activity); traffic characteristics (number and type of trains), fuel consumption, time distribution according to the schedules; fuel consumption and operating hours for shunting activities.

- Maritime and inland traffic on Waterways (SNAP Code 0803 and 0804)
  Source definition: Rivers and channels as line sources, harbours as area sources
  Traffic flow: by analogy with road and rail traffic; characteristics and emission factors following the recommendations of the MEET-Deliverable No. 19 "Emissions from ships".

<u>Data needed:</u> Following the data structure for road and railway traffic
- Air traffic (SNAP Code 0805)

    The necessary activity data must be determined by analogy with the above mentioned mobile emission sources, following the recommendations of the MEET deliverable No. 18 "Emissions from Air traffic".

    <u>Data needed:</u> Following the data structure for road and railway traffic

Allocation of the emission sources

High or low spatial resolution of emission data is one of the most important dimensions which characterize emission inventories. To provide a micro-scale or a local scale assessment, an inventory with high spatial resolution is needed. To estimate background concentration, inventories with a lower spatial resolution are sufficient.

According to their source strength or to the degree of detailing of the inventory, emitters can be defined as point, line or area sources. In general, they can be grouped into stationary and mobile sources. Mobile sources on the main road network can be treated as line sources, while many small emitters (e.g. small stacks from residential combustion) can be treated as area sources. The co-ordinates must fit to the national reference co-ordinates or the UTM system.

An accurate allocation of the emission sources is very important for urban (micro-scale) emission inventories.

<u>Point sources</u> must be assigned via their geographical co-ordinates (longitude, latitude and altitude) and height above ground (stack height).

<u>Line sources</u> must be divided into sections according to certain requirements. An intersection (knot-point) must be set
- for a second line source branch of (crossings),
- when the emission situation changes due to certain conditions (e.g. changes in vehicle speed, changes in road/rail gradient etc.),
- when the location of a simplified line source differs more than a certain value (1%o from the scale used) from the geographical location.

<u>Area sources</u> are used when the activity data or the emission factor is a function of an area (e.g. emission in g/m^2 or activity in persons/m^2). The area source must be defined as a closed polygon line according to the location where the emission takes place (e.g. fugitive emissions on a waste disposal site) including the emission height. The accuracy of the boundaries of the area source should be 1 %o from the scale of the map used.

Distinction between point and area sources

In addition to physical area sources like landfills, larger collections of point sources are often combined to form area sources. This would appear to make sense particularly when a lot of small chimneys are emitting within the "canopy layer", as is the case, for example, with domestic heating. The situation with regard to industrial emissions is a similar one where a large number of diffuse sources emit directly above roof (e.g. dust from hall ventilation).

Another distinguishing feature can be found in the degree of detailing of the emission inventory. The greater the degree of detail, the more emission sources are indicated as point sources. The question as to whether point or area source is thus, in addition to the physical description, also a question of survey (in situ information) and available data material.

In connection with the creation of emission inventories, the following definition characteristics evolve for distinguishing between point and area sources.

- Basic and line type of emitter
    - point source: definable single source allowing an exact indication of the emission site related to a chimney
    - area source: diffuse emission from a certain area
- Distinction on the basis of survey precision
  Specifying a threshold value allows us to define a distinguishing feature as to whether an emitter can be described as a single emitter or point source. This threshold value can refer to the thermal output of a plant, material throughput or a particular volume of emission. Following the SNAP code, thermal plants with a capacity of 50 MW, for example, must of course be individually surveyed. This threshold value is of course far too high for urban emission sources. Threshold values should be selected as dependent upon the size of the survey area and the expected total emission volume in such a way that the majority of the emissions (> 90 %) from all activities – except Traffic (SNAP code 07/08), Agriculture and forestry (SNAP code 10), Nature (SNAP code 11) and activities connected with private households – are described with the aid of point sources.
- Distinction on the basis of activity data
  For some activities – such as domestic heating – it is necessary to use statistical data. These data generally exclusively refer to territorial units (e.g. statistical district). Such emitters are described as area sources.

3.2.1.4 Top down

Spatial disaggregation from CORINAIR level

Information on emissions is very often being derived on the scale of an administrative unit. For this administrative unit the respective statistical and socio-economical data is available. Within international inventories, administrative units used are countries. Starting from national figures, a top-down approach allows attributing the emissions to smaller units (for example, in CORINAIR inventory national data are finally broken down to the level of NUTS-III regions, which have the size of a British county or a French département). Parameters used for such a disaggregation are listed in Table 3.1, detailed information has been given by Orthofer and Winiwarter (1998).

Table 3.1. Examples for parameters used for disaggregation to smaller administrative units

SOURCE CATEGORY	DISAGGREGATION PARAMETER
Industrial fuel use	Employment statistics: industrial employment Industrial statistics: nominal production, energy consumption, fuel consumption, etc.
Industrial processes	industrial employment by branch industrial statistics: nominal capacity of production, raw material consumption
Domestic combustion	population or heating equipment and specific heat demand, domestic fuel consumption or sales
Domestic solvent use	Population
Regional road traffic	car registrations; population; fuel consumption or sales
Nature and agriculture	Statistics: total area or specific for forest uses of the administrative units

Furthermore, disaggregation may be performed using land use information rather than administrative boundaries. Features that are relevant with respect to emissions are:

- directly emitting area, emissions related to area: forests
- directly emitting area, emissions related to activity: roads
- area with emissions related to activities that can be derived from statistical data: settlement area, built-up area

Allocation / limitations

The top-down approach described above is characterized by the fact that the emissions are derived for the largest unit, with uncertainty increasing the further down emissions are broken. Note that, despite increasing uncertainty, specific emission attribution within an area still increases the overall accuracy for that area with respect to using just averages. Nevertheless such approaches lack of including local data, as they use typical country-wide behavioural patterns which may not be reflected by an urban area to be considered specifically.

With respect to bottom-up emission assessments, a trade-off between effort spent on assessing single sources and accuracy of specific information has to be performed in all cases. Notably, this trade-off between top-down and bottom-up approaches is quite different for application to different scales.

From the previous, it becomes clear that the widely available national CORINAIR figures might not be useful for urban inventories. Even the emissions on NUTS 3 level are usually derived from the national totals by disaggregation and therefore do not carry much additional information. A resolution needed for urban plume characterization is on the scale of several km^2, an urban flow model needs input in the order of a few 100 m resolution.

Temporal disaggregation

The general idea of a temporal disaggregation adheres to a similar philosophy as the spatial disaggregation. The basis for the disaggregation is the sum of annual total emissions. These numbers are relatively easily available in many countries, because underlying statistics usually refer to annual time period, too.

The use of standardized fixed emission quotas for similar source categories is a very straightforward approach to approximate the temporal emission patterns. The method aims at modelling the "typical" rather than the "actual" emission dynamics using surrogate data. While in a kind of bottom-up approach also specific data might be used, the standardized fixed quota approach applies surrogate parameters that in general are not directly or not at all linked with the emission source categories.

The annual emissions are broken down to monthly emissions. The monthly quota is derived as the contribution of the respective month in relation to the average monthly contribution. This means that the monthly emissions can be calculated by multiplying the annual emissions with the monthly quota and divide by the number of months by year. Similar quota can be obtained for each day of the week (for anthropogenic activities only), and for each hour in a day. The emission for one certain hour within a year now can be computed from the annual total emissions by:

$$E_h = E_a * f_a * f_w * f_d / 8760 \qquad \text{(3.1.)}$$

with:

E_h Emission for a certain hour
E_a Annual total emissions
f_a Emission quota for annual cycle
f_w Emission quota for the weekly cycle
f_d Emission quota for the diurnal cycle
8760   Number of hours per year

The emission quota (f_a, f_w, f_d) are unities for constant emissions and vary with respect to the emission variability.

The most important step in creating a useful temporal disaggregation is the selection of adequate surrogate data from socio-economic, statistical or meteorological information. Table 3.2. shows an example of surrogate statistics used for a temporal disaggregation of different source groups.

It should be noted that in some cases also regional differences need to be considered. In a study for Austria particular regional patterns for domestic heating and for traffic (Winiwarter 1993) were noted. Heating activities depend on the climate patterns which are very different in the Alps and in the lower regions. Traffic patterns, on the other hand, were attributable to consumer behaviour.

Table 3.2. Surrogate data used for the definition of standardized fixed emission quota for the disaggregation of NMVOC emissions in Austria (Winiwarter 1993)

SOURCE GROUP	SURROGATE DATA
Domestic/Industrial heat demand *	Gas consumption, degree-days
Traffic – exhaust **	Traffic counts
Traffic – evaporation	Hourly traffic counts, one hr delayed; fuel volatility & monthly temperature
Traffic – motorbikes	Traffic counts
Solvents	Model for 2 working shifts in construction industry
Other industry	Uniform emissions (no dynamics)
Stubble burning	Model for one month burning in summer, 12 hours per day
Forests	Monthly mean temperatures; diurnal temperature variations

*   three regionally different quota   ** two regionally different quota

In addition to the parameters shown above, which may be difficult to obtain in some countries, other derived parameters can be used for disaggregation.

The main advantages of the standardized fixed quota approach are the wide range of surrogate data which can be used and the easy introduction into larger models. However, it must be noted that the standardized fixed quota cannot consider specific "untypical" patterns of emission sources or irregularities of individual emission sources. For instance, the weekly quota will not consider national holidays other than Sundays. Also, some important stochastic influences of climate patterns on emissions (such as the influence of sunshine or rain on weekend traffic patterns) are usually neglected in the fixed quota system. Thus the outcome of such disaggregation is the pattern of a "typical" rather than a "real" situation of an episode. Nevertheless, as "real" information and source-specific quota frequently will not be available anyway for a great number of emission sources, this standardized quota approach has to be employed within all emission inventories at least to some extent, if temporal disaggregation is needed.

3.2.1.5 Combined method

Exclusive use of "bottom-up" methods fails due to lack in available input data, while exclusive use of "top-down" methods will only lead to an undesirable level of accuracy for urban emission inventories. For this reason, both methodical approaches are always combined.

In order to achieve the necessary level of accuracy required for urban air modelling, it is recommended to use the following methodical approaches:

Mobile sources:

Road traffic
- Bottom up: Traffic on the main road network
- Top-down: Traffic on the secondary road network (area sources)
                 Cold start emissions and evaporation losses
                 Activity data based on statistical information like fleet distribution per
                 model year (share of VMT model year)

Off-road traffic
Top-down approach based on statistical information

Railway traffic
- Bottom-up: Regular traffic on the rail network
- Top-down: Shunting activities

Maritime and inland traffic on waterways
- Bottom-up: Scheduled traffic for ferry boats, etc.
- Top-down: Harbour activities

Air traffic
   Methodology in analogy to railway and waterway traffic

Stationary sources

Bottom-up approaches are applied for point sources while area sources are mostly treated with top-down methodologies.

If the activity data is given on a statistical basis (e.g. for residential combustion), a top-down methodology must be applied, e.g. for SNAP sectors 2 (heat demand), 6 (solvents), 10 (agriculture) and 11 (nature).

If enough information on location, activity data and emission factors is available it is possible to use a bottom-up approach instead of a top-down method. E.g.: if detailed information on land use cover and location is available for emission sources from nature and agriculture, and specific meteorological data and emission factors can be used the approach can be considered as bottom-up.

3.2.1.6 Quality assurance

Quality assurance is imperative for urban emission inventories at all and especially for those which should eventually be used for urban air modelling. The following section contains information from Cirillo et al. (1996) extended and adjusted for urban air modelling.

General principles

All information compiled under the heading "Quality control / quality assurance" needs to be made available at least within the respective project in order to warrant the conditions of its collection. A quality management person must be appointed within each project to be responsible for assessing the fulfilling of the guidelines given below.

Emissions

Emission inventories are compiled from information on the activity of any given source category, and its emission factor. Any emission inventory that may undergo quality checks needs to be well documented in both terms. Additionally, the assumptions and functions used for temporal and spatial disaggregation as well as for the species split (e.g. NMVOC's) need to be documented (see the ANNEX 1).

Completeness analysis (Checklist)

A completeness analysis shall ensure that all activities - anthropogenic and natural - which generate emissions of a certain pollutant have been considered, or at least the emission sources not considered are negligible with respect to considered time and spatial scale (see the ANNEX 2). The check list shall follow the source description including pollutants covered and resolution requirements given in the previous sections.

3.2.1.7 Uncertainty

Essential for any quality evaluation is an assessment of uncertainties. The uncertainty of an emission inventory with respect to each of its elements is related to one of the following[1]:

    i)   uncertainty related to the choice of the indicators;
    ii) uncertainty related to the quantitative value of the indicators;
    iii) uncertainty related to the emission factors;
    iv) uncertainty related to the representativity of the emission factor for the sources it is applied to;
    v) uncertainty related to a complex structure of emissions estimate models (e.g. in the case of traffic or natural emissions).
    vi) uncertainty related to the temporal and spatial attribution of emissions (esp. for top-down inventories) or the species split.

While (ii) and (iii) can be described quantitatively by the variance of the inputs which may be used to quantify the "random error" associated with the inventory, (i), (iv) through (vi) describe a "systematic error" which cannot be quantified.

---

[1] list has been adapted from Cirillo et al. (1996)

All of these points must be considered for each of the input parameters used for the inventory. An uncertainty matrix must be created (cf. ANNEX 3) and used as a checklist, such that all considerations are documented.

3.2.1.8 Retrospective assessment

In addition to the considerations towards the elements of the inventory, an overall assessment must be performed and documented according to the following items:

i.   analysis of the relative weight of different activities/emission sources typologies, with respect to the total emission;
ii. sensitivity analysis;
iii. overall uncertainty analysis;

   While (i) and (iii) are mandatory for any urban emission inventory, (ii) is optional. In this context, realistic upper and lower limit inputs are to be applied on emission models in a Monte-Carlo type assessment.

3.2.1.9 Validation

Emission inventories are based on a certain amount of assumptions, best guesses and engineers' judgments. Especially urban emission inventories, while assembling a considerable amount of detail and determined to be very accurate, may lack precision due to the fact that emission factors that may be applicable to averages may be not so for individual sources.

   This is where the need arises to compare the results of an inventory to an independent approach. In the following, examples for such comparisons are given. For an urban emission inventory, at least one of such validation approaches will have to be taken, most of which are connected with measurements.

Alternative emission assessment

While not giving sufficient detail for an urban inventory, for most European cities there exist data from the CORINAIR exercise (regionalized to NUTS III level). These top down data (or data from other inventory) must be compared in detail with the emission totals from the urban emission inventory. The method will not be valid when the UEI is derived closely from the CORINAIR inventory, or when it is not possible to discuss possible reasons for discrepancy (e.g. due to insufficient documentation of an inventory).

Emission trends vs. ambient air concentration trends

If the emission inventory comprises data of several years, data on air quality from routine monitoring sites may be employed to compare trends. Due to different atmospheric dispersion at different times of a year, only annual averages should be compared. As a minimum, the trends for three components need to be assessed (SO_2, NO_x, CO).

An improvement of this method would rely on a dispersion model to simulate the concentration to be expected from the emissions and emission changes. This way the meteorological fluctuations would be taken care of.

In both cases, trend analysis of emission vs. concentration allows to conclude on the validity of the emission changes, but not on the absolute magnitude of the emissions.

Flux estimations from downwind measurements

On the scale of a city, the fact that urban emission densities are at least an order of magnitude larger than in the surroundings may be taken advantage of. An urban plume therefore can easily be discerned from the background air. Measurements have shown that under favourable conditions the vertical concentration profile in such an urban plume is relatively uniform in some distance from the individual sources (Winiwarter et al. 1997). The mean plume increment (plume – background concentrations) is derived from airborne measurements along horizontal cross-sections of the plume.

Obviously, for reactive species the atmospheric transformation must be considered even for transport times of as little as one hour.

Atmospheric models

Atmospheric models may be applied to account for the processes occurring between emission and receptor. Basically such models are able to handle also complex transport conditions and thus have a wide range of applicability. Also, due to the widely used concept of modelling the vertical structure of the atmosphere in different layers, results are available also for ground levels and can therefore be directly compared to ground-based measurements.

Simple dispersion models that are designed to be applied to point sources will lose performance when applied to a multitude of sources, as is available in cities, however. On an even larger scale, emission inventories have been validated using complex atmospheric models (Pulles et al. 1996). Such approaches, however, yield less relevant results for urban conditions. While a direct comparison to measurement data is the main benefit of this approach, discrepancies observed cannot clearly be attributed to either the emission inventory or the simulation of transport and transformation or the measurement.

Fingerprints and receptor modelling

An additional validation method is available which may be applied to any scale. This method takes advantage of the relative contribution of individual compounds in the plume. Using the ratios, a certain "fingerprint" of pollution can be assessed, and downwind measurements (also ground-based) may be compared with an emission profile.

The quality of conclusions derived depends upon number and quality of the species for which both measurements and emission data are available. Ideally, discrepancies can be attributed to source classes using receptor models and this way identified. No quantification in absolute terms is possible, however.

3.2.1.10 Documentation

Documentation is a main issue in context with quality assurance. The documentation has to cover the methods used, emission factors, sources for activity data and notes of the QA-procedure itself.

Emission inventories need to be carefully documented with respect to:

- the input data used and the emissions derived
- the completeness of the inventory (checklist)
- the uncertainty associated with each of the inputs
- a subsequent analysis with respect to the overall contributions and uncertainties
- and their validation.

3.2.1.11 Acknowledgments

These guidelines were prepared with the help of colleagues within the framework of the EUROTRAC 2 subprojects SATURN and GENEMIS. The authors want to thank H. ApSimon: Imperial College London, T. Böhler: NILU, P.O.Box 100, N-2007 Kjeller, M. Lopes, C. Borrego: Departamento de Ambiente e Ordenamento, Universidade de Aveiro, P-3810 Aveiro, C. Mensink: Centre for remote sensing and atmospheric processes, VITO, B-2400 Mol, P. Blank, R. Friedrich: Institut für Energiewirtschaft und Rationelle Energieanwendung, Universität Stuttgart, Heßbrühlstraße 49a, D-70565 Stuttgart, M.Volta, G.Finzi : Dipartimento di Elettronica per lÁutomazione, Universita´degli Studi die Brecia, I-25123 Brescia, and N. Moussiopoulos: LHT, Aristotle University Thessaloniki, GR- Thessaloniki for their contributions.

3.2.2. Statistical uncertainty analyses

J. Kühlwein, J. Theloke, R. Friedrich

With the results of statistical uncertainty calculations it is possible to assess emission data's qualities and to filter out those model input data whose errors are contributing most to the error of total emissions. Therefore it is possible to find out the weak points of the used emission model and to do efficient model improvements in future.

3.2.2.1 Methods of error estimation

The quantification of extensive pollutant emissions by measurements requires high efforts. So, usually emissions are not measured directly (except for large point sources) but determined by model calculations based on results from mediating observations. The emission E of a pollutant can be expressed as a function of a set of input parameters $x_1, x_2, ..x_n$. These input parameters are emission factors and activity parameters like traffic flows, vehicle fleet compositions, driving pattern mixes, road specific parameters, national energy balances, solvent consumption etc. (Kühlwein and Friedrich 2000).

Usually the values of parameters used for emission modelling of a definite case are not known exactly. This is mainly because of the necessity to transfer measurement results from other cases (e.g. from random samples) to the current case of interest. Additional errors arise from temporal and spatial disaggregation of more global data and from measurement uncertainties. About some important influential parameters reliable measurement data is not available at all. Such data gaps have to be filled temporarily by highly uncertain judgements.

Systematic errors have to be distinguished from statistical ones. Quantifying them is a much more difficult task than calculating statistical errors, because independent data sets are necessary to discover them. Such additional data sets are not available in each case.

Ideally, the probability density functions (PDF) of all model input parameters are well known. The statistical distributions of all input quantities can then be combined by the use of the Monte Carlo method which is based on performing multiple model runs with randomly selected model inputs. This method results in a PDF of the calculated emissions as model output. The emission PDF can then be converted easily to commonly used uncertainty quantities like standard errors or variation coefficients.

In reality, the input PDFs are normally not known in detail. Rather standard errors can only be roughly calculated or estimated. Some input quantities cannot be described with simple PDFs because of their complex physical structure (e.g. vehicle fleet compositions). This makes it difficult or even impossible to use the Monte Carlo method.

An alternative method of combining statistical errors of model input data to get the calculated emission's uncertainty is the use of the principle of error propagation. By variation of a single input parameter x_i in the calculation model, the sensitivity $\partial E/\partial x_i$ can be determined on the basis of partial derivation of the emission E with respect to the chosen input parameter. The absolute error of emission ΔE_i related to the error of the input parameter x_i results from the multiplication of the sensitivity $\partial E/\partial x_i$ with the calculated or estimated absolute error dx_i or Δx_i.

With the help of the error propagation law on conditions that
- the error estimations of the input parameters are correct by approximation,
- the input parameters are statistically independent from each other and
- there is a high degree of linearity between E and x_i ($\partial E/\partial x_i \approx$ const.) or the errors of the input parameters Δx_i are small compared with x_i,

it is possible to calculate the mean total error $\overline{\Delta E}$ of the emission E for any functional relations between the input parameters x_i according to Eq. (3.2).

$$\overline{\Delta E} = \sqrt{\sum_{i=1}^{n}\left(\frac{\partial E}{\partial x_i}\Delta x_i\right)^2} = \sqrt{\sum_{i=1}^{n}(\Delta E_i)^2} \qquad (3.2)$$

with:

n:	number of input parameters
$\partial E/\partial x_i$:	sensitivity of calculated total emission related to the input parameter x_i
Δx_i:	error of input parameter x_i (e.g. standard deviation)
ΔE_i:	single error of calculated total emission E related to Δx_i

Input parameters with asymmetric distributions can be handled by a bilateral approach. Positive and negative limits of the chosen confidence intervals of all input quantities are regarded separately. This approach results in a positive and a negative emission error quantity with different amounts.

The following results of statistical calculations are given as standard deviations resp. variation coefficients. This means that the estimated error ranges describe a 68.3 % confidence interval. In case of emission modelling usually no higher degree of confidence (e.g. 95 %) is reported because of the high general level of errors of calculated data.

3.2.2.2 Results for emissions from road transport

The results presented in the following are related to the high resolution emission factors and the associated traffic data which are available for the area of West Germany. The investigations are restricted to the pollutants NMHC (Non Methane Hydrocarbons) and NO_x (NO and NO_2) as they take main responsibility for creation of atmospheric oxidants. Some systematic errors have been found in commonly used input data sets. These results can be used to correct modelled emission data from road transport based on these input data sets.

The underlying emission calculation model was developed at the IER, university of Stuttgart (John 1999), and is similar in structure to emission models used in European scale (MEET 1999), but more detailed. The model is based on current parameters for regional emission modelling with high temporal and spatial resolution. Emissions are calculated for all segments of non urban roads and urban areas.

Error ranges for emissions from different road categories

Error estimations of calculated annual emissions have been carried out for different road categories according to the method described. In addition to the statistical errors, systematic errors have been quantified and documented separately, as far as data for the estimation of these errors were available. The error calculations have been carried out for the year 1994 and the area of West Germany.

The statistical error ranges of the five main input parameters have been quantified as follows:
- Mean gradients for individual road sections are available in gradient classes of 1 % steps from the road databases, resulting in a maximum error of ± 0.5 %.
- Errors in traffic flow data are caused by manual count errors and by errors due to interpolation from count samples to annual values. They have been estimated on a random basis by comparisons of the traffic flow data used for modelling with annual traffic flow data from automatic permanent counting stations.
- Usually, information about the driving pattern mixes of individual road sections is missing, because section specific data about speed limits, road courses etc. are not available. Plausible assumptions have been made to estimate possible ranges of driving patterns for each road category.
- Fleet compositions vary from one region to another. The standard errors and the effects of these spatial variations on the emission rates have been quantified by evaluating vehicle stock data of different administrative districts in West Germany.
- Statistical errors of commonly used emission factors have been investigated by the TÜV Rheinland (Hassel et al. 1998a, 1998b). Standard errors from the single values of extensive dynamometer measurements have been calculated per vehicle layer. These highly resolved interim results have been aggregated to standard errors for the most important driving patterns.

Emissions from vehicles with warm engine outside towns

The results of the error estimations for the road categories motorway and federal road are presented in Fig. 3.1. The results are related to road sections on the following conditions:

- gradient class: 0-1 %,
- full extent counts at the official German federal traffic census in 1990 and 1995 (8 counting days, 36 counting hours),
- estimated (not locally recorded) section specific route situation (speed limit, bends etc.)
- estimated (not measured) section specific traffic data (driving pattern mix and fleet composition).

The determined statistical errors (variation coefficients) of calculated annual emissions have been found in the range of 21 to 26 % for NMHC and 16 to 22 % for NO_x for both road categories. Especially on motorway sections a negative error (emissions are calculated too high) is more probable than a positive one because of asymmetrical effects of the errors of the driving pattern mix.

The emission factors represent the most important source of error especially for NMHC. Since the frequency distribution of the driving pattern mix is not normal distributed, the statistical error due to driving pattern mix is not symmetrical.

In addition to the statistical errors, two important sources of systematic errors (fleet composition and emission factors) have been found. The term "fleet composition" describes the further differentiation of traffic flow data which are already differentiated to vehicle categories by manual traffic counts. The traffic flow is now further distributed among vehicle layers which are defined by the criterions type of engine, exhaust gas purification system, legislative exhaust gas limit, age, cubic capacity and vehicle mass. These highly resolved traffic flow data are available e.g. from the "UBA-Handbook" (UBA 1999) for the three road categories motorways, rural and urban, related to the German situation. Comparing vehicle stock data (Kraftfahrt-Bundesamt 1994) that are projected to the dynamic fleet composition by concept- and size-related projection factors (Steven 1995), with the data from the "UBA-Handbook" shows, that shares of vehicles without or with older exhaust gas purification systems are much higher in the projected stock data.

Systematic deviations of emission factors can be partially deduced by interpreting investigations made by TÜV Rheinland (Hassel et al. 1994). Emission factors for different driving cycles usually are constructed by first dividing the test driving cycle and the corresponding emissions into parts (modal analysis of e.g. one second each) that are defined by present speed (v) and acceleration (a) of the tested vehicle. The measured emissions in high temporal resolution are classified by a v-a-grid. To calculate the integral emission factor of any driving pattern, it is necessary to add up the mean emission values of the different grid cells by the specific frequency distribution of driving conditions. These added up emission factors are published in large databases (UBA 1999) and are used for emission modelling. Some of these modal measured emission factors have been exemplarily compared to integral values (bag analysis), measured over the whole driving cycle. Occurring deviations are caused by systematic errors of the current method used for determining emission factors with dynamometer measurements.

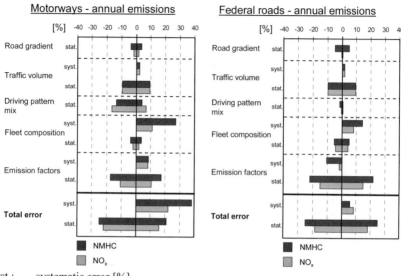

syst.:   systematic error [%]
stat.:   statistical error: variation coefficient [%]

Fig. 3.1. Error ranges of annual emissions on road sections in West Germany, 1994

The resulting systematic errors are conspicuous. Whereas these two error groups have different signs at the federal road sections, which means that they compensate each other, at motorway sections the same signs lead to systematic deviations of about + 38 % (NMHC) and + 22 % (NO_x). That means that there is a systematic underestimation of emissions on motorways when using the usual input data sets.

Emissions from vehicles with warm engine inside towns

In contrast to non urban traffic no detailed standardized traffic flow data are available for individual road sections inside towns. (Of course, in many towns traffic counts have been made - but these are not standardized on a regional or national level, so it takes too much effort to use them.) Total traffic flows on town level are available for some towns only. So additional infrastructure data (e.g. number of inhabitants) have to be evaluated. With it, total urban traffic flows on federal or regional level are distributed to the individual towns.

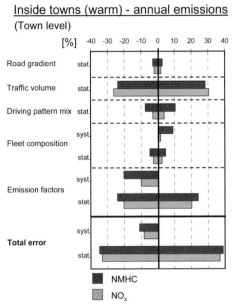

Fig. 3.2. Error ranges of annual emissions with warm engine inside towns in West Germany, 1994

As shown in Fig. 3.2, the statistical total errors have been determined to be approx. 37 % (NMHC) and 35 % (NO_x). Thus, they are considerably higher than the errors of road sections outside towns. There is an additional systematic error of about - 11 % for NMHC and - 8 % for NO_x. The two input parameters contributing most to the total error are the traffic flow and the emission factors. The step of distributing total traffic flows to town level using the number of inhabitants per town involve a variation coefficient of about 23 % (Schmitz S et al. 1997). Compared to this, the determination of the total traffic flow inside towns for all motor vehicles (e.g. for the old West German states) is quite reliably (variation coefficient: approx. 11 %) and the errors of differentiation of the traffic flow to different vehicle categories are relatively small, too (3.8 % in case of NMHC and 11.9 % in case of NO_x). The uncertainties of emission factors for urban transport are slightly higher than those for non urban road sections.

Cold start emissions

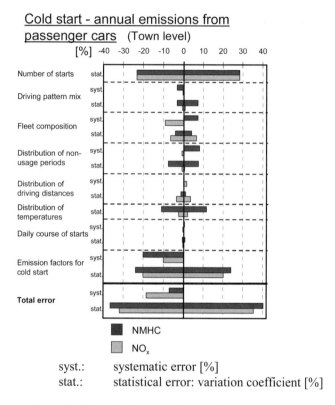

Fig. 3.3. Error ranges of annual additional cold start emissions from passenger cars in West Germany, 1994

A calculation model, which contains mean distributions for the driving pattern mix, the fleet composition, parking times before start, driving distances, outdoor temperatures and the daily course of starting frequencies (Umweltbundesamt 1999), lays the foundations for the error estimations of additional cold start emissions. The number of starts per town is deduced from traffic flow and population data. The errors of additional cold start emissions are mainly marked by the spatial differentiation of the number of starts on town level and the cold start emission factors. Error contributions of more than 10 % have been determined for the number of drives related to traffic flow and the annual distribution of temperatures (only NMHC).

Similar to the traffic flow in case of determining emissions with warm engine inside towns, necessary information about the number of starts is not available on town level. So the total number of starting processes is determined from the total driving performance of passenger cars (in West Germany) and a mean driving distance. Then the total number of starts is distributed to the individual towns according to the numbers of inhabitants. The statistical uncertainties of the spatial distribution of starts correspond with those of the spatial distribution of traffic flow. They amount to 23 % with a slight asymmetry (Fig. 3.3).

The statistical total errors up to now amount to about 39 % in case of NMHC and 33 % in case of NO_x. Systematic deviations have been determined from the available data as about - 7 % for NMHC and - 19 % for NO_x.

3.2.2.3 Uncertainties of temporal resolution of traffic emissions

In order to model temporally highly resolved emission data, mean annual, weekly and daily courses of traffic flow per road category are applied. Temporally highly resolved data (hourly intervals) are available from traffic counts from automatic permanent counting stations. Comparing hourly shares in annual traffic flows, deviation ranges between different road sections of the same road category can be derived.

The deviations of variation coefficients caused by the daily courses can be roughly separated into two fields: 1. late morning till evening, 2. night till early morning. Besides the deviations caused by the daily courses there are significant differences between the weekdays (Monday-Friday, Saturday, Sunday).

Variation coefficients amount from 10 to 70 % for light duty vehicles and from 15 to 100 % for heavy duty vehicles. Low error ranges occur during the day hours with high traffic flow, and higher error ranges occur during the night hours with low traffic flow. In general the variation coefficient increases with decreasing traffic flow. So, an increase with decreasing order of road category (motorway → state road) can be seen. The deviations on weekends are higher than those on weekdays. An exception was found for LDV during the night hours where variation coefficients stay unchanged respectively decrease slightly on weekends. For HDV in all examined cases higher deviation ranges occur compared to LDV.

3.2.2.4 Conclusions

The results of statistical error analyses carried out for annual emissions in 1994 on road sections show variation coefficients (68,3 % confidence interval) between 21 and 26 % in case of NMHC and between 16 and 22 % in case of NO_x. In addition systematic errors of commonly used input data sets have been identified especially affecting emissions on motorway sections with + 38 % for NMHC and + 22 % for NO_x. The statistical uncertainties of annual emissions with warm engine inside towns on town level amount to 37 % (NMHC) and 35 % (NO_x); they are much higher than the errors outside towns. Error ranges of additional cold start emissions determined so far are in the same order of magnitude.

For two reasons all determined total errors have to be regarded as lower limit of the real total errors:

1. Because of the incomplete data situation the uncertainties of some input parameters could be quantified only partially. There are considerable data gaps especially for emission factors (vehicles with newer technology, cold start, road gradients, heavy duty vehicles, motorcycles, running losses), the

driving patterns on ordinary roads and inside towns and the traffic flow data. Additional investigations are necessary to close these gaps.

2. In most cases it is not possible to calculate the true total statistical errors taking into account possible correlations because current available data is not sufficient to do essential quantifications of covariances.

Significant systematic errors could be recognized for the input parameters emission factors and fleet composition by including additional data sets. To clarify these systematic deviations further dynamometric measurements are necessary. As the errors of the emission factors quantified so far are caused mainly by the method of modal analysis (Hassel et al. 1994) the methodology of deriving emission factors from dynamometer testings should be improved. Experimental determinations of dynamic fleet compositions should be carried out on different road categories (e.g. by license plate number evaluations).

To avoid the peaks of uncertainty caused by temporal resolution, additional traffic flow data related to the individual road sections of interest are required in correspondingly high temporal resolution.

3.2.2.5 Results for emissions from solvent use

The method for calculation of emissions from solvent use is described in Chap. 2.2 (Theloke and Friedrich 2003).

Uncertainty ranges of calculated emission data have been analysed for this sector. as well. With respect to the large number of different types of solvent applications and due to the very inhomogeneous structure of corresponding input data for the estimation of solvent emissions, it is not possible to perform a detailed statistical error analysis in the same way as for the road traffic sector.

For different solvent use sectors statistical error ranges in the order between 30% and 50% have to be expected. The uncertainties of calculated emissions from solvent use are quite difficult to quantify. Annual emissions from solvent use typically contain an uncertainty of ± 30% according to our assessment (Kühlwein et al. 1999). However, there are substantially greater uncertainties in the analysis of individual source sectors. The reasons are the heterogenic structure of the sources and the quality of available input data. A large variety of sources and emission activities exist with significant uncertainties in the assignment of production, consumption, application and resulting emissions. A chance for the assessment of plausibility is the analysis of surveys, emission declarations, expert surveys (for example from industry associations) and the solvent bilance (see Table 2.15). The stated annual amounts of solvent use in different industries varies significantly even within the same source group (see Fig. 3.4). Another example of input data uncertainties is the wide distribution of solvent content of different paint application systems (see Fig. 3.5).

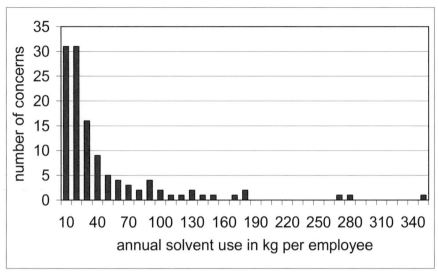

Fig. 3.4. Annual solvent use in different concerns of mechanical engineering (Kühlwein et al. 1999)

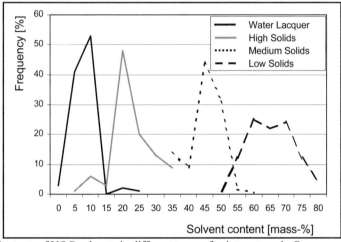

Fig. 3.5. Contents of VOC-solvents in different types of paint systems in Germany (Kühlwein et al. 1999)

Different approaches for assessment and calculation of uncertainties of NMVOC emissions from solvent use

Different approaches are possible for the assessment and calculation of uncertainties from NMVOC emissions. First a differentiation has to be made between uncertainty analysis methods and possibilities for verification of the calculated emissions. In uncertainty analysis a further distinction between qualitative, semi-quantitative and quantitative methods can be made. As quantitative methods, either error propagation or Monte Carlo simulation can be applied. For verification purposes plausibility checks for example with data from solvent industry on domestic solvent consumption can be used (Table 2.15). In addition to that, experiments, for instance for a whole city can be conducted with a comparison between modelled and measured ambient air concentrations. In qualitative and semi-quantitative approaches uncertainties are typically classified into five qualitative classes, then bandwiths of the confidence intervals are assigned in percent (see Fig. 3.6). The distributions are mostly non-symmetrical.

Table 3.3. Qualitative approach for assessment of uncertainties of emissions from solvent use (Kühlwein et al. 1999)

Source groups	Uncertainty bandwith			
	Yearly emissions	Spatial resolution	Temporal resolution	VOC speciation
Paint application	2-3	3	3-4	3
Domestic solvent use	3	2	5	2
Printing processes	2	2	3	2
Synthetics processing	4	4	4	3
Metal degreasing	4	4	4	4
Other source groups	3	3	3-4	4
Solvent use Complete assessment	3	3	4	3

One possibility to calculate quantitative uncertainties is the application of error propagation. Error propagation has the following requirements for application: Firstly, the input parameter must be subject to a Gaussian distribution with a standard deviation in the order of one percent of the mean value.

However, error propagation methods have the following disadvantages:

- Other distributions as Gaussian distributions cannot be properly assessed with the error propagation law.

- The input data required for the calculation of emissions from solvent use are usually not subject to Gaussian distributions and furthermore often correlate.

- Thirdly error ranges of more than 30 % of input data can be treated badly with error propagation. But error ranges of more than 30 % are not unusual for the input data of the solvent emission model.

Hence it can be deduced that error propagation is not a suitable method for the quantification of uncertainties of emissions from solvent use. Fig. 3.6. shows results from a Monte Carlo simulation of the applied calculation model for emissions from paint application in Germany in 2000. For this purpose, the commercial software *RISK 4.5* for EXCEL was applied. The average mean resulting from Monte Carlo simulation (10 000 iterations) is 231 kilotonnes. The 90%-confidence interval is 231 kilotonnes ±10 %.

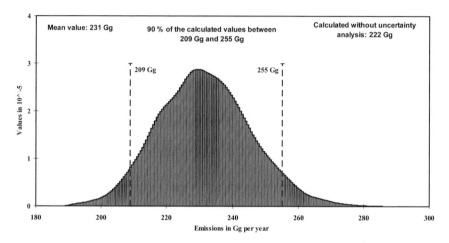

Fig. 3.6. Result of a Monte Carlo simulation with the model for emissions from paint application

Fig. 3.7 shows the results of two city experiments in Paris (Vautard et. al. 2003), France, and Augsburg, Germany (Kühlwein et al. 2002a) (conducted in 1998 in Augsburg and 1999 in Paris)[18]. On the x-Axis the relation of modelled to measured concentrations of selected hydrocarbons are displayed vs. the measured and modelled ambient air concentrations of Carbon Monoxide (CO). The results in Fig. 3.7 show a relatively good agreement. The deviations between modelled and measured concentrations deviate at a maximum factor of 2 apart for Propane in the Paris experiment. One of the reasons for the underestimation could be the under representation of Liquified Petrol Gas (LPG) heaters in the model. Fig. 3.7 indicates much larger deviations between measured and modelled ambient air concentrations. In Augsburg rather large deviations for some species could be noticed. These originated primarily from solvent use and here in particular from white spirits (main component: n-Decane). The reasons for these significant differences could not be fully analysed and explained yet, hence verification experiments are necessary.

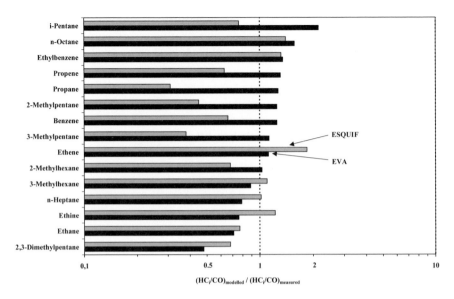

Fig. 3.7. Results of City experiments in Augsburg (EVA), Germany and Paris, France (ESQUIF) (Kühlwein et al. 2002a; Vautard et al. 2003)

Conclusions

Annual emissions from solvent use have an uncertainty bandwith of ± 30% according to our assessment with a semi quantitative approach. Results from a quantitative uncertainty analysis using Monte Carlo simulation (10 000 iterations) for our modelled emissions from paint application in Germany in 2000 show an average mean of 231 kilotons. The 90%-confidence interval is 231 kilotons ±23 kilotons. A plausibility check resulted in good agreement between the solvent consumptions for 1994 estimated with the product based approach and the same using the solvent based approach (with data from the European solvent industry as a basis). Furthermore, the modelled emissions were evaluated by two city experiments in Paris, France and Augsburg, Germany. The comparison between modelled and measured ambient air concentrations of single species was partial in good agreement for some individual species. However, in some cases significant discrepancies were identified. In the case of Augsburg, particularly large differences for some NMVOC species typically from solvent use and here primarily from white spirits (main component: n-decane) were found. The reasons for these differences between measured and modelled ambient air concentrations are yet to be analysed in detail. Thus, further verification experiments are necessary.

3.3 Comparisons of emissions

3.3.1 Comparison of emission inventories

W. Winiwarter

3.3.1.1 Introduction

An important aspect of quality considerations is the comparison with independent sets of data. Ideally, this would be ambient measurement data, as such ambient data also allow for a closure of cause (emissions) to effects (pollution). Ambient measurement data in sufficient quantity and quality are however only available for very specific campaigns. But there are also cases when emission inventories can be compared. Especially this is possible for inventories that have been assessed independently for the same area. A number of methods have been developed to describe the similarities and differences between such inventories. If discrepancies are flagged, it is possible to discuss and potentially decide upon an ideal approach. Even if such a decision is not possible, a comparison allows for an understanding of the reliability and the uncertainty of the input used to make up the inventory.

A necessary precondition for this concept is that the two inventories are independent which are compared in such a way. It is usually quite difficult to determine full independence, as inventories at least in their concept contain similar approaches and consequently also similar input data. The largest difference in concept is between bottom-up inventories (made up from information about a multitude of individual sources) and top-down inventories which combine large scale data (as national totals) with surrogate information for disaggregation. Here it is possible to identify conceptual discrepancies, but not the detailed methodologies. The latter only can be performed when a closer concept is being compared, and when additionally the input information is made available for the comparison.

There is several levels at which emission inventories can be compared. The following sections will give a more detailed account of this. Basically, comparisons are possible along the following paths:
- Emission total (by chemical species)
- Sector emission total (by chemical species)
- Spatial distribution
- Temporal distribution

The respective pathways of differentiation reflect the diverging needs of users of emission inventories. While modellers are first of all interested in emission totals and the spatial and temporal resolution of these emissions, which all are basic input parameters, policy makers require information on individual sources and source groups and therefore also on the confidence of the respective information. The attribution of emissions to different source groups, however, is a common cause of discrepancy between two inventories. While this is not relevant for atmospheric chemistry, policy decisions are based upon such information and consequently the topic is extremely relevant.

3.3.1.2 The Milan emission inventory: A case study

Probably the most intensely investigated area in terms of emissions is the Milan province, Italy. A number of emission inventories are available for the area, which derive from different institutions. It is not possible however to fully trace to what extent emission inventories are different. Indications are that some of the inventories have been developed from the same origins. Nevertheless it is interesting to note what actually exists. Some characteristics of those Milan province emission inventories we know of are listed in Table 3.4.

Table 3.4. Some emission inventories available for the province of Milan, Italy

No.	Institution	Project	Extension and spatial resolution	Spatial resolution	Source/contact person
1	JRC	Auto-Oil	Lombardy region	unknown	Angelino E., A. Skouloudis*
2	University of Brescia	SATURN	100 x 100 km²	2 km	Catenacci et al. (1999)
3	Regione Lombardia	LOOP	141 x 162 km²	3 km	G. Maffeis, TerrAria s.r.l.*
4	ENEL - CRAM, Milano		Lombardy region	4 km	G. Brusasca*
5	University of Stuttgart	GENEMIS	141 x 162 km²	3 km	Schwarz et al. 1999
6	ARC Seibersdorf	IMPRE-SAREO	Milano province	1 km	Hayman et al. 2002

* no publication available

   In the following an example for comparing inventories is presented. For this purpose it was irrelevant which to compare, as long as the geometric coverage and resolution was sufficiently identical, and the emission data were made available at sufficient detail. This was the case between the LOOP inventory and the GENEMIS inventory, as the GENEMIS inventory was specifically derived to match the coordinates of the LOOP requirements. As an additional advantage, these two data sets also were widely independent by their concept already: LOOP basically compiled the inventory from individual data to a total (in the following termed the bottom-up inventory), GENEMIS applied national and provincial data and redistributed it according to surrogate statistics such as population data (the top-down inventory). A more detailed description of the two inventories has been given by Winiwarter et al. (1999) and by Maffeis et al. (1999) or Schwarz et al. (1999) respectively. For this study, the gridded emission data were made available, but not the input information.

3.3.1.3 Comparison of regional emission totals and individual source sectors

In many cases, emission figures are given as annual totals as this is most compatible to underlying statistics. In the examples given here, emissions have been calculated for a specific period (compatible with a campaign of the EUROTRAC subproject LOOP) in late spring for the bottom-up inventory. The respective top-down inventory was fully adapted to the time period and spatial resolution, even if based on an older state of emission data (1994 vs. 1998).

For a comparison of totals, the average for the full period was calculated. This is a full week for the top-down inventory, but only three days in bottom-up. The weekend differences were accounted for by using information from Lombardian traffic authorities that average weekly traffic accounts for about 90 % of weekday traffic when considering weekends. Industry was considered to emit at a constant level even on weekends for this purpose.

Such mean values for each of the source groups were determined to assess the discrepancy in emission levels (Fig. 3.8). The source groups used by different inventories were also not identical – especially the group "point sources" which covered just 12 individual sources in the bottom-up inventory, but SNAP category 1 (power plants and energy conversion) in the top-down approach. At least for the total, the different assignments do not influence the results. As an additional check also CORINAIR '90 data are presented in Fig. 3.8: Annual data for Lombardy, 1990 were simply averaged for 1 s and one grid (9 km²). While this approach of course is just a rough approximation, the result shows where the top-down data actually derive from, as the agreement is evident (compare columns 1 and 3). The only minor differences occur in the sector "other" for CO, where CORINAIR includes heating emissions as it is an annual average. The differences in industry/CO and point sources/NO_x most probably are due to the differences in the area included.

The more interesting comparison is between the top-down and the bottom-up inventory. While there exists very good agreement for total NO_x, the split into source sectors proves that this is a mere coincident. The higher traffic emissions in the top-down inventory may be related to the technological improvements (introduction of catalytic converters has been effective at least for Italy as a whole between the different base years). But the difference for industrial emissions cannot even be explained by adding the "point sources" which show up only in the top-down inventory. Likewise there are large discrepancies also for NMVOC and CO which are also evident from the total. For traffic, CO emissions are larger in the (older) top-down inventory, consistent with the explanation given for NO_x. This is not the case for NMVOC. Compared to other countries and also inventory methods (see Ntziachristos et al. 2001), the NMVOC/NO_x ratios and specifically also the NMVOC emissions are unusually high for Italian traffic in any case, which is possibly influenced by an accounting of evaporation at the high end of possible estimation.

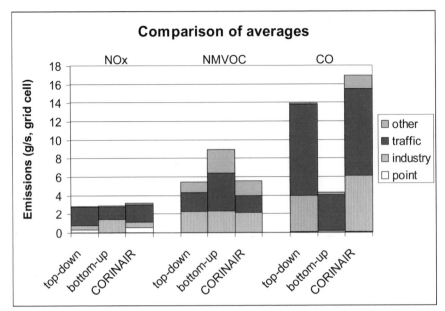

Fig. 3.8. Comparison of averages for the grid area. Note that the value denoted as CORINAIR refers in fact to the Lombardy region rather than the grid area, and time base is derived from the annual average

Even more than the traffic emissions, the emissions from industry exhibit important differences. The bottom-up inventory lacks of CO emissions almost completely. This may either be an omission, or it may be that an important source of CO (like steel industry) has been attributed as part of the area by the top-down inventory which in reality is not there. In contrast, it is surprising that the difference is not larger for NMVOC, as here usually differences exist between applying a top-down solvent balance vs. a bottom-up account of solvents used. In the Italian inventory, the latter method seems to have been used also for the top-down inventory, as data is available on a detailed source split.

3.3.1.4 Approaches to assess the spatial variability

In the absence of underlying information, a possibility to look into more detail of the inventory is an assessment of the spatial variability. At the same time this spatial variation is a quality of the emission inventory on its own, which also requires specific consideration. Winiwarter et al. (2002) have performed a detailed assessment of methods which can be applied towards such spatial comparisons. Here the most important of these will be highlighted, making use of the emission inventories discussed above.

The visual inspection of an inventory usually plays an important role already in the phase of creating at least a bottom-up inventory. Especially the inspection of continuity (along roads, settlement areas), but also the agreement of the emission distribution with known patterns (population agglomerations, industrial zones, unpopulated mountain ranges contrasting the valleys in between) is important. Fig. 3.9 allows performing such an inspection with both inventories (here shown for NO_x only).

The main features to be seen are the high emission density in the city and surroundings of Milan. Elevated emissions along major highways agree with a digital road network (provided by ESRI, Redlands, CA). Clearly visible is also the decrease of emissions close to zero in the uninhabited mountain areas. In the top-down inventory (Fig. 3.9 – left panel), also a marked difference between lower levels in the western and especially in the northern part of the grid may be discerned. These lower levels clearly coincide with the Swiss border in the north of the domain, and the border to the neighbouring Italian region of Piedmont (Regione Piemonte) in the west. The top-down inventory, treating each of the three areas separately, therefore was not able to yield consistent emission values across the borders.

In a similar way, the comparison of the discrepancies can provide relevant input. The discrepancies – the two different approaches are displayed in Fig. 3.10 – can be calculated either on an absolute scale (here the difference is calculated) or on a relative scale (displayed as the values of the top-down inventory divided by the values of the bottom-up inventory). What can be discerned in the example of NO_x (the behaviour of CO and NMVOC is largely identical) are large relative discrepancies in the northern and western parts that easily go to a factor of 10 and beyond, which however are not important on the absolute scale. By contrast, discrepancies are relatively small in the direct vicinity of the urban centre, indicating general agreement of the underlying factors and parameters for emission calculation. Just in these cases the absolute differences are considerable. The largest of these differences coincide with the highest absolute values for the top-down approach. In fact, these may be point sources that ran out of operation since the establishment of the original inventory (high grid cell emissions are caused by overproportionally high values for SNAP categories 1 and 3), or for which the activities indicated in the respective inventories are assumed totally different. Totally there are 6 such grid cells, two of which are dominated by a point source in both inventories.

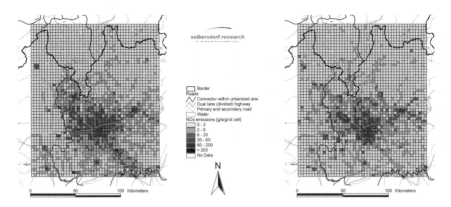

Fig. 3.9. NO$_x$ emissions calculated for the identical grid using a top-down approach (left panel) vs. a bottom-up approach. The scales are identical.

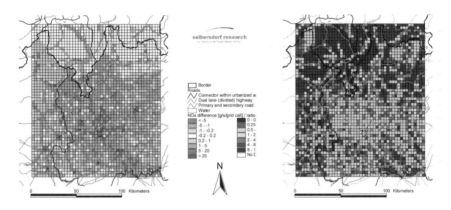

Fig. 3.10. Difference and ratio of the two inventories[2].

In addition to map display, the extent of an agreement between two grids can also be expressed in terms of numerical indicators. The simplest of these indicators is the coefficient of determination derived from a regression analysis, but Winiwarter et al. (2002) also describe the application of the Moran coefficient, and of an applicability criterion. Here we will limit the analysis to the regression analysis, which is presented together with the temporal aspect in the following section.

---

[2] This figure requires colour to be fully comprehensible. Please contact the author to obtain access to the original coloured figure.

3.3.1.5 Comparison of emissions - the temporal aspect

Distributing emissions (originally as annual emissions) over the course of a year, a week and a day has become a standard procedure in emission inventorying (see the contributions collected by Ebel et al. 1997). Disaggregation is the more common procedure, there are however also a few attempts to assess the temporal emission behaviour bottom-up for very specific campaigns using specific traffic counts, power plant logs and questionnaires to industry to assess the individual temporal activity distribution. Generally little information is available on temporal patterns for different activities, and patterns are often country dependent. Consequently, it is difficult to compare distribution functions.

In the following, we instead take a look at the combined spatial and temporal resolution, and we determine the correlation between the two Milan grids used for different hours of a day, and for the average. In the further processing, four distinctive time periods were defined to base the comparison on. These were a Tuesday (= second day for both inventory periods) at 2-3 a.m., local time (Central European daylight saving time CEST = 0-1 a.m. UTC) denoting the lowest night level, at 8-9 a.m. to characterize the morning peak and at 2-3 p.m. for representing afternoon emissions.

The attempt to identify agreement or disagreement of the two inventories using regression analysis is presented in Table 3.5. The extremely low values seen for NO_x do not indicate total disagreement but rather poor applicability of the method.

Table 3.5. Coefficient of determination (R^2) between bottom-up and top-down inventories

NO_x	Average	2:00 - 3:00	8:00 - 9:00	14:00 - 15:00
Total	0.11	0.02	0.15	0.15
Traffic	0.56	0.50	0.58	0.54
Point sources	0.00	0.00	0.31	0.32
Industry	0.05	0.04	0.06	0.06

VOC	Average	2:00 - 3:00	8:00 - 9:00	14:00 - 15:00
Total	0.40	0.08	0.43	0.58
Traffic	0.62	0.50	0.61	0.62
Point sources	0.08	0.03	0.00	0.01
Industry	0.41	0.22	0.38	0.49

CO	Average	2:00 - 3:00	8:00 - 9:00	14:00 - 15:00
Total	0.54	0.15	0.57	0.57
Traffic	0.57	0.44	0.59	0.59
Point sources	0.02	0.00	0.00	0.34
Industry	0.14	0.14	0.13	0.14

The general difficulty with using the coefficient of determination is that it is characterized almost solely by the extremely high emission values (see Winiwarter et al. 2002). Both the total and the point sources are affected, only traffic is less prone to this problem. As CO and VOC are also not as much affected by point sources, the values of R^2 may be discussed for these compounds.

In all cases the best correlation is seen with traffic emissions. This proves that the road network is sufficiently well known and is covered adequately in different digital databases, and that there is not even a significant geographical shift. Also the traffic emissions show most clearly, that there is poor agreement during night, but daytime agreement is similar as for time average. For industry, daytime agreement is even better. No clear pattern can be derived for point sources. This may to some extent be due to the different definition of point sources in the two inventories, but it also indicates that the temporal pattern is not adequately established. As long as the application of the inventory concentrates on daytime (due to the much more relevant emissions during day, approx. one order of magnitude) the considerably poorer correlation during nighttime does not pose a problem.

3.3.1.6 Discussion and conclusions

The results presented for the test case show the value of intercomparing different emission inventories. While the quality of the two approaches presented certainly is very different, also with respect of the amount of effort that went into the specific inventory, it is not in all cases clear that discrepancies only occur for one inventory. Therefore it should be generally assumed that differences that occur between the emission inventories presented here are representative of differences occurring in emission inventories that are used in practice.

While no efforts were made to establish a right and a wrong version of a specific topic, it is clear from what has been shown that the reason for a number of the discrepancies observed is clearly to be regarded an error. Source sectors are missing, emission factors attributed incorrectly, and assessments made for different administrative units in incompatible ways have been assembled using assumptions which later on prove to be not applicable.

The fact that such discrepancies do not only occur here is clearly underlined by results from the LOOP project (see Neftel et al. 2002, for details) which provides an independent assessment of the LOOP emission inventory shown here using field measurements. The fact that the field measurements exhibit a distinctively higher ration of CO/NO_x is consistent with the missing CO from industry seen in this analysis. Discrepancies between inventories and measurements as well as between inventories do occur (see also Mannschreck et al. 2002a), and at the current state of knowledge part of such discrepancies are simply errors.

An account of what emission inventories can in fact deliver has been given by Winiwarter (2002). With current methods it is not possible to go beyond these limitations, but techniques have been developed (see above) which allow intercomparisons and therefore at least will be able in the future to improve the results. Currently, emission inventories can only be considered valid when adequately compared to an independent set of data. For this reason, such a requirement was also entered into the Urban Air emission inventory guidelines (see Sect. 3.2.1)

Such a validation concept should, ideally, not be limited to the comparison, but also include an assessment of the underlying data and a correction / improvement procedure. But the comparison alone allows for a better understanding of the uncertainties involved in inventory assessment.

3.3.2 Comparison with measurements: road transport

3.3.2.1 Tunnel studies

J. Staehelin and P. Sturm

Introduction

Measurements in road tunnels are suitable tools to quantify the emissions of collectives of road vehicles. The information from road tunnel studies is complementary to dynamometric test results of single vehicles, which are used as input of current road traffic emission models. Road tunnel studies can be used for the following purposes:

1. Tunnel studies can be used to evaluate road traffic emission models based on a collective of vehicles ("real-world emissions"). However, for conclusive answers which respect to possible deficits of the emission models the uncertainties of both, tunnel measurements and emission models, need to be evaluated.

2. Tunnel studies performed in the same tunnel after several years are valuable to quantify the temporal evolution of the road traffic exhaust emissions. By such measurements, the success of the introduction of new technologies (e.g. catalytic converters for gasoline powered vehicles) can be evaluated. Similar information can be obtained from monitoring measurements performed close to busy streets, however, in case of such measurements the contribution of other emission sources needs to be taken into account as well.

3. Measurements of tunnel studies were used to quantify the VOC-split, i.e. the relative contributions of single compounds to the organic exhaust gases. This task is important because measurements of single VOC compounds in the exhaust of road vehicles are limited from dynamometric tests (the analysis of speciated organic exhaust compounds is not routinely performed in dynamometric test facilities because special analytical requirements necessary to measure single organic compounds, e.g. using gas chromatography). However, single organic compounds are very important in atmospheric chemistry and air pollution research, because different compounds contribute very differently to the formation of photooxidants because of their distinct differences in reactivities in ambient air.

Tunnel studies also have their limitations. The driving pattern might not be identical as when driving outside of tunnels and cold start emissions are usually not covered by tunnel measurements, because these studies are usually performed in high-way tunnels outside of cities.

Measurements and analysis of road tunnel studies

In particular when attempting to compare the results of tunnel studies with numerical emissions models the following requirements need to be fulfilled.

1. Measurements need to include the air pollutant concentrations inside the tunnel as well that of fresh (inlet) air.

2. The air flow through the tunnel must be known precisely, requiring detailed quantitative information of the ventilation system of the tunnel.

3. The number of the vehicles and the composition of the fleet driving through the tunnel and the vehicle speed need to be known.

4. Other variables such as road gradient are important. In case of changes in road gradient in the tunnel an additional uncertainty is introduced making the quantitative comparison more difficult.

The results of the tunnel studies can be compared with the emission models by two approaches:

(a) Direct comparison: In this approach of the results tunnel measurements and an emission model are compared for the respective time intervals, e.g. 15 min or 30 min. The numerical model predicts the total emission of the collective driving through the tunnel using the respective information of the fleet composition. These results are compared with the measurements in the respective period. The comparison of results for different fleet compositions (e.g. presence or absence of heavy duty vehicles) can be used to obtain more information for the interpretation (see e.g. Sturm et al. 2001).

(b) Statistical analysis. The tunnel measurements performed during e.g. one week can be analysed first by a statistical (regression) model which allows to deduce the EF of categories of vehicles. Most often the categories of light duty vehicles (LDV) and heavy duty vehicles (HDV) are used. This classification is usually based on measurements of loop detectors (limits of the gross weights of 3.5 t) yielding LDV (including passenger cars and vans) and HDV (including trucks and buses) (see e.g. Staehelin et al. 1997). By this approach the EF calculated for the collective of vehicles is linearised by the proportion of one class of vehicles using the following formula:

$$EF = \alpha + \beta\ pHDV + \varepsilon \tag{3.3}$$

where:

EF:	calculated for the emissions of the entire collective of vehicles driving through the road tunnel during a specific time interval
α:	EF of the light duty vehicles (mainly passenger cars)
$\alpha + \beta$:	EF of the heavy duty vehicles
ε:	random error

The statistical approach has also limitations. The proportion of HDV typically fluctuates during a week between 0 and 30% in European road tunnels. Therefore, the EF of LDV can be determined much more precisely by the given formula than the EF of HDV, which needs an interpolation from approximately 30 to 100%. NO_x emissions of HDV can be calculated by equation (3.3)because the EF of HDV (diesel vehicles) are much larger than those of LDV (which mainly consist of gasoline driven vehicles), whereas EF of HDV of CO and t-NMVOC can not be obtained by statistical analysis, because EF of gasoline engines are larger than those of diesel vehicles.

Subsequently the comparison of the EF of the classes can be compared with the results of the emission model for the same classes (John et al. 1999). This approach allows the attribution of the emissions to classes, which can provide useful additional information particularly in case of possible discrepancies between numerical models and tunnel studies.

Results of recent European tunnel studies

<u>Long-term changes in road traffic emissions</u>
Tunnel measurements were performed at least twice in the Tauren tunnel (Austria), namely in 1988 and 1997 (Schmid et al. 2001) and 7 times in the Gubrist tunnel in Switzerland (near Zürich) between 1990 and 2000 (Steinemann and Zumsteg 2001). The results of both studies document the dramatic decrease in CO, t-VOC and NO_x emission of gasoline driven vehicles (at high vehicle speed and hot engines) since the introduction of catalytic converters which started in Europe in the second half of the 1980s (see Table 3.6). The decreases in the Austrian study are larger (decrease in averaged EF of CO and t-NMVOC by a factor of 10) than in the Swiss study, because the proportion of cars with catalytic converters was virtually zero in the first tunnel study in Austria in 1988 while in the first Gubrist tunnel study in Switzerland (1990) already 40% of the gasoline driven vehicles were equipped with controlled converters. The measurements of the Gubrist tunnel (Switzerland) also indicate, that the NO_x emissions of gasoline vehicles decreased by a factor of approximately three during the 1990s. The NO_x emissions of heavy duty (diesel) vehicles decreased only insignificantly, reflecting the fact that no new technology reducing vehicle exhaust comparable to catalytic converters was introduced for diesel vehicles in Europe at the present (Steinemann and Zumsteg 2001).

Table 3.6. Results of European tunnel studies performed in the same tunnels (compare text). The factors of decrease in EF are approximate numbers because uncertainties in the derived EF by statistical analysis are considerable.

Year	prop. catalyt. convert.	Factor of decrease in EF		NO_x	
		CO LDV	t-NMVOC LDV	LDV	HDV
Tauren tunnel, Austria					
1988	appr. 0				
1997	60 %	appr. 10	appr. 10		
Gubrist tunnel, Switzerland [1]					
1990	40% [2]				
2000	95%	3	6 (1991 to 2000)	3	insignificant

[1] Measurements were performed in 1990 (excluding t-VOC measurements), 1991, 1992,1993, 1994, 1995, 1997 and 2000.

[2] Proportion of catalytic converters in cars

Comparison of road tunnel studies with road traffic emission models
In the EUROTRAC subproject GENEMIS extended measurements were performed at the Gubrist tunnel in September 1993 (Staehelin et al. 1995). They were first analysed by a statistical modelling. The subsequent comparison showed good agreement between the road traffic emission model currently used in Germany, Switzerland and Austria (UBA 1995) and the tunnel measurements (high speed driving and hot engines) for carbon monoxide and t-NMVOC. Also the agreement between the measured EF of single VOCs and the numerical prediction was acceptable in most cases (Staehelin et al. 1998). For NO_x emissions the agreement between tunnel measurements and model prediction was satisfactory for gasoline vehicles (LDV, in the Swiss fleet diesel driven vehicles only contributed insignificantly to LDV) whereas the model underpredicted the NO_x emissions of the heavy duty vehicles by approximately by factor of two (John et al. 1999). In additional sensitivity tests, the possible effect of the tail wind which could cause lower engine loadings in a tunnel environment was simulated. The results indicated, that the tail wind effect did not alter the main conclusion of the study.

In the year 1999 a large tunnel study was performed in the Plabutsch tunnel near Graz in Austria (Sturm et al. 2001). The tunnel ventilation system is different as in the Gubrist tunnel and the authors used approach (a) for the comparison of the measurements with the emission model based on the "Handbook". They found that the numerical model predicted the emissions of CO and NO_x of the entire fleet well during weekend when truck (HDV) driving is restricted in Austria.

However, the Hanbdook (Keller et al. 1998) substantially unpredicted the NO_x-emissions during workdays when also HDV emissions substantially contribute to the fleet emissions.

Based on these tunnel studies, a large effort was initiated to investigate HDV NO_x-emissions in the EU-project ARTEMIS, in which the emission models of types of traffic are improved. In this effort more dynamometric tests were performed in particular for HDV. The revision of the emission model for NO_x emissions of diesel vehicles, which is based on dynamometric test data, is presently in progress (Hausberger et al. 2002). Preliminary results show, that NO_x EF for HDV in the handbook are too low (Hausberger et al. 2002). Further tunnel studies were conducted in spring 2001 in Lundby tunnel near Goteborg in Sweden and a second study in the Plabutsch tunnel in fall 2001. The first tentative results from the Plabutsch tunnel suggest, that not only the underprediction of the EF of NO_x but also other problems, such as possibly the used driving pattern in the tunnel and the loading factors of HDV, might contribute to the underprediction of HDV NO_x emissions in the presently used version of the Handbook. In addition it has been shown, that humidity plays a non negligible role when calculating HDV NO_x emissions (Hausberger et al. 2002).

Conclusions and outlook

Tunnel studies performed at the same tunnel sites in European tunnels impressively demonstrated the success of the introduction of more stringent emission standards in passenger cars since the middle of the 1980s in the European road traffic fleet. However, this is not the case for NO_x emissions of HDV. The measurements performed do not reflect emission reductions as one could expect when considering the much more stringent emission standards. Furthermore, tunnel studies were used to demonstrate, that the emission models currently used in the German speaking countries in Europe (Germany, Austria and Switzerland) ("Handbook") strongly underestimates the real world NO_x emissions of HDV vehicles. This problem is currently under investigation and needs further attention.

Further decreases in the emissions of road traffic emissions are predicted because of the introduction of new technologies (EURO 4 and 5). However, the success of the introduction of these new technologies needs to be evaluated carefully. For this purpose further tunnel studies in the same tunnels seems most suitable.

3.2.2.2 Real-world emissions from gasoline passenger cars measured by on-road optical remote sensing

Å. Sjödin

Introduction

Air quality is improving in many areas of the world, partly as a result of stricter emission standards for gasoline passenger cars, requiring the use of effective closed-loop three-way catalyst systems (TWCs). For instance, the growth in the number of TWC-cars is believed to be the main cause behind the observed strong decrease in ambient urban levels of both NO_2 and benzene in Sweden since the early   90's   (Sjödin   *et*   *al.*   1996; *http://www.ivl.se/en/miljo/projects/urban/trendvoc.asp*).

This demonstrates the great importance of LDV emissions for urban air quality. However, the experience from the US is that air quality is not improving as fast as is predicted from the growth in the number of TWC-cars. It has been claimed that this may partly be due to existing inspection and maintenance programs failing to maintain the optimum vehicle emission performance as vehicles age (Lawson 1993 and 1995; Bishop et al. 1996; Stedman et al. 1997). Furthermore - in general - knowledge of the actual emissions from in-use LDV fleets is still limited, since most available emission data arise from e.g. FTP-measurements on small vehicle samples that may be unrepresentative for the in-use fleet.

One approach to possibly overcome the problem with test fleet representativity in conventional road vehicle emission measurements is to use the on-road optical remote sensing device originally developed by the University of Denver in the late 80's (Bishop et al. 1989). As part of the GENEMIS-2 project a remote sensing measurement campaign were conducted in 1998, as a follow-up to earlier measurements in 1995 and 1991 at the same freeway ramp site in Gothenburg, Sweden, with the main objective to evaluate the change in emission performance over time of Swedish gasoline passenger cars. For details on these measurements see Sjödin and Andréasson (2000) and Sjödin (1994).

Fleet average emissions

In each year - 1991, 1995 and 1998 - emissions from around 10,000 gasoline passenger cars were measured by the remote sensor at the exit of a sharp, 270 degrees curve of an uphill freeway interchange-ramp, considered to carry only warmed-up vehicles. Measurements in 1991 involved only CO and HC, whereas measurements in 1995 and 1998 involved all three parameters CO, HC and NO. The remote sensor detector and filter applied for the CO measurements were the same in all three years. The HC detector applied in 1991 utilised a filter which is sensitive to positive liquid water interference (Sjödin 1994; Guenther et al. 1995), whereas the HC detector applied in 1995 and 1998 utilized a filter less responsive to water. NO measurements in 1995 were made by means of non-dispersive UV (Zhang et al. 1996), whereas NO measurements in 1998 were conducted by a newly developed high-speed UV diode array spectrometer, with a much lower level of noise at the same time less sensitive to positive HC interference (Popp et al. 1998). Overall fleet average tailpipe emissions for the gasoline passenger cars on the freeway ramp in 1991, 1995 and 1998 are given by Table 3.7.

Table 3.7. Overall fleet average tailpipe emissions (in volume-%) for gasoline passenger cars on the freeway ramp in 1991, 1995 and 1998 according to the remote sensing measurements. Errors are one standard deviation according to gaussian distributions, calculated by means of the methodology used by Bishop et al. (1993) for the gamma distributed remote sensing data.

Year	%CO	%HC (propane)	%NO
1991	0.71±0.10	0.059±0.005*	-
1995	0.73±0.13	0.022±0.004	0.064±0.018*
1998	0.57±0.08	0.037±0.006	0.055±0.009

*) different detectors were used the first year compared to the succeeding year(s).

According to Table 3.7 there was no significant change in the average tailpipe CO concentration for the gasoline passenger car fleet on the freeway ramp between 1991 and 1995. On the contrary, a slight, still however very significant ($p<0.001$ according to a student-t distribution) decrease of about 20% in CO emissions was observed between 1995 and 1998. To a large extent these observations can be explained by the increase in average fleet age of the gasoline passenger cars on the freeway ramp from 5.5 years in 1991 to 7.4 years in 1995, remaining nearly constant, or 7.3 years, in 1998. The increase in fleet age between 1991 and 1995 reflects the general poor sales of new cars in Sweden in the early 90's due to the recession in the economy, and, in terms of fleet average emissions, counterbalance the increase in the fraction of TWC-cars, estimated to 49% in 1991, 61% in the 1995 and 77% in 1998 from the observed model year distributions on the freeway ramp.

Average HC emissions dropped markedly from 1991 to 1995, and then increased again in 1998. However, any firm conclusions regarding possible changes in the overall HC emissions from the Swedish light-duty gasoline fleet from the remote sensing measurements cannot be made, since there was evidence for the accuracy of the remote sensing HC measurements being poor both in 1991 (Sjödin 1994) and in 1998 (see below).

Average NO dropped significantly ($p<0.05$ according to a student-t distribution) by 15% between 1995 and 1998. The NO average for 1995 reported in Table 3.7 is calculated after removing all data where HC exceeded 0.05% to eliminate the reported positive HC interference of the non-dispersive NO measurement technique used in 1995 see further below. This decrease makes sense in light of the 20% drop observed for CO during the same period, mainly caused by the increase in the fraction of low-emitting three-way catalyst cars from

61% in 1995 to 77% in 1998. It also indicates that threeway catalysts are somewhat more efficient in reducing CO emissions than NO emissions.

Average emissions by model year

Fig. 3.11 through Fig. 3.13 display average tailpipe exhaust concentrations for the three gases CO, HC and NO by model year for the gasoline passenger cars on the freeway ramp in 1991, 1995 and 1998. The sharp drop in average emissions from pre-1987 to post-1987 model years, especially for CO, reflects the introduction of new emission standards, requiring closed-loop threeway catalyst systems, compulsory from model year 1989, as observed already in the 1991 measurements (Sjödin 1994). By comparing the average emissions by model year curves for CO for all three years, it can clearly be seen from that there is both a significant deterioration of emissions as cars grow older, as well as an improvement in emission performance of new cars with time. From the data yearly rates of deterioration in CO emissions were estimated to 17-30% for TWC-cars, and 5-8% for non-catalyst cars. The decreasing slope of the CO emission by model year curve with time for the most recent model years of non-catalyst cars (1981-1986), indicates that their deterioration rate actually decreases with age. In 1998 this part of the emission by model year curve is nearly horizontal, indicating that any deterioration of non-catalyst cars CO emissions no longer takes place.

More than 500 unique TWC-cars and about as many non-catalyst cars appearing in two consecutive measurement campaigns yielded additional opportunities to study the effect of ageing on CO emissions. In all cases, the average CO emissions of fleets consisting of the same car individuals exhibited higher values in the succeeding campaign (1995 or 1998) compared to the preceding campaign (1991 or 1995). TWC-cars emissions were consistently higher in the second campaign compared to the first campaign about a factor of 2, whereas non-catalyst cars emissions were higher about 15-20%. It was also found that the main reason to the observed increase in TWC-cars CO emissions was an increase in a small fraction of CO high-emitters.

A significant improvement in emission performance due to improved emission control technology for new cars is indicated by the fact that CO emissions of the newest cars, i.e. 0-3 years of age, in 1995 were roughly only half of those of the cars of corresponding age in 1991. However, taking into consideration also the 1998 measurements, results are contradictory, since according to the remote sensing data truly new cars (i.e. 0-1 year old, model year 1998) in 1998 had about twice as high emissions as truly new cars in 1995. Several explanations for this observation was sought for, i.e. differences in driving pattern or in fleet composition of new cars in 1995 compared to new cars in 1998 with regard to make, model or weight, but none of these could account for the observed differences. However, CO emissions of new cars, i.e. 0-3 years old, in 1998 were still lower than cars of the corresponding age in 1991.

The average emission by model year curves for HC given by Fig. 3.12 are much harder to interpret, due to the already mentioned less accurate or erroneous HC measurements in one or more years. In the 1991 measurements it was found that particularly catalyst cars HC emissions were erroneously high due to water interference from the HC detector used at that time (Sjödin 1994). A careful analysis of the 1998 HC data indicated a severe HC detector problem in the 1998 measurements, with a detector exhibiting excessive noise and non-linearity. This calls for the introduction of a new standard for field calibration procedures for the remote sensor, involving daily multi- instead of single-point calibrations.

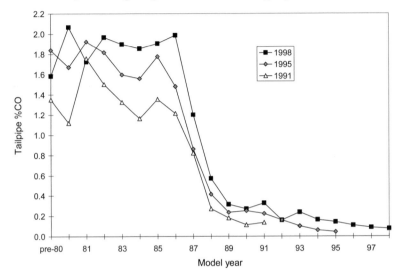

Fig. 3.11. Average tailpipe CO concentrations (by volume-%) by model year for gasoline passenger cars on the freeway ramp in 1991, 1995 and 1998 according to the remote sensing measurements.

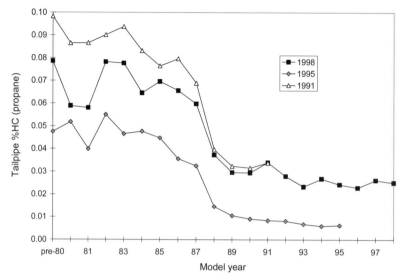

Fig. 3.12. Average tailpipe HC concentrations (by volume-% as propane) for gasoline passenger cars on the freeway ramp in 1991, 1995 and 1998 according to the remote sensing
measurements (measurements particularly in 1991 influenced from water interference and
in 1998 from excess HC detector noise and non-linearity problems).

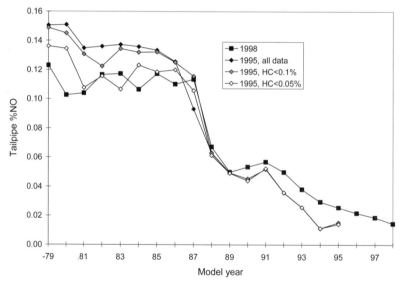

Fig. 3.13. Average tailpipe NO concentrations (by volume-%) for gasoline passenger cars
on the freeway ramp in 1995 and 1998 according to the remote sensing measurements.

As a means of eliminating the effect of the posted HC interference on the NO
detector used in 1995, three curves are presented for the 1995 NO data in Fig. 3.13:

All data with no discrimination, data after excluding NO values when HC exceeded 0.1% (propane) and finally data after excluding NO values when HC exceeded 0.05%. It can be seen from Fig. 3.13 that the average NO emissions actually decrease with increasing discrimination of records with high HC values, however for the non-TWC fraction of the fleet only. With the highest level of HC discrimination applied on the 1995 data, the observation is that NO emissions of catalyst cars of all model years (except model year 1989) present in the 1995 measurements exhibit a significant increase, whereas emissions of non-catalyst cars do not appear to have changed at all from 1995 to 1998. However, if all records in 1995 with high HC emissions, which might be synonymous with all cars operating rich, i.e. those cars for which NO emissions may be suppressed due operational factors, are excluded systematically, it cannot be ruled out that by such an approach non-catalyst cars NO emissions in 1995 to some extent may be underestimated, possibly hiding an actual "negative" deterioration effect for NO for the non-catalyst fraction of the fleet. Note that the relative deterioration tends to be higher for later model years than for earlier model years of catalyst cars. Note also that new cars NO emissions in 1998 are not significantly lower than new cars emissions in 1995. It thus seems as the big leap in TWC technology as regards improved emission performance for NOx was from model year 1991 to 1994. However, further evidence is needed to proof these statements.

Emission distributions

As observed in many other remote sensing studies, emission distributions were skewed, particularly for the TWC-fraction of the fleet and for CO. For the latest model years, CO emissions were below 1% CO for more than 99% of the fleet, and even for the oldest model years of catalyst cars, and for non-catalyst cars, the fraction of cars with very high emissions was low, although contributing typically as much as 50% of the overall fleet CO emissions. Lower percentiles, median and higher percentiles for tailpipe %CO all increased with increasing age, indicating that the deterioration in CO emissions observed has two components, a slower deterioration which might be synonymous with a slow deterioration of the catalyst itself, and one resulting from an increase in the fraction of high-emitters within each model year, which may be due to malfunctions of other parts of the emission control system than the catalyst, i.e. the EGR or the lambda-sond or other critical components of the emission control system.

The fraction of TWC-cars exceeding any given cutpoint for CO appeared to be a linear function of vehicle age, as seen in Fig. 3.14. A very similar pattern was observed for NO. Such observations have also been made for US fleets from remote sensing measurements (Stephens, 1994). Therefore, linear relationships could be established by means of linear regression analysis of the plots represented by Fig. 3.14, to estimate the fleet fractions exceeding a given cutpoint i as a function of vehicle age, according to the formula:

$$\mathbf{F_i = ax + b} \qquad\qquad (3.4)$$

where Fi is the fraction of the fleet exceeding cutpoint i, x is the fleet age and a and b are the coefficients derived from the linear regression analysis. These coefficients were calculated for various cutpoints for CO and NO for TWC-cars for the measurements in 1991, 1995 and 1998. It was found that regardless of what year the measurements were made, the relationships, particularly represented by the slope coefficients (a), for a given pollutant (CO or NO) and a given cutpoint, were fairly uniform, indicating that for the Swedish TWC-fleet vehicle age rather than model year is the most important factor determining CO and NO emissions. For non-catalyst cars the fraction exceeding a given cutpoint tended to be independent of age, cf. Fig. 3.14.

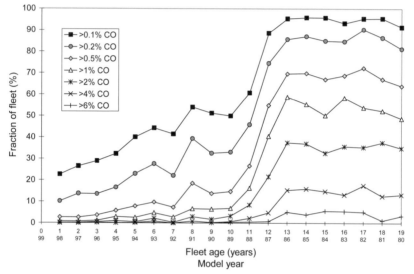

Fig. 3.14. Percent fraction of the overall gasoline passenger car fleet of a given age exceeding various cutpoints for tailpipe %CO on the freeway ramp in 1998.

3.3.2.3 Summarizing analyses of tunnel and open road studies

J. Kühlwein, R. Friedrich

Statistical approaches are not satisfying in every case when aiming at quantifying differences between modelled and real emission rates, for two reasons: 1. Statistical error analyses result in confidence intervals which represent position (emission) probabilities. But there is no information given about the real position of true emission values and therefore amount and direction of deviation are not known. 2. Systematic errors cannot be derived from statistical analyses if no additional independent data sets are available (Kühlwein 2003).

Instead, real-world emissions have to be derived in case of road traffic from tunnel measurements and from open roadway experiments and compared with modelled emission rates.

Principles

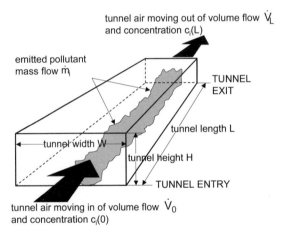

emitted pollutant mass flow $\dot{m}_i$

tunnel air moving out of volume flow $\dot{V}_L$ and concentration $c_i(L)$

TUNNEL EXIT

tunnel length L

tunnel width W

tunnel height H

TUNNEL ENTRY

tunnel air moving in of volume flow $\dot{V}_0$ and concentration $c_i(0)$

Fig. 3.15. Principle of pollutant mass balances from tunnel experiments

The principle sketch of pollutant mass balances from tunnel experiments is given at Fig. 3.15. An ideal tunnel with rectangular cross-section and length L is shown. Tunnel air is moving in at the tunnel entry with a specific volume flow and background pollutant concentration. The tunnel air is leaving at the tunnel exit with a volume flow, that should be identical with the tunnel entry, and with a pollutant concentration that should be considerably higher as at the entry because of the pollutant mass flow emitted by vehicles passing the tunnel.

The principle of calculation and the necessary input data to derive real world emission factors from tunnel studies are given at Eq. (3.5).

$$EF_i = \frac{(c_i(L) - c_i(0)) \cdot A_T \cdot u_T}{\dot{n} \cdot L} \qquad (3.5)$$

with:

EF_i: mean emission factor per vehicle and pollutant i [g/m]
$c_i(L)$:   concentration of pollutant i at tunnel exit [g/m³]
$c_i(0)$:   concentration of pollutant i at tunnel entry [g/m³]
A_T: tunnel cross-sectional area [m²] ($= W \cdot H$)
L: tunnel length [m]
u_T: tunnel wind speed [m/s] ($= \dot{V} / A_T$)
$\dot{n}$:   traffic flow [1/s]

The principle of mass balances at open roadways is shown in Fig. 3.16. An imaginary box includes the road section on the ground and the total exhaust plume. There are two vertical measuring lines, one upwind and one downwind of the road. The pollutant mass flows entering at one side (x_1) and leaving at the opposite side (x_2) have to be determined. Because the wind speed perpendicular to the road and pollutant concentrations are not expected to be homogenous with height z, vertical wind and pollutant concentration profiles have to be measured. The mathematical description on how to derive the emission factors from measurement data is given with Eq. (3.6).

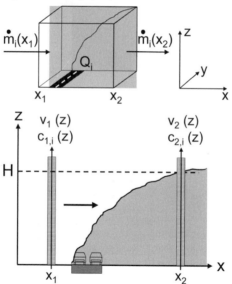

Fig. 3.16. Principle of pollutant mass balances from open roadway experiments

$$EF_i = \frac{\int_0^H v_2(z) \cdot c_{2,i}(z)dz - \int_0^H v_1(z) \cdot c_{1,i}(z)dz}{\dot{n}} \qquad (3.6)$$

with:

EF_i: mean emission factor per vehicle and pollutant i [g/m]
$v_1(z)$: upwind wind speed at height z [m/s]
$v_2(z)$: downwind wind speed at height z [m/s]
$c_{1,i}(z)$: upwind concentration of pollutant i at height z [g/m^3]
$c_{2,i}(z)$: downwind concentration of pollutant i at height z [g/m^3]
H: height of exhaust plume [m]
$\dot{n}$: traffic flow [1/s]

Table 3.8 gives a comparison of advantages and disadvantages between tunnel and open roadway experiments.

Table 3.8. Advantages and disadvantages of tunnel experiments compared with open roadway experiments

Advantages of tunnels	Advantages of open roadways
• no dependencies on meteorological influences (wind direction, wind speed, rain, atmospheric stability)	• much larger selection of possible road sections (different driving patterns, fleet compositions and road gradients)
• homogeneous concentration profiles (turbulent mixing) and small temporal variations	• no wall effects (adsorptive or catalytic effects)
• constant flow direction and flow speed of pollutants (external ventilation)	• no „tail wind effects" and necessary corrections of model emissions (at perpendicular streaming)
• comparatively small efforts in measuring pollutant concentrations and wind speeds (no vertical profiles)	• representative driving conditions (tunnel driving behaviour is more passive for psychological reasons)
• accumulation of pollutant concentrations (factor 5-20)	
• no photochemical reactions because of sun irradiation	

Additional input data are necessary to facilitate comparisons of measured real-world emission factors with model emission factors derived from test stand measurements. This concerns traffic data like driving patterns, fleet compositions and road data. Therefore it is essential to include the following objects into the measurement program to achieve a comprehensive and useful data set:

• measurements of speeds and accelerations, separated for different vehicle categories

- manual (or automatic) counts to disaggregate total vehicle flows into different vehicle categories
- license plate number evaluations to do a further fleet differentiation into vehicle layers
- altitude measurements to determine road gradients of selected road sections

Results

Up to now, eight different experiments from the German-speaking countries have been evaluated: five tunnel studies (Gubrist tunnel, Zürich 1993; Airport tunnel Tegel, Berlin 1994; Heslach tunnel, Stuttgart 1994; Arisdorf tunnel, Switzerland 1999; Plabutsch tunnel, Graz 1999), two open motorway studies (A 3, Frankfurt-Köln 1987-1989; A 656, Heidelberg-Mannheim 1997) and one study about measurements in two closed car parks in München 1998. Further details about these studies are given in (Kühlwein 2003).

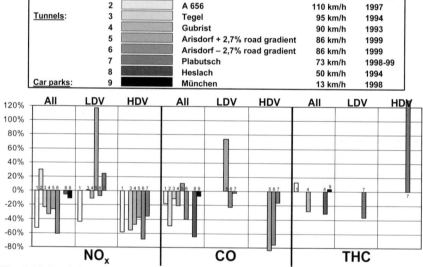

Fig. 3.17. Deviation of model emission factors (test stands - UBA) from real world emission factors

The experimental real world emission factors have been compared with the model emission factors from the emission factor handbook, that is published by the German Umweltbundesamt and the Swiss BUWAL (UBA 1999). Fig. 3.17 shows the results of these comparisons. Most of the model emission factors from the handbook are lower than the experimental ones.

According to all studies with separated emission factors for heavy-duty vehicles (HDV) and light-duty vehicles (LDV), NO_x emission factors for HDV from the model are between 35 % and 65 % lower than those from real world observations. The deviation seems to be much lower for LDV, except for road sections with considerable ascending gradient. Here the Arisdorftunnel study shows a distinct model overestimation.

The differences between the results for CO are partly caused by different data quality. But the main reasons for the occurring deviations have to be seen in the different characteristics concerning different driving behaviour, fleet compositions (what is also indicated by the different study ages) and road gradients. It is conspicuous that at those road sections with high engine loads because of accelerations or ascending road gradients model underestimations of 40 % and more occur. The differentiation between LDV and HDV is more difficult comparing to NO_x because LDV and HDV emission factors for CO are much closer to each other. Because of that, the differentiated results have to be interpreted warily. Anyway, the results of one individual experiment are not transferable to other road sections with different characteristics, which cannot be taken into account when using current available data bases. Rather, more detailed reliable emission factors which also cover situations with high engine loads are required.

Total hydrocarbons (THC) agree well for the total vehicle fleet within a range from + 10 % to - 30 %. Up to now, only one study is available with separated THC emission factors for LDV and HDV (Plabutschtunnel). This study indicates a large overestimation for HDV and an underestimation of about 40 % for LDV.

Because of the complex structure of parameters influencing vehicles' emission behaviour, the available results have to be consolidated and broaden by additional verification experiments. The whole spectrum of driving conditions, fleet compositions and road gradients should be covered when planning future studies and selecting suitable tunnels and open road sections.

3.3.3 Comparison with measurements: The EVA city experiment

3.3.3.1 Overview

F. Slemr

Photochemical models require as input actual emissions of photooxidant precursors such as NO_x and VOCs with high temporal (usually 1 h) and spatial resolutions (e.g. 1 x 1 km^2). Because of widely different reactivities, the VOC emissions have to be resolved to groups of compounds with similar reactivity or even to individual compounds, depending on the model. Since the measurement of the emission data with this resolution is impractical, the emission data have to be calculated using complex emission models. A typical emission model proceeds in two steps. In the first step, broad emission categories are estimated by various methods on an annual or other temporal aggregate basis for a country or other aggregate spatial basis. VOC emissions are usually estimated as a sum from each source (e.g. gasoline evaporation), rarely as an emission of a specific compound (e.g. of some solvents). In the second step, the aggregated emissions are „disaggregated" to the required spatial, temporal resolution of individual compounds or their classes using a combination of techniques such as „source profiles" (emissions of a source resolved to individual compounds), typical temporal allocation functions (e.g. diurnal variation of traffic activities), and surrogate spatial allocation functions (e.g. distribution of emissions from domestic heating by population density) (Placet et al. 2000).

The uncertainty of highly resolved emissions accrues both from uncertainties in the input data for the emission models and from approximations made to disaggregate the aggregated emissions. Many uncertainties are associated with the quality of the input data for the emission models such as the reliance on obsolete emission factors, problems with missing data or varying data quality from different sources (Placet et al. 2000; Sawyer et al. 2000). The process of disaggregation suffers also from uncertainties in input data, e.g. in source profiles, and from assumptions made when constructing the temporal and spatial allocation functions (Placet et al. 2000).

To assess the uncertainties of the highly resolved emissions estimated by an emission model, an experiment called EVA (Evaluation of Highly Resolved Emission Inventories) was executed in 1998. An overview of this experiment and of the major results is given by Slemr et al. (2002b). A detailed account of the emission model to be evaluated (Kühlwein et al. 2002a), of the different techniques used to determine the emissions and their uncertainties (Kalthoff et al. 2002; Klemp et al. 2002; Möllmann-Coers et al. 2002a; Panitz et al. 2002), and of the comparison of measured with calculated emissions (Kühlwein et al. 2002a; Mannschreck et al. 2002a; Möllmann-Coers et al. 2002c) is presented in a supplement of Atmospheric Environment.

The EVA project was specifically designed to evaluate the emission model by Kühlwein et al. (2002a) which provides the highly resolved emissions for photochemical model, and to address the four major questions:

- Do the highly resolved CO and NO_x emissions calculated by this emission model agree with the measurements?

- Do the measured emission ratios such as VOC/NO_x, VOC/CO, and VOC/sum of VOC agree with the calculated ones?

- What are the possible reasons of differences when observed?

- Is the emission model complete or is there an indication of hitherto unconsidered sources?

The city of Augsburg was chosen to address these questions. Augsburg is a medium sized city with a population of about 255,000 in southern Germany in a predominantly rural area of almost flat terrain. Other reasons for the choice of Augsburg were the relatively small size of the city which enables the use of the tracer technique and the dense flight measurements needed for the determination of the mass balance of the city. A sophisticated traffic flow model was also available. The evaluation proceeded in three steps. In the first step preliminary emission inventories (level 0) were calculated as input for photochemical model simulations used to assess the feasibility of emission measurements and to design the experimental activities. In a second step emission data (level 1) have been calculated with the methods usually applied for regional modelling, i.e. the determination of emissions for larger areas like a region with several communities or a country. To get some insight into possible reasons for deviations between modelled and measured emissions, a more detailed emission inventory (level 3) was generated using traffic counts, a more sophisticated traffic flow model and other data collected during the measurement campaign.

Emissions of the city were determined by three independent techniques: absolute emissions of CO and NO_x using a mass balance and a tracer techniques, and NMHCi/CO, NMHCi/sum NMHC (with NMHCi being the individual NMHC compound), and sum $NMHC/NO_x$ emission ratios from concentration ratios at downwind receptor sites and from aircraft measurements. The uncertainty of each independently determined emission of CO and NO_x was estimated and the measured emissions were compared. In most cases the emissions measured by mass balance and tracer techniques agreed within their combined estimated uncertainties. The measured CO and NO_x emissions and emission ratios were then compared with the level 1 emissions calculated by a regular emission model. To aid the interpretation of the observed differences, level 3 emission data were used.

The measurements were designed to provide data necessary for the determination of the emissions and, within budget limits, to get as much information about the meteorology and air chemistry of the investigated area as possible. Joint efforts of the institutions listed below were necessary to execute the extensive measurements made within the EVA project:

- Abteilung Arbeitssicherheit und Strahlenschutz (ASS), Forschungszentrum Jülich, tracer measurements,
- Bayerisches Landesamt für Umweltschutz (LfU), Augsburg, ground based monitoring stations and support,
- Fraunhofer Institut für Atmosphärische Umweltforschung (IFU), Garmisch-Partenkirchen, coordination, mass balance and receptor measurements,
- Institut für Chemie der Belasteten Atmosphäre (ICG-2), Forschungszentrum Jülich, receptor measurements,
- Institut für Energiewirtschaft und Rationelle Energieanwendung (IER), Universität Stuttgart, emission model, modelled emission, traffic counts,
- Institut für Meteorologie und Klimaforschung (IMK), Universität Karlsruhe/Forschungszentrum Karlsruhe, mass balance, meteorological measurements, photochemical model simulations,
- Institut für Verfahrenstechnik und Dampfkesselwesen (IVD), Universität Stuttgart, spatial sounding of meteorological and chemical parameters using tethered balloons and an airship.

The measurements were further supported by the city council of Augsburg, by Bayerisches Institut für Abfallforschung (BifA), and by many individuals who allowed us to use their property for the measurements.

Fig. 3.18 shows a simplified map of the city with the positions of the ground stations, the flight track used for the mass balance determinations, and the location of the tracer emission and receptor sites. An overview of the meteorological and chemical measurements made on different sites and using different platforms is given in Table 3.9 and Table 3.10, respectively. In addition, air samples for NMHC analysis were taken into electropolished stainless steel canisters onboard aircraft, at Stätzling site during the intensive measurements, at Leitershofen site (near Radegundis) during westerly winds, and using adsorption tubes onboard of a tethered balloon.

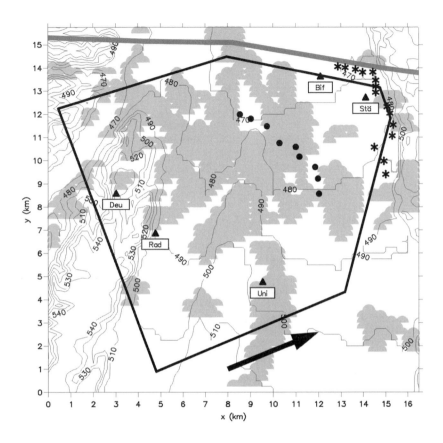

Fig. 3.18. Augsburg and its surrounding with the projection of the flight pattern (black line) and the position of the ground measurement stations (Bif – Bayerisches Institut für Abfallforschung, Deu – Deuringen, Rad – Radegundis, Stä – Stätzling, Uni – University). The sealed areas and the motorway are shaded in grey colour. The tracer release points are marked by circles, the tracer collection points by stars. The arrow indicates the wind direction prevailing during the special observation periods.

Table 3.9. Meteorological measurements made during the EVA campaigns. Temperature is indicated by T, relative humidity by Rf, pressure p, wind speed by v, wind direction by dir, net radiation by Q, soil heat flux by B, and sensible heat flux by H. z_0 defines the lowest measurement height in m above ground level and Δz the vertical resolution.

Site	Institute	System	Parameter	Time resolution	Height above ground
BifA (Bif)	IMK	Radiosonde	T, Rf, p, v, dir	5 s	up to 6 km
BifA (Bif)	IMK	Energy balance	T, Rf, p, v, dir, Q, B, H	10 min	4 m
University (Uni)	IMK	Sodar	V,dir	15 min	$z_0 = 50$ m $\Delta z = 25$ m
Radegundis (Rad)	IFU	Ground station	T, Rf, p, v, dir	10 s	8 m
Deuringen (Deu)	IMK	Wind profiler	T, v, dir	30 min	$z_0 = 60$ m $\Delta z = 45$ m
Leitershofen (Lei)	IVD	Tethered balloon	T, Rf, p, v, dir	10 s	50 – 620 m
Stätzling (Stä)	IVD	Tethered balloon	T, Rf, p, v, dir	10 s	0 – 460 m
Stätzling (Stä)	IVD	Airship	T, p, v, dir	10 s	0 – 425 m
Stätzling (Stä)	ICG-2	Ground station	T, Rf, p, v, dir	10 s	10 m (T, Rf: 4.5 m)
Dornier 128	IMK	Aircraft	T, Rf, p, v, dir	2 s	
Partenavia 68	IFU	Aircraft	T, Rf, p, v, dir	2 s	

The measurements were carried out during two four week long campaigns in March and October, 1998. These months were selected to avoid intensive photochemistry in summer and to detect a possible emission difference due to residential heating which is expected to be more intensive in March than in October. Altogether six tracer experiments were carried out (on March 18 and 27 and October 10, 21, 22, 23) of which four fulfilled the assumption of nearly stationary conditions in the course of the experiment. Four mass balance experiments were executed on October 10, 21, 22, and 23. October 10 was Saturday, all other days were working days. Emission ratios from measurements of concentration ratios at receptor sites could be determined on about half of the almost 60 days of the campaigns. The determinations on the other days were hampered by meteorological conditions changing in the course of the experiment.

The following chapters deal with different aspects of the EVA experiment. The first chapter describes the measures taken to ensure that the data quality is known and adequate for the purposes of the experiment. The following chapters describe the mass balance and source-receptor techniques used to determine the CO and NO_x emissions of the city, and the long term receptor measurements of HCi/CO and HCi/ sum HCi emission ratios. A special chapter is devoted to an intercomparison of two techniques for measurement of formaldehyde and to the formaldehyde/NO_x emission ratio. The last two chapters summarize the comparison of modelled and measured CO and NO_x absolute emission and HCi emission ratios. In the last chapter conclusions are made.

Table 3.10. Chemical measurements during the EVA campaigns.

Site	Institute	System	Parameter	Time resolution	Height above ground
BifA (Bif)	IMK	Ground station	CO, NO_x, O_3	10 min	4 m
University (Uni)	IMK	Ground station	CO, NO_x, O_3	10 min	4 m
Radegundis (Rad)	IFU	Ground station	CO, NO_x, NOy, O_3, CH_2O, NMHC	10 min	4 m
Leitershofen (Lei)	IVD	Ground station	CO, CO_2, SO_2, NO_x	10 min	4 m
Stätzling (Stä)	ICG-2	Ground station	CO, NO_x, NO_y, O_3, CH_2O, NMHC	10 min	10 m
Stätzling (Stä)	IVD	Airship	NO_x, O_3	10 s	0– 425 m
Stätzling (Stä)	IVD	Tethered balloon	NO_x, O_3, NMHC	10 s	0– 460 m
Frechholzhausen (Fre)	LfU	Ground station	CO, CO_2, SO_2, NO_x, BTX	30 min	3.5 m
Bourges Platz (Bou)	LfU	Ground station	CO, NO_x, SO_2	30 min	3.5 m
Königsplatz (Kön)	LfU	Ground station	CO, NO_x, SO_2	30 min	3.5 m
Haunstetten (Hau)	LfU	Ground station	O_3, SO_2	30 min	3.5 m
Tracer sampling line	ASS	Ground sites	SF_6, CO	20 min	1 m
Ultralight aircraft	IFU	Sampling	SF_6, CO		
Dornier 128	IMK	Aircraft	CO, NO_x, O_3, NMHC	2 s	150– 2000 m
Partenavia 68	IFU	Aircraft	CO, NO_x, O_3, NMHC	2 s	150– 2000 m

3.3.3.2 Quality assurance and quality control (QA/QC)

F. Slemr

As listed above seven institutions participated in the EVA experiment and six of them were involved in the experimental work. The institutions used sometimes different techniques to measure the same species and different calibration gases to calibrate their instruments. Consequently, a stringent quality assurance and quality control (QA/QC) was necessary to obtain data of known and adequate quality for the purposes of the EVA experiment. The QA/QC work was made as a part of the overall QA/QC work in Tropospheric Research Programme (TFS). The review of the TFS QA/QC work for inorganic and organic compounds is given by Kanter et al. (2002) and Volz-Thomas et al. (2002), respectively. The following paragraphs are limited to EVA specific QA/QC work.

The quality of the measurements was assessed by intercomparisons before and after each measurement campaign. The intercomparison before the March campaign was made at the airport of Mindelheim to enable the intercomparison of the aircraft measurement systems with ground measurements. Unfortunately, the air was very clean during these intercomparisons, preventing the intercomparison of NMHC and NO_x measurements in ambient air. Ozone, NO_x, and CO measurements were intercompared successfully using synthetic gas mixtures.

The intercomparisons after the March campaign, and before and after the October campaign were all carried out at the Stätzling site. CO, O_3, NO, NO_x, NOy, and NMHC in ambient air were measured during these intercomparisons for a period of 24 hours and the results were intercompared. An excellent agreement was achieved for almost all NMHC compounds measured by two different on-line instruments of ICG-2 and IFU, and in air samples taken into stainless steel canisters which were analysed at IFU laboratory by a third instrument (Volz-Thomas et al. 2002). All other continuous measurements were found to be in agreement within the margins of the data quality objectives defined for the EVA experiment.

The quality of the aircraft data was assessed before the October campaign by a flight with both aircraft flying side by side for preselected legs of the flight. An excellent agreement was found for the CO measurements, which provide the basis for the determination of the mass balances (Kanter et al. 2002). Both aircraft sampled air into the stainless steel canisters provided by IFU. The air samples were analysed by IFU and, as mentioned above, the IFU analyses were found comparable with the on-line NMHC measurements made by IFU and ICG-2.

In summary, the CO, NO, NOy, and NMHC measurements on the ground and onboard aircraft during both EVA campaigns were found to be within the data quality objectives defined for EVA purposes.

As a part of the QA/QC work we further investigated the question whether the emission ratios measured at the Stätzling site are representative for the whole area of Augsburg. This was investigated by intercomparison of the emissions ratios determined at Stätzling with emission ratios derived from aircraft measurements made during the mass balance experiments in October 1998. During these flights air samples were taken into electropolished stainless steel canisters randomly along the flight path and analysed for CO and NMHC content at IFU laboratory. The intercomparisons of the HC_i/CO emission ratios measured at Stätzling site and onboard aircraft on October 10, 1998, is shown in Fig. 3.19.

Fig. 3.19 shows the individual emission ratios, each of which was calculated as a slope of a linear regression of individual HC versus CO measurements. The bars represent the standard deviations of the HCi versus CO slopes of about 10 aircraft and about 8 ground based pairs of measurements made approximately at the same time (the air samples were collected during a 3 h long flight, the on-ground measurement cover a period of about 8 h). The regression line ($R^2 = 0.945$) indicates that in average the measurements at the Stätzling site agreed with measurements onboard aircraft (the agreement is better than 10 %) which are representative for the whole area of Augsburg. To the best of our knowledge, no comparable intercomparison has been published yet.

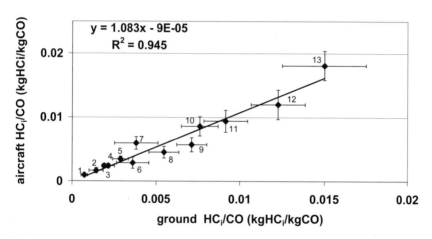

HC$_i$/CO emission ratios: aircraft vs. ground

Fig. 3.19. HCi/CO emission ratios of the city of Augsburg from measurements onboard aircraft versus those measured downwind of the city at the receptor site in Stätzling on October 10, 1998. The bars show the 1σ standard deviations of the regression analyses of both measurements. The compounds are. 1: cyclohexane; 2: heptane; 3: hexane; 4: 2,2-dimethylbutane; 5: 2-methylpentane; 6: i-butane; 7: benzene; 8: n-pentane; 9: n-butane; 10: ethyne; 11: ethene; 12: i-pentane; 13: toluene. The HCi/CO emission ratios are based on mass equivalents (kgHC$_i$/kgCO). Only compounds measured both onboard aircraft and at the Stätzling site are shown. About 20 other compounds were measured at the Stätzling site.

3.3.3.3 Determination of CO and NO_x emissions by the mass balance technique

For determination of CO and NO_x emissions from field measurements, meteorological parameters, CO and NO_x were measured onboard 2 aircraft along a path in Fig. 3.19 at as many altitudes as possible within a 3 hour period in the early afternoon (12 a.m. – 3 p.m., local time). Model simulations by Panitz et al. (2002) have shown that the requirements of nearly constant meteorological conditions and almost constant CO and NO_x emissions are most likely fulfilled at this daytime.

Under the assumption of stationary meteorological conditions, constant CO and NO_x emissions, and the neglect of vertical fluxes at the capping inversion, the emission of the substance X, E(X) expressed in kg/h, within the investigated area would be (Klemm and Schaller 1994):

$$E(X) = 3.6 \ 10^{-9} \int_z \int_s \bar{v}(s,z) \cdot \bar{n}(s) \cdot \frac{M(X)}{M_A} \cdot \rho(z) \, \mu(X,s,z) \, ds \, dz \qquad (3.7)$$

where $\bar{v}(s,z)$ is the measured wind vector (m/s) in the plane along the flight path s around the city, $\bar{n}(s)$ the unit vector normal to the flight path, $\rho(z)$ the air density (kg/m³), $\mu(X,s,z)$ the measured mixing ratio (ppb) of substance X, and ds dz the area element (m²) in the plane of measurement. M(X) and M_A denote the molecular weights of species X and dry air, respectively. The factor 3.6 converts the result into kg/h. Measurement were made only at 4 – 14 altitude levels and the emissions were calculated by IMK and by IFU in two slightly different procedures. These procedures are described in detail by Kalthoff et al. (2002).

IMK considered aircraft, airship, tethersonde and surface measurements to interpolate the NO_x and CO mixing ratios for grid cells of 25 x 25 m² on the envelope of the control volume confined by the flight pattern. To generate the horizontal wind vector on the entire envelope of the control box, aircraft and, below the lowest flight level, sodar data were used. The mass fluxes of CO and NO_x were calculated from the wind speed and the mixing ratios for each grid cell of the entire envelope. It was observed that a substantial part of the flux through the envelope is due to subsiding movement of the air within the measurement area. The mean downward vertical wind speed at the top of the control volume was calculated from the continuity equation and the incoming mean mixing ratios of NO_x and CO were derived from measurements above the mixing layer. The CO and NO_x import through the lid of the measured volume was then calculated from the mean vertical wind speed and the CO and NO_x mixing ratios, respectively. The air import through the lid contribute about 40 % of the CO and, due to its lower background concentration, about 20 % of the NO_x mass balance. The large contribution of the small subsiding movement of only about 2 cm/s to the CO and NO_x mass balance is due to the large lid area (130 km²) of the control volume in comparison to the area of its vertical envelope of only about 30 km².

IFU assumed the measurements at one flight level to be representative for an altitude interval halfway between the flight levels below and above. The CO and

NO_x fluxes along the flight trajectory were calculated from the measured CO and NO_x mixing ratios, respectively, and from the wind vector provided by wind profiler for the altitude of the measurements which was considered to be representative for the entire control volume. The use of the same wind profile in the entire control volume corrects for the import term in the mass balance in a very similar way to the IMK approach. The fluxes were integrated along the flight circle. The circle flux integrals were then multiplied by the altitude interval and summed up from the ground to the inversion layer. For safety reasons, the aircraft was not allowed to fly below 300 m above ground. The flux in the lowest layer above ground was, therefore, corrected for the decrease of wind speed towards the ground using sonar data.

The uncertainties were estimated using the error propagation analysis of all input parameters. Uncertainty from the limited density of measurements was considered by IMK as deviations of measurements from the interpolated values and by IFU by comparing the original emissions with those obtained by omitting individual flight levels.

The results are summarized in Table 3.11. The CO emissions determined by IMK and IFU agree within the stated uncertainties of typically about 20 %. The uncertainties of the NO_x emissions of typically about 40 % are mostly due to larger uncertainty of the NO_x measurements and could be reduced by using more accurate measuring techniques (Kanter et al. 2002). The NO_x emissions determined by IMK and IFU also agree within the stated larger uncertainties.

Table 3.11. Inversion height z_i, NO_x, and CO emissions of Augsburg, calculated for the special observation periods by IMK and IFU. The given uncertainties are estimated from uncertainty propagation and correspond to 1σ.

Day	z_i (m asl)	IMK CO emissions (kg h^{-1})	IFU CO emissions (kg h^{-1})	IMK NO_x emissions (kg h^{-1})	IFU NO_x emissions (kg h^{-1})
October 10	950	2261 ± 474	2110 ± 365	126 ± 46	170 ± 51
October 21	950	1843 ± 608	2370 ± 495	434 ± 213	530 ± 174
October 22[a]	1050	1600 ± 304	1540 ± 370	301 ± 117	380 ± 125
October 22[b]	1050	1223 ± 220	1010 ± 251	231 ± 99	240 ± 106

[a] 12:21 – 13:56 local time (10:21 - 11:56 UTC)
[b] 13:45 – 15:00 local time (11:45 – 13:00 UTC)

3.3.3.4 Experimental determination of absolute emissions from concentration ratios by means of tracer studies

M. Möllmann-Coers, D. Klemp, K. Mannschreck

Introduction

In the frame of the joint research project EVA (Slemr et al. 2002b) which is focused on the evaluation of an emission model, the absolute CO emission rates of the city of Augsburg were determined by means of the source-tracer-ratio (STR) method. Applying this method several atmospheric tracer experiments were performed within two field campaigns in March and October 1998. Sulphur hexafluoride (SF_6) was used as tracer which was released in Augsburg to mark the city plume. The emission rates of CO are then calculated by means of the known emission rates of SF_6 and the measured SF_6 and CO mixing ratios downwind of Augsburg. From correlation analyses between CO and other substances (NO_x and NMHCs) in the plume absolute emission rates of these species were also determined.

Experimental

Principle of the source-tracer-ratio (STR) method

Tracers are widely used to study advection and dispersion processes (Pasquill 1974). For this purpose, suitable species have to be chemically inert and not toxic and have not to be present in the normal environment. SF_6 fulfils these requirements to a large degree and it can be measured in the low ppt range by electron capture detector technique (ECD). A brief overview of the experimental technique is given by Zeuner and Möllmann-Coers (1994).

When using the STR method the unknown emissions of a species (in our case CO) are marked by a tracer (in our case SF_6) released with a known emission rate. Providing a comparable distribution of CO and SF_6 sources and assuming identical atmospheric transport and dispersion of SF_6 and CO, the emission rate of CO can be calculated from the concentration ratio and the known SF_6 emission rate according to equation (3.8):

$$\dot{q}_{CO} = \frac{c_{CO}}{c_{SF_6}} \dot{q}_{SF_6} \qquad (3.8)$$

In this equation $\dot{q}_{CO}$ is the unknown emission rate of CO, $\dot{q}_{SF_6}$ is the known emission rate of SF_6, and c_{CO} and c_{SF_6} are the measured concentrations of CO and SF_6, respectively.

Tracer release and sampling

The experimental site and the locations of the tracer release and receptor sites are shown in Fig. 3.20. Under southwesterly wind conditions SF_6 is released at position 1 to 9 in the city with a known emission rate. The large number of release positions provides for a homogeneous build-up of the SF_6 plume comparable to that of the respective CO emission plume. Air samples were taken along the reception line downwind of Augsburg. The reception line consists of 16 individual sampling positions.

Conducting the experiments, pure SF_6 is released directly from gas cylinders with a constant emission rate of 1 g/s. The release height is 4m above ground.

The samplers for up to 9 sampling bags (aluminium coated plastic bags Linde Plastigas®) were designed and built by Research Centre Jülich. The air samples are taken about 1 m above ground. The time base for the individual samplers is given by a radio controlled clock and the begin and the end of the sampling period can be set individually for each sample.

Gas analysis

The SF_6 concentrations in the air samples were analysed using a gas chromatograph (Siemens, Sichromat 3) equipped with an ECD. The samples were analysed for CO using the fast-response resonance fluorescence method described by Gerbig et al. (1996, 1999) and Volz-Thomas and Gerbig (1998). More details of the analysis technique are given in Möllmann-Coers et al. (2002a).

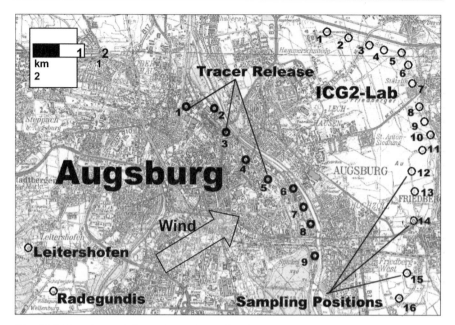

Fig. 3.20. Experimental site and experimental design. The downwind positions of the sampling units are shown on the right in the figure, as well as the upwind positions at Radegundis and Leitershofen on the left side. The tracer release positions are depicted in the centre of the figure. In addition, the position of the ICG II mobile laboratory is marked. Position 7 of the reception line is identical with the position of the ICG II mobile laboratory. By means of this laboratory a wide range of hydrocarbons as well as CO and NO_x have been measured (Klemp et al. 2002).

Application of the STR method in Augsburg

Using CO as a reference trace gas

For the application of the STR method to evaluate modelled emission inventories CO was chosen as reference substance for trace gas emissions from Augsburg. The use of CO has several advantages:

1. Apart from CO_2, CO is by far the most abundant species released from anthropogenic sources.
2. CO is nearly exclusively emitted by combustion processes and especially by vehicle exhaust (Kühlwein et al. 2002a). Within the Augsburg area there are only very small CO-emitting point sources (Kühlwein et al. 2002b). Consequently the assumption of an area source for CO (as requested for the STR method) is met with good approximation. The emission rates of other mainly traffic-related species like NO_x and some NMHCs (i.e. benzene, ethyne, ethene) can be calculated from their concentration ratios to NO_x.

3. Within the advection time from the release to the sampling position (0.5 h), CO can be assumed to be inert.

The main disadvantage of CO is its relatively high atmospheric mean CO background, which exceeds the impact of the city emissions downwind of Augsburg.

Experimental design

The experimental design is depicted in Fig. 3.20. For the chosen conditions the city plume of Augsburg can be approximated by a line of point sources through the city for southwesterly winds. This is shown by detailed model studies (Möllmann-Coers et al. 2002a). The tracer sampling line is established about 3 km easterly of the city. The area between the city border and sampling line consists mainly of agricultural land without significant CO sources. The distance of three km is sufficient for a vertically well mixed city plume within the atmospheric mixing layer. The latter assumption is supported by the results of model studies mentioned above.

The tracer experiments

To ensure a clear source receptor relationship on the one hand and to receive a significant increase of the CO concentration within the city plume on the other hand, the conduction of tracer experiments was restricted to wind speeds between 2.5 and 7.5 m/s at 10m height. In addition, the experimental design restricts the experiments to southwesterly winds (cf. Fig. 3.20)

The tracer experiments last four hours from 11 to 15 local time. In order to get a stationary tracer plume, the tracer release started 20 minutes before the begin of sampling. Nine 20 minute samples were taken at each position of the receptor line. Four of six experiments could be analysed. Two were not analysed due to an unexpected wind shift during the experiment.

Data analysis

The evaluation of the experimental data requires the knowledge of the background CO and SF_6 concentration:

1. From test samples in Augsburg a mean SF_6 concentration of 7.2 ppt was found which is about 7% of the mean experimental tracer concentration at the reception line.
2. Due to its lifetime of more than 2 months background concentrations of CO may vary ± 30 ppb depending on the history of the air masses (Mannschreck 2000) for different experiments. Further, even in the city plume the CO background contributes to more than 70% (Klemp et al. 2002) to total CO levels. Consequently, the CO background has to be determined for each experiment individually using two additional sampling positions upwind of Augsburg.

3. The experimental results show that several samples are contaminated by local events e.g. occasional harvest traffic on the fields close to an individual sampling position. On the other hand, due to variations in the wind direction the city plume may not cover the whole reception line all the time. In any case, for this type of event the differences of the concentrations between two neighbouring positions exceed the 1σ-range. Within a sampling interval these samples were excluded from the evaluation by a median-based filter algorithm:

$$\left| C^{CO}_{Median} - C^{CO}_i \right| \le f\ C^{CO}_{Median}\ ,\ \text{with } f = 0.3^3.\qquad (3.9)$$

The mean CO/SF_6 emission ratio is calculated by orthonormal regression analysis (York 1966) using the remaining pairs of CO and SF_6 mixing ratios after filtering and the respective background measurements performed in Radegundis and Leitershofen. The 1σ-uncertainty of the slope ($\approx 30\%$) is denoted as uncertainty of the calculated CO/SF_6 emission ratio. More details about the application of the STR method are discussed in Möllmann-Coers et al. (2002a).

Results and discussion

Table 3.12 summarizes the absolute CO emission rates derived from the application of the STR method. Within the range of uncertainty, the results agree with correspondent results from a mass balance method (Kalthoff et al. 2002). It is shown that the experimentally determined emission rates also agree with the results of the emission model (Kühlwein et al. 2002a). The same is valid for the NO_x emission rates listed in Table 3.12. These values were calculated from the CO values (Table 3.12) according to Eq. (3.8) and from the NO_x/CO concentration ratio determined from the measurements of the ICG II mobile laboratory (cf. Fig. 3.20). The uncertainties of NO_x are calculated by quadratic additions of the individual uncertainties of the CO/SF_6- and NO_x/CO- correlations, respectively.

Table 3.12. Absolute CO and NO_x emission rates together with their respective uncertainties

Date:	27.3.	10.10.	22.10.	23.10.
CO emission rate (t/h)	3.0±1.0	1.8±0.4	0.98±0.24	0.9±0.2
NO_x emission rate (t/h)	0.35±0.12	0.25±0.06	0.24±0.07	0.22±0.05

The CO emission rate on 27th of March (a working day) is significantly higher than the CO emissions for working days in October. This is probably due to residence heating, not present in autumn.

---

[3] Sensitivity tests indicated that the choice of f within a wide range (0.1 to 3) primarily affects the uncertainties and has only weak influence on the calculated emission rates.

The ICG II laboratory at Stätzling was integrated in the STR reception line (cf. Fig. 3.20). Consequently, the NMHC/CO concentration ratios (Klemp et al. 2002) determined at Stätzling can be combined with the CO emission rates derived by STR method. Based on Eq. (3.8) the emissions rates of individual hydrocarbons can be calculated in the same way as the NO_x emission rates. Table 3.13 shows the result for NMHCs which can be specified individually from the emission model results (Mannschreck 2000).

Several conclusions can be drawn from the comparison of the experimental and model results (cf. Table 3.13):

- Since the measured and modelled CO emission rates compare well within their uncertainties, significant differences between modelled and measured hydrocarbon emission rates have to be attributed to the differences in hydrocarbon emissions. Comparing the experimentally derived NMHC emission rates with the model results deviations up to a factor of 5 can be found for some species (i.e. ethane, propane). These deviations exceed by far the uncertainty of the STR method which is about 30%. For some other species (e.g. toluene), the agreement between both emission rates is quite good.

- The measurements suggest that the emission rates of most NMHCs are about twice as large in March than in October. But the model predicts - apart from the butenes and the C_4-C_5 alkanes – the same emission rates in spring and autumn.

- The modelled emission rates of C_4 and C_5 alkanes are in October by 30% - 40% higher than in March. This is mostly due to its higher share of evaporation processes in October in comparison to the March campaign. This behaviour, however, was not observed in the experimental data, suggesting that the increase of evaporation-related emissions with increasing temperature is only of minor importance. Emission rates of C_4 and C_5 alkanes in October were smaller than those in March suggesting to combustion related emissions as the major source of C_4 and C_5 alkanes. This is in line with the conclusions of Mannschreck et al. (2002a) and Möllmann-Coers et al. (2002c).

Table 3.13. Absolute emission rates for a number of hydrocarbons calculated from their $NMHC_i$ /CO ratios together with their 1-s-standard deviations (Experiment) and the corresponding emissions calculated with the emission model (Model) (EVA-Database, 2000)

Date	27.3.		10.10.		22.10.		23.10.	
	Emission rate		Emission rate		Emission rate		Emission rate	
	[kg/h]	[kg/h]	[kg/h]	[kg/h]	[kg/h]	[kg/h]	[kg/h]	[kg/h]
Species	Experiment	Model	Experiment	Model	Experiment	Model	Experiment	Model
Ethene	37.5±18.6	14.7	12.3±4.4	15.7	10.8±3.55	19.4	10.0±3.1	18.7
Ethyne	24.0±9.0	7.9	17.0±5.8	7.2	10.5±3.0	7.4	9.7±2.7	7.7
Ethane	13.9±7.3	2.8	8.8±3.3	5.7	7.1±3.3	11.3	6.6±3.0	6.2
Benzene	16.5±6.2	9.1	14.9±5.1	10.1	9.0±2.3	10.5	8.3±2.0	10.8
Propene	10.8±5.1	6	11.8±4.8	6.1	5.5±8.5	6.7	5.1±7.8	6.9
Propane	18.8±11.9	3.8	12.3±4.7	4.2	5.6±2.4	5.8	5.2±2.2	4.8
Butene	7.9±3.7	4.6	6.9±2.7	6.6	4.9±1.7	6.7	4.5±1.5	8.7
i-Butane	13.7±5.6	6.5	7.1±2.4	9.6	7.5±2.9	9	6.9±2.6	13.6
n-Butane	23.1±11.3	14.8	12.7±4.2	22.8	9.2±3.3	21.4	8.5±2.9	32.5
i-Pentane	25.2±9.9	29.5	17.9±6.1	42.8	7.8±2.1	40.3	7.3±1.9	59.3
n-Pentane	19.9±9.5	9	8.7±3.0	13	10.8±6.1	12.1	10.0±5.6	17.6
Toluene	39.0±18.3	38	27.6±9.0	30.2	26.7±7.8	35.6	24.7±6.9	34.7

Summary and Conclusions

The STR method was successfully applied to determine the absolute CO emission rates of Augsburg. The results obtained with the STR method compare favourably with those obtained by the mass balance method (Kalthoff et al. 2002; Kühlwein et al. 2002a)

The overall accuracy of the STR method applied here is in the range of 30%-40%. Within the range of uncertainty the experimentally derived CO and NO_x emission rates agree with the results of the emission model. But for some hydrocarbons the experimentally determined emission rates exceed from the modelled ones up to a factor of 5.

3.3.3.5 Determination of anthropogenic emission ratios in the Augsburg area by long-term concentration measurements downwind of the city

D. Klemp, K. Mannschreck, F. Slemr

Introduction

Air quality models require temporally and spatially highly resolved emission inventories of NO_x, CO and individual organic compounds as one of the most important input parameters. Since the quality of the output of the air quality models depends decisively on the quality of the input emission data (Placet et al. 2000), the latter has to be assessed.

In a joint project "**EVA**luation of calculated highly resolved emissions of a city" (**EVA**) within the German Tropospheric Research Programme (TFS), several institutions determined the emissions of the city of Augsburg by three independent techniques and compared the results with the calculated emissions (Slemr et al. 2002a). One of the techniques was the determination of the emission ratios from the concentration ratios measured at a ground receptor site downwind of the city and onboard aircraft. Special attention is devoted to the speciation of the hydrocarbon emissions. The question of representativity of emission measurements in Augsburg is also addressed.

Experimental

Augsburg is a medium-sized city (250,000 inhabitants) in Bavaria in Southern Germany. CO, $C_2 - C_{10}$-hydrocarbons, NO and NO_2 at the receptor site were continuously measured throughout two 4 week long campaigns in March and in October 1998. For this purpose the mobile laboratory of Forschungszentrum Jülich was placed north-east of Augsburg in Stätzling in a distance of about 5 km from the city centre.

During intensives, grab samples were taken onboard aircraft circling around the city of Augsburg and analysed at IFU Garmisch for NMHC content. The samples were taken randomly upwind and downwind of the city at altitudes varying from about 150 m above ground up to the top of the mixing layer and a few samples were taken in the lower free troposphere.

Analytical techniques

The analytical instruments were installed in the air conditioned mobile laboratory of Forschungszentrum Jülich (Klemp et al. 2002). The inlets and meteorological instruments were mounted on a pneumatic mast 10 m above the surface. Briefly, CO was measured using a commercial infrared filter correlation absorption spectrometer (Thermo Environmental Instruments, Modell TE48). Interferences by atmospheric water vapour were eliminated by drying the sampled air using Nafion® drier. Problems with zero drift were substantially reduced by passing ambient air through a Hopcalite® scrubber for 5 minutes every hour.

NO_x and NO_y were measured using chemiluminescence detectors (ECO Physics, CLD 770AL ppt) and a photolytic converter (ECO Physics PLC 760) to convert NO_2 to NO. Due to the fact that the conversion efficiency of the PLC in ambient air is affected by the mixing ratios of O_3 and NO_2 a photochemical box model is used to simulate the respective conditions in the PLC chamber (Pätz et al. 2000) and to correct the measured NO_2 raw signals.

Different to the other analytical equipment the NMHC measurement technique is based on a home-built GC-system, designed for the trace gas analysis in the lower ppt-level. $C_2 - C_{10}$ hydrocarbons were measured using a gas chromatograph (HP 5890A) with a flame ionisation detector (FID) and a specially designed sampling device (Schmitz T et al. 1997). Water from sampled air was removed by a cooling trap kept at –25 °C. NMHCs from a sample volume of 500 cm^3 were then preconcentrated in a liquid nitrogen cooled sample loop (20 cm long, 2 mm diameter) filled with glass beads. The flow of sampled air was kept at 50 ml/min resulting in 10 min sampling interval. When sampling is completed, the sample loop is heated up to 80 °C and the hydrocarbons are injected on the capillary column (DB-1, 90 m x 0.32 mm ID, film thickness: 3μm). After injection the column is kept isothermal at –50 °C for 5 min and then heated up to 200 °C with a rate of 5 °C/min. Subsequently the column temperature is kept at 200°C for 15 min. The complete analysis took about 80 minutes.

Individual peaks in the chromatogram were identified via injection of the pure species and the identification was confirmed using the 70 component gas mixture (Slemr 1999). The mean detection limits for measured hydrocarbons varied between 10 ppt (C_3-compounds) and 3 ppt for hydrocarbons > C_8. From error propagation analysis the experimental uncertainties ($\Delta\mu_{HCi}$) are calculated to account to less than 15% to the respective mixing ratios μ_{HCi} of HC_i.

With respect to the NMHC grab samples taken onboard aircraft, air samples were drawn by two Metal Bellow pumps in series into 1 l electropolished stainless steel canisters. The canisters where then shipped to the laboratory and analysed: Briefly, 400 ml of air samples were directed through CO_2 and H_2O scrubber (columns packed by NaOH and $Mg(ClO_4)_2$) and the $C_2 - C_{10}$ NMHCs were trapped at -25 °C on Carbotrap (Chrompack). The enriched NMHCs were then desorbed at 120 °C and cryofocussed on the initial part of the separation column at -120 °C. The NMHCs were separated on Al_2O_3/Na_2SO_4-PLOT column (50 m long, 0.53 mm diameter, 10 μm film thickness) using a temperature program (5 min at 50 °C, 8 °C/min up to 170 °C, 6 °C/min to 200 °C and hold for 45 min). Total uncertainty of the NMHC measurements as given by the reproducibility of the analyses including sampling and sample storage, and by the quoted uncertainty of the calibration gas mixture which was in total about 8% for alkanes, 12% for aromatic compounds, and 20% for alkenes.

The quality of ground based measurements and of the analyses of air samples in canisters was assessed by intercomparisons before and after each EVA measurement campaign. The results of the intercomparisons with synthetic gas mixtures and ambient air measurements are described in detail by Kanter et al. (2002) and Volz-Thomas et al. (2002). It is shown in Volz-Thomas et al. (2002) that the agreement between the ground and canister NMHC measurements was usually better than 20%.

Determination of emission ratios from concentration ratios

The concentration of a pollutant measured at a receptor site downwind of a city results from the concentration upwind of the city (background) and from the mixed in and diluted emissions within the city area. For time periods of constant background concentrations and in the absence of chemical reactions, the concentration ratio of two co-emitted species remains constant, because both species are subjected to the same atmospheric mixing processes. During those conditions the concentration ratio at the receptor site of two co-emitting pollutants is determined solely by their emission ratio. The emission ratio of two pollutants is thus calculated as slope of an orthonormal correlation (York 1996) of synchronously measured concentration pairs over a period of constant background concentrations.

For a calculation of emission ratios, $E(A)/E(B)$, of a city from experimentally derived concentration ratios, $\mu(A)/\mu(B)$, downwind of the source several requirements have to be fulfilled:

- The chosen measurement site has to be located in the main wind direction downwind of the city.
- Contamination from sources near to the receptor site have to be negligible.
- The distance between the city and the receptor site has to be long enough for sufficient mixing.
- The transport time should be short enough to prevent significant photochemical removal.

As an appropriate measurement site we have selected a location 5 km northeast of Augsburg centre next to the small village Stätzling. Photochemical removal during transport is minimized by choosing time periods of lower photochemical activity (March and October). Local contamination can be neglected for the considered species since the area between Augsburg and Stätzling is used exclusively for agricultural purposes.

Emission ratios from NMHC and CO measurements in air samples taken onboard aircraft were calculated formally in the same way as described above as slopes of the orthonormal correlations. In contrast to the measurements on the receptor site at the ground, these correlations contain measurements upwind and downwind of the city area. Since the air samples were taken randomly along the flight track, all emissions within the area confined by the flight track are considered.

Results and discussion

Filter criteria for the ground-based measurements

Given by the analytical system with the lowest time resolution (GC-system, sampling interval: 10 min), all other data were averaged over 10 min intervals. The data from Stätzling were only considered when the site was downwind of the city. In order to ensure stable wind flow conditions, only data with wind speeds ≥ 3 m/s were used. 45 and 50 % of the data from March and October, respectively, met both the wind direction and speed criteria.

Characteristic trace gas ratios for urban air masses

Since the contribution of the background may substantially exceed the impact of the city emissions (October campaign: Mean CO concentration: 201 ppb), it is indispensable to subtract the background concentrations[4]. This is achieved by considering concentration ratios and by calculating the emission ratios via correlation analyses. Table 3.14 lists the characteristic emission ratios of $\Sigma HC/NO_x$, the mean reactivity $\langle k_{OH} \rangle$ and mean number of C-atoms $\langle C \rangle$ together with their uncertainties and regression coefficients (R^2) for the October campaign. The mean reactivity $\langle k_{OH} \rangle$ and the mean chain length $\langle C \rangle$ of the hydrocarbons are calculated as follows:

$$\langle \alpha_i \rangle = \frac{\sum_i \alpha_i \cdot HC_i}{\sum_i HC_i}, \text{ with } \alpha_i = k^i_{OH} \text{ and } \alpha_i = C^i. \tag{3.10}$$

The CO/NO$_x$ emission ratios on weekends were more than twice as high as on working days. This is probably due to the absence of heavy duty vehicles on weekends, which are strong NO$_x$ emitters (Hassel et al. 1995). The mean reactivity $\langle k_{OH} \rangle$ and the mean number of C-atoms $\langle C \rangle$ were significantly higher on weekdays than on weekends. It has been shown by Mannschreck et al. (2002a) that the mean NMHC reactivity on weekdays is substantially higher due to a larger contribution of alkenes and aromatics relatively to that of the alkanes.

---

[1] Mean CO-background values (October campaign: 165 ppb) were calculated using regression of CO against concurrently measured NO$_x$ and extrapolating CO background concentrations to a NO$_x$ concentration of 1 ppb (Mannschreck et al. 2002)

Table 3.14. Characteristic parameters of urban air masses, relative uncertainties (1σ) and respective correlation coefficients (R^2) for the October campaign (Augsburg, urban sector).

Characteristic parameters	Weekdays	Saturdays	Sundays
$\Sigma HC_i/NO_x$ [ppbC/ppb]	2.9±0.2 (0.71)	2.7±0.5 (0.47)	3.8±0.5 (0.67)
$\langle k_{OH} \rangle$ [cm^{3}s^{-1}10^{-12}]	7.4±0.1 (0.96)	4.4±0.5 (0.78)	5.8±0.4 (0.87)
Number of C atoms $\langle C \rangle$	4.2±0.04 (0.99)	3.5±0.1 (0.97)	3.6±0.1 (0.98)
CO/NO_x [ppb/ppb] (all data)	6.8±0.2 (0.82)	17.5±1.0 (0.67)	14.2±0.8 (0.59)

Comparison of emission ratios derived from ground based measurements with those from measurements onboard aircraft

The receptor site and the air-borne measurements are based on different fetches: Whereas the correlations from ground-based measurements result from averaging of time series of several weeks duration, the correlations from air-borne investigations yield spatial averages due to the circling of the aircraft around the city of Augsburg. It is thus important to compare the results obtained by these different techniques.

Fig. 3.21 shows a plot of average $HC_i/sum(HC)$ emission ratios on working days from air-borne measurements versus mean emission ratios measured at Stätzling on working days in October 1998. Emission ratios determined by the two different techniques correlate well ($R^2 = 0.93$). This result suggests that long-term receptor measurements of the emission ratios yield a stable hydrocarbon pattern, which can be treated as representative for the emissions of the whole city.

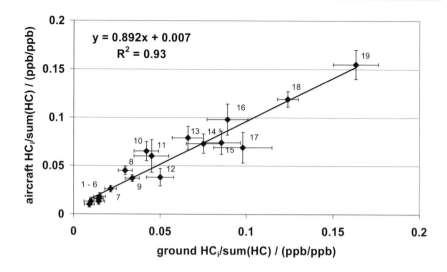

Fig. 3.21. HC$_i$/sum(HC) emission ratios from aircraft measurements (mean ratios for the intensives performed on weekdays) versus HC$_i$/sum(HC) emission ratios measured downwind of the city on the receptor site in Stätzling (mean ratios for the October campaign, weekdays). The bars represent the 1σ standard deviations from the regression analyses for the aircraft measurements as well as for the ground based measurements. 1: n-hexane; 2: cyclohexane; 3: heptane; 4: 2,3-methylhexane; 5: 3-methylpentane; 6: 2-dimethylbutane; 7: 2-methylpentane; 8: propene; 9: butenes; 10: i-butane; 11: n-pentane; 12: benzene; 13: n-butane; 14: propane; 15: toluene; 16: i-pentane; 17: ethane; 18: ethyne; 19: ethene. The HC$_i$/sum(HC)-ratios are based on mixing ratios (ppb/ppb).

Representativity of the emission ratios measured in Augsburg

An estimation for the representativity of the Augsburg hydrocarbon pattern for German city conditions can be obtained by comparison with an averaged NMHC emission pattern for urban conditions.

This pattern was constructed from measurements in German cities between 1982 and 1995 (Schmitz T et al. 1997; Abraham et al. 1994; Bayerisches Landesamt für Umweltschutz 1994; Bruckmann et al. 1983; Ellermann et al. 1995; Thijsse and van Oss 1997) and is given in the "Database for Volatile Organic Compounds" (Mannschreck et al. 2002b). Fig. 3.22 shows the mean hydrocarbon distribution for the Augsburg measurements and the averaged urban emission scenario taken from Mannschreck et al. (2002b).The hydrocarbon composition of both campaigns in March and October are very similar to the averaged urban emissions for German cities suggesting that the HC-composition measured in Augsburg is representative for urban scenarios in Germany.

Summary

Long-term ground based receptor emission measurements represent a well-suited method to get a representative picture of characteristic emission ratios from a whole city.

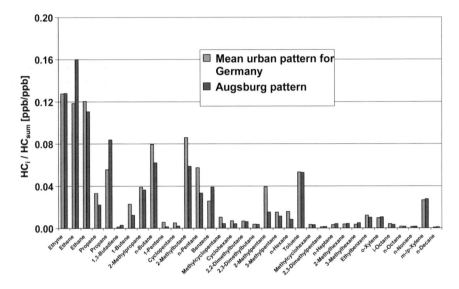

Fig. 3.22. Results of ground-based measurements performed during EVA (October and March) and a average HC-composition for German urban conditions (Thijsse and van Oss 1997)), original references: Schmitz T et al. (1997); Abraham et al. (1994); Bayerisches Landesamt für Umweltschutz (1994); Bruckmann et al. (1983); Ellermann et al. (1995); Thijsse and van Oss (1997)

The choice of the measurement site 5 km downwind of the city centre turned out as a reasonable compromise between the requirements of sufficient mixing of the individual sources of the city and of sufficiently high concentration fluctuations needed for correlations. A good agreement between emission ratios derived from aircraft and from receptor site measurements shows that the emitted pollutants were sufficiently mixed at the receptor site to represent the emissions of the whole city. Thus, it has been shown that this experimental condition is fulfilled, which is necessary for the comparison of measured concentration ratios from long-term measurements with those from the emission model (Mannschreck et al., this issue).

The good agreement of our HC_i/HC_{sum} emission ratios with the average hydrocarbon composition for German city conditions (Mannschreck et al. 2002b) suggests that the Augsburg emission pattern can be considered as typical for German cities.

3.3.3.6 Comparison of two different HCHO Measurement Techniques: TDLAS and a commercial Hantzsch Monitor – Results from Long-term Measurements in a City Plume during the EVA Experiment.

D. Klemp, K. Mannschreck, B. Mittermaier

Introduction

For the evaluation of an emission model two field campaigns have been performed in the Augsburg area in March and in October 1998. As part of an integrated concept (Slemr et al. 2002b) of ground-based and air-borne measurements as well as tracer experiments, the long-term measurements were applied to characterize the composition of the city plume of Augsburg. For this purpose, a mobile laboratory was placed 5 km downwind of Augsburg in Stätzling, equipped with analytical instruments for the measurement of NO, NO_2, NO_y, CO, $C_2 - C_{10}$-hydrocarbons, HCHO, O_3 and meteorological parameters. In this study special emphasis is placed on the results of atmospheric HCHO concentrations, measured by two different measurement techniques: Tuneable Diode Laser Absorption Spectroscopy (TDLAS) and a commercial Hantzsch monitor.

Experimental

Formaldehyde and NO, NO_2 and NO_y were continuously measured over a period of more than 4 weeks at Stätzling in a distance of about 5 km from city centre in March 1998. Analytical instruments for the measurement of HCHO, NO_x and NO_y are part of the equipment, which is installed in the air conditioned mobile laboratory of Forschungszentrum Jülich (Klemp et al. 2002; Mannschreck et al. 2002a). The inlets were mounted on a pneumatic mast 10 m above the surface. For all connections PFA tubing (¼ inch OD) were used. In order to compensate for the different time constants (10%- to 90%-increase) of the analytical instruments (TDLAS: $\approx$ 10 s, Hantzsch: $\approx$ 150 s, NO_x, NO_y: $\approx$ 10 s) 10-minutes averages were calculated from the base data set.

Formaldehyde: The commercial *Hantzsch monitor* detects HCHO by fluorescence of 3,5-diacetyl-1,4-dihydrolutidin (Dasgupta et al. 1988), which is produced from the reaction of HCHO with 2,4-pentadione and NH_3. Atmospheric HCHO is transferred quantitatively into the liquid phase using a stripping coil, which was kept at a constant temperature of 15 °C. Zero signals were determined by scrubbing HCHO from ambient air using a Hopcalite® filter. The instrument was calibrated by a relative and an absolute procedure. Relative calibrations were performed daily using the internal HCHO permeation device. Absolute calibrations were made once a week using a series of liquid standards ($2 * 10^{-8} - 2 * 10^{-7}$ M). With respect to the 10 minutes intervals an average detection limit of 100 ppt (2-σ) and a precision of 10 % at ambient air levels above 1 ppb were achieved during the EVA campaign.

The *TDLAS technique* is based on the monitoring of the specific absorption from a single rotational-vibrational line in the middle infrared spectrum of the molecule of interest. In order to achieve detection limits which allow monitoring atmospheric concentrations of HCHO (detection at 1605 cm^{-1}), a White cell (total light path $\approx$ 100 m) at reduced pressure (30 hPa) is used. Due to the reduced pressure the probability of overlap between absorption lines from other atmospheric species is nearly excluded by reducing the pressure broadening of the rotational lines. In combination with 2-f-derivative detection technique (modulation frequency: 12.5 kHz) optical densities of less than 10^{-5} can be measured. The time resolution of this technique is only limited by the exchange time (< 10 s) of the ambient air, which is pumped at the reduced pressure through the White cell. In order to monitor slowly variating background structures which limit the attainable detection limit of the TDLAS, background spectra were monitored every 5 – 10 minutes by adding N_2 instead of ambient air at the inlet system. Calibrations were performed every 30 minutes by adding a known mass flow of HCHO from a home-built thermostated permeation device to a N_2 gas stream at the inlet. Ambient air concentrations were calculated from the comparison of the ambient air spectrum with that from the calibration cycle by fitting Voigt-profiles to the respective data sets. The HCHO mass flow of the permeation device was absolutely calibrated by two independent methods: by gravimetry and by colorimetry as described in Harris et al. (1989). For the TDLAS technique a detection limit of 200 ppt (2-σ, 10-minutes averages) is achieved throughout the EVA campaign and the precision of the TDLAS measurements at more than 2 ppb is calculated to be 10 %.

NO_x and NO_y were measured using two chemiluminescence detectors (ECO Physics, CLD 770AL ppt), a photolytic converter (ECO Physics PLC 760) to convert NO_2 to NO, and a home-built gold converter to convert NO_y (i.e. NO_x = NO + NO_2 and its oxidation products, mainly HNO_3) to NO on a hot gold surface (Fahey et al. 1985), which was directly placed on top of our pneumatic mast in order to avoid NO_y losses. NO calibrations were performed daily by adding a known flow of calibration gas (10 ppm NO in N_2) into the NO_x-free air stream. The conversion efficiencies of the PLC and the Au-converter for NO_2 were also determined every night. The conversion efficiency of the Au-converter for HNO_3 was checked weekly using a home-built permeation source. For NO a precision of 8 % and a detection limit of 20 ppt (2-σ) were achieved for an integration time of 10 minutes. The corresponding data for NO_2 and NO_y are 9 % and 30 ppt, and 8 % and 200 ppt, respectively.

Results

Comparison of concurrently measured HCHO concentrations

In Fig. 3.23 more than 1000 concurrently measured 10-min averages are plotted. Only data from the city sector are considered. For most pairs of measurements (i. e. HCHO < 2 ppb) both detection techniques yield comparable mixing ratios.

For higher mixing ratios, however, some Hantzsch measurements exceed those of TDLAS by up to a factor of two.

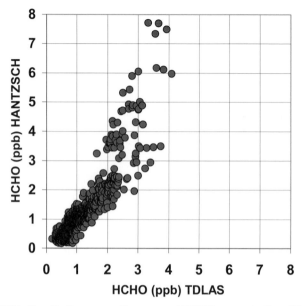

Fig. 3.23.. Result of concurrently measured HCHO concentrations (Hantzsch monitor and TDLAS, 10 minutes averages) during the EVA campaign (2. 3. – 31. 3. 1998) downwind of Augsburg.

As shown in Fig. 3.24, both subsets of data can be successfully separated by using the NO_x concentrations and the respective NO_y/NO_x-ratios as filter criteria. The NO_y/NO_x-ratio is used as a "photochemical clock" in order to calculate the effect of photochemistry on the measured air masses which are transported from Augsburg to the receptor site. Changes in the NO_y/NO_x-ratio of an air mass are mainly caused via OH attack of NO_2 to produce HNO_3 (which is the most abundant of the NO_y species and which is not emitted directly). That means, the NO_y/NO_x- ratio of an air mass is a measure for its photochemical exposure and for freshly polluted air masses the respective NO_y/NO_x-ratio is always[5] close to 1.

---

[5] The NO_y/NO_x-ratios in a city plume can be understood as the result of mixing of aged air masses of background air (containing high NO_y/NO_x-ratios but low concentration levels) with fresh NO_x emissions. Nevertheless, due to the much stronger NO_x increase caused by the impact from the city the resulting NO_y/NO_x-ratios concentration levels yield values of only slightly higher than 1.

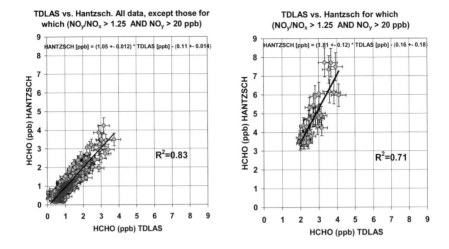

Fig. 3.24. Result of correlation analysis (orthonormal regression) between Hantzsch and TDLAS using ($NO_y/NO_x > 1.25$ and $NO_y > 20$ ppb) as separation criteria. The error bars denote the respective 2-σ errors of the individual 10-minutes averages.

A quite good agreement between TDLAS and Hantzsch in the Augsburg plume is observed in fresh emissions ($NO_y/NO_x < 1.25$) and for low and moderately polluted conditions i.e. $NO_y < 20$ ppb, ($NO_y^{Mean(EVA-1)} = 8.36$ ppb). The orthonormal regression analysis yields a slope of $HCHO^{HANTZSCH}/HCHO^{TDLAS} = 1.05$ ($R^2 = 0.83$). During heavily polluted events ($NO_y > 20$ ppb) with photochemically active conditions ($NO_y/NO_x > 1.25$) the slope is $HCHO^{HANTZSCH}/HCHO^{TDLAS} = 1.81$ ($R^2 = 0.71$).

If photochemically active conditions connected with high precursor concentrations are excluded from the long-term data set, the HCHO measurements of Augsburg yield an agreement within 5% between Hantzsch and TDLAS. Moreover, from the good accordance between both methods outside photochemically active conditions and from the stability of calibration signals, calibration errors can be excluded to be the reason for the observed deviations.

In principle, with respect to the observed deviations between both HCHO measurement techniques during photochemically active periods two explanations are possible: a positive interference of the Hantzsch technique caused by unknown species or a negative interference, which reduces the TDLAS absorption signal. The latter explanation seems to be rather unlikely. Any interference by an additional species at the same spectral range would even enhance the measured absorption signal of the TDLAS system and could not explain HCHO mixing ratios, which are up to a factor of two smaller than those measured by Hantzsch. Saturation effects of the TDLAS system, which may decrease the sensitivity of the system, can also be excluded: An excellent linearity ($R^2 > 0.99$) of the TDLAS system is observed between 0 and 15 ppb (Klemp et al. 1994).

Comparison with the results of a box model study

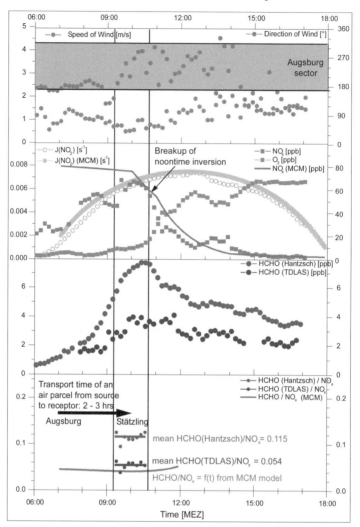

Fig. 3.25. Box model study (MCM, (Derwent et al. 1998)) for the 31. 3. 1998, Augsburg, EVA-1 campaign. Boundary conditions for the MCM box model run: NMHC/NO_x: 2.5 ppbC/ppb; HCHO/NO_x: 0.035 ppb/ppb; HCHO background concentration: 0.8 ppb; Mixing height below the persistent inversion layer: 300 m; $K_{dep}(NO_2)$: 0.2 cm/s; $K_{dep}(HCHO)$: 0.4 cm/s; a split of 45 individual NMHC was considered.

For a period of summer smog which occurs in the last days of the EVA-1 campaign with substantial deviations between both techniques, a box model study was performed using the MCM model (Derwent et al. 1998). This type of model study leads to reasonable results as long as exchange of polluted air masses from

Augsburg with those from free troposphere is hindered by a stable inversion layer during the morning hours. As shown in Fig. 3.25, air masses from Augsburg are advected to Stätzling from the early morning hours up to 14:00 with low wind speeds of around (1.0 ± 0.5) m/s (c.f. upper part of Fig. 3.25). The day is characterized by clear-sky conditions. The break-up of the inversion layer took place at around 10:45, c.f. crossing of the plotted O_3 and NO_x-lines. Modelled dilution was adjusted in a way that modelled NO_x-values (start: 80 ppb at 7:00) follow the temporal behaviour of measured values (c.f. second part of Fig. 3.25). HCHO concentrations measured by the Hantzsch system and by TDLAS both show highest levels at around 10:00, like they are also observed for NO_x, but Hantzsch levels were up to twice at high as the ones measured by TDLAS (c.f. third part of Fig. 3.25). Modelled HCHO/NO_x-ratios were compared with HCHO(TDLAS)/NO_x-ratios and HCHO(Hantzsch)/NO_x-ratios only between 9:15 (minimal transport time, calculated from wind speed and distance from the city) and 10:45 (break-up of the inversion layer). In case of the Hantzsch measurements modelled and measured ratios deviate by more than a factor of two, whereas the respective ratios from TDLAS measurements were close together with the modelled ratios. From the scatter of the individual datapoints it is apparent that the differences between measured ratios are significant.

Mean HCHO/NO_x-source ratios

For the calculation of mean HCHO/NO_x-source ratios from the whole data set only data with wind speeds ≥ 3 m/s are used. This approach guarantees for transport times of less than 30 min from the city to the receptor place and it also ensures for neglectable photochemical removal during transport.

Fig. 3.26 shows mean formaldehyde concentrations (based upon 10-minutes averages, measured by a TDLAS system) as a function of NO_x concentrations. 10-minutes averages of HCHO were grouped by different classes of NO_x (ΔNO_x = 2ppb) and mean formaldehyde concentrations, 25- and 75-percentiles were calculated for each individual NO_x-class.

Fig. 3.26 suggest two conclusions: i) The regression of HCHO with NO_x yields an HCHO/NO_x emission ratio of around $(3.7 \pm 0.5) \cdot 10^{-2}$ ppb/ppb for fresh emissions. ii) The extrapolation of formaldehyde to the seasonal NO_x background concentration of 1 ppb (Mannschreck 2001) yields a background concentration of HCHO of (0.6 ± 0.1) ppb. This result is in good agreement with the background HCHO concentration of (0.7 ± 0.3) ppb observed by us in Jülich-Mersch in March 1995 (Schmitz T et al. 1997).

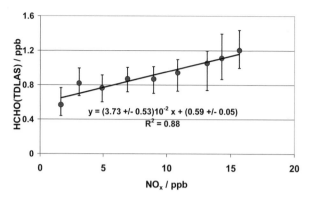

Fig. 3.26. Mean formaldehyde concentrations (TDLAS measurements, based upon 10 minutes averages, March campaign) for different NO_x-classes ($\Delta NO_x = 2$ ppb). The error bars indicate the 25-and 75-percentiles of formaldehyde for each NO_x-class.

The same type of investigation is performed for the HCHO mixing ratios measured by the Hantzsch monitor. Table 3.15 shows that both the HCHO/NO_x ratios and the HCHO background concentrations are in good agreement for the two independent measurement techniques. Also listed in Table 3.15 is the mean emission ratio of HCHO/NO_x = 3.3 $\cdot 10^{-2}$ ppb/ppb for March 1998 from the emission model provided by IER Stuttgart (Kühlwein et al. 2002a). Within their error limits, experimentally derived HCHO/NO_x-ratios agree with those from the emission model.

Table 3.15. Comparison of experimentally derived HCHO/NO_x ratios with those from the emission model, provided by IER, Stuttgart (Kühlwein et al. 2002a) for the March campaign. HCHO measurements were clustered by NO_x intervals and correlation analyses were performed. Two different HCHO measurement methods were used (a commercial Hantzsch system (AL4) and a TDLAS system (TDL)).

	HCHO/NO_x (ppb/ppb)	HCHO-Background (NO_x=1 ppb) (ppb)
AL4	$(3.28 \pm 0.48) \cdot 10^{-2}$ R^2 =0.82	0.49 ±0.07
TDL	$(3.73 \pm 0.53) \cdot 10^{-2}$ R^2 =0.88	0.61 ±0.05
Emission model	$3.3 \cdot 10^{-2}$	- -

Conclusions

1. An agreement of better than 5 % between Hantzsch and TDLAS is observed for low and moderately polluted conditions.

2. For heavily polluted events under photochemically active conditions the Hantzsch system shows higher values of up to a factor of two. The reasons for this behaviour are still unknown, but there are two arguments which suggest that the Hantzsch HCHO levels are biased by an interference under those conditions:

- The observed $HCHO^{Hantzsch}/NO_x$-ratios yield values of around 11%, whereas the $HCHO^{TDLAS}/NO_x$-ratios at around 4% (as it is observed as average throughout both EVA-campaigns).
- Additional production of HCHO from chemistry can be excluded from a box model study (initialised by a comprehensive set of measured species) to cause differences of up to 4 ppb during transport from the city to the receptor place 5 km downwind of the city.

Within their error limits, experimentally derived $HCHO/NO_x$-ratios agree with those derived from the emission model provided by IER.

3.3.3.7 Mass balancing by means of vertical and horizontal profiles determined with tethered balloon and airship soundings

G. Baumbach, U. Vogt, P. Bauerle, K. Glaser

Introduction

The main target of the joint project EVA (Evaluation of Highly Resolved Emission Inventories) was to determine the mass flow of emissions originating from the whole city containing different sources with the help of ambient air measurements by a combination of ground level, tethered balloon, airship and aircraft measurements.

Here the mass flow calculations windward and leeward of a German city are described, based on results of tethered balloon (vertical distribution) and airship (horizontal distribution) measurements. These experimentally determined results are compared to results of emission modelling.

Measurement strategy

The emission mass flow of the city of Augsburg should be gained by calculating the difference between the incoming flow of air pollutants windward and the outgoing flow leeward of the city. The principle of the method can be seen in Fig. 3.27. The idea was to set up "virtual windows" windward and leeward of the city in right angles to the prevailing wind direction and to measure the vertical and horizontal distribution of all necessary parameters within these two windows. With tethered balloon (vertical distribution) and airship (horizontal distribution) soundings and additional help of ground level monitoring stations and aircraft flights (horizontal distribution) the windward and leeward concentration fields for NO_x and 9 different volatile organic compounds (VOC) and the wind fields could be determined and thus the flows for NO_x and VOC could be calculated. Whereas the aircraft had the opportunity of catching the total plume *(virtual window: large)*, the combination of tethered balloons and airship could catch the plume only partially by setting up a *smaller virtual window* with a horizontal width of 4 km and a vertical extension of 300 m. The advantages of the method with the small *virtual window* were:

- Availability of data directly from ground level up to a height of 300 m above ground
- A high temporal and spatial density of data.

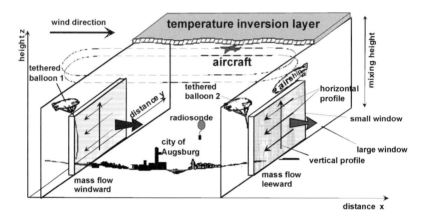

Fig. 3.27. Principle of the mass flow calculation method for air pollutants incoming and outgoing of Augsburg and the used measurement equipment.

The air analysed at the windward side took up emissions over the city and passed during the selected investigation day at an average wind speed of 6 m/s the leeward side about 30 minutes later. For the vertical windward soundings one tethered balloon was positioned southwest of Augsburg. Leeward of the city in the northwest in "Stätzling", two tethered balloon measurement systems were applied to measure the vertical profiles of all the parameters in detail. To determine horizontal profiles an airship on the leeward side operated in different heights. Two aircraft measured the horizontal distribution of the parameters around the city at different height levels, windward as well as leeward (Kalthoff et al. 2001).

Measurement techniques of tethered balloon and airship

With the tethered balloon systems vertical profiles of the meteorological parameters wind speed, wind direction, temperature and humidity were determined continuously, as well as the vertical profiles of NO_x, NO_2 and O_3. VOCs were sampled discontinuously at different sounding heights (leeward side, with a second balloon system) or as integral value (windward side) throughout an ascent and a descent. The samples were taken on activated carbon tubes and subsequently extracted and analysed in the laboratory (Steisslinger and Baumbach 1990; Baumbach and Vogt 1995; Glaser 2001). Our institute fulfilled the quality assurance criteria of the EVA project (Slemr 2001) for about 20 volatile organic compounds (VOCs).

The helium-filled airship is remote-controlled and solar-driven. It was constructed by the Institute for Statistics and Dynamics of Aerospace Structures (ISD) of the University of Stuttgart for research measuring and observation flights. Its maximum operation height in Augsburg was 300 m above ground. The components NO_2, O_3 and VOC as well as wind speed, wind direction and the current position (by GPS) were recorded during the flights in different heights.

Meteorological Conditions at the selected investigation day

Measurements were carried out at several days. But on 23rd October 1998 there were the most appropriate meteorological conditions for evaluating the mass balancing experiment. From 10 UTC onward the morning inversion layers were dispelled completely and a nearly homogenous vertical structure existed around noontime. During this time the prevailing wind direction was exactly south-west over the whole measurement height and the wind speed was around 6 m/s. Thus, all preconditions for mass balancing were fulfilled on this day from 10 to 13 UTC.

NO_x mass flow

The NO_x mass flow of the city of Augsburg was determined by balancing the outgoing and the incoming flows over a certain time, using the measurement windows shown Fig. 3.27. The NO_x mass flows through these *windows* were determined by integrating and interpolating the product of the measured NO_x concentration and the wind speed perpendicular projected to these *virtual windows*. The horizontal extension of each window amounts to 4000 m (abscissa), the vertical extension to 300 m above ground respectively 500 to 800 m above sea level, according to the operation ranges of the airship and the tethered balloons. The results of the measurements are shown in Fig. 3.28. The points in the diagram depict the places where data from the balloon and airship measurements were recorded. The isopleths were gained by interpolation between these points using the Kriging Method (Zongmin 1986). The depicted windward distribution results from the interpolation of aircraft measurement results. Under the precondition of a nearly homogenous pollutant distribution the mass flows could be extrapolated to the whole plume of the city (outgoing air) and the whole windward incoming air, respectively. In Fig. 3.28 it can be seen that the NO_x concentrations on the windward side are distributed very well. On the leeward side the NO_x was also nearly homogenous on this day.

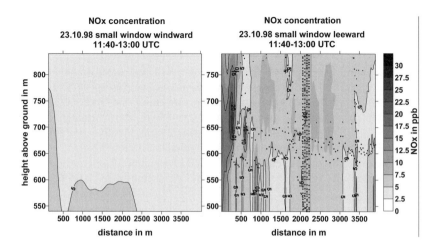

Fig. 3.28. Spatial isopleth diagrams of NO_x concentrations distribution windward (left diagram) and leeward (right diagram) of Augsburg on October 23rd, 1998, from 11:40 to 13:00 UTC. (Small points in the right diagram indicate the location of measurement values. The areas between the points were filled with help of an interpolation calculation).

With these NO_x concentration fields and the corresponding wind fields, also determined by airship and tethered balloon measurements, the outgoing and incoming NO_x mass flows through these "windows" were calculated. To determine the mass flow from the whole city an extrapolation from the flows through these *small windows* to the whole city plume had to be carried out. For this purpose the horizontal extension of the city's plume was estimated using the city's map. The width of the plume determined in this way was 9 km. The height was assumed to be the atmospheric mixing height of 500 m above ground, determined by the vertical profiles from the tethered balloon measurements and the radio soundings carried out regularly during the project (Kalthoff 2001).

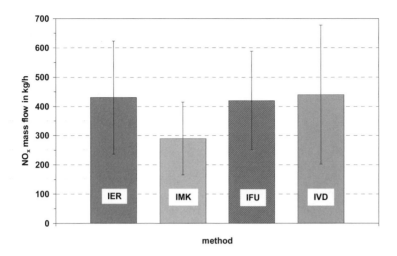

Fig. 3.29. NO_x mass flows from the city of Augsburg with estimated uncertainty ranges as results of different mass balancing methods, October 23rd, 1998, 10 to 13 UTC. IER: emission simulations (Kühlwein 2001), IMK and IfU: mass balancing from aircraft measurements (Kalthoff 2001), IVD: results of this contribution.

Thus, the plume's cross-section was calculated to be 4.5 km². The cross-section of our *small measurement windows* covered an area of 1.23 km². Thus, the size of the whole plume's cross-section was calculated to be 3.346 times larger than the size of our small window. With this factor of 3.346 the NO_x mass flow was extrapolated from the *small measurement window* up to the whole city's plume. Thus, an average NO_x mass flow of about 440 kg/h was calculated over the period from 10 to 13 UTC (on October 23, 1998).

In Fig. 3.29 this calculated NO_x mass flow is compared to the NO_x mass flow from IER emission modelling (Kühlwein 2001) and to the mass flow balance calculations from IMK and IFU aircraft measurements (Kalthoff 2001). The estimated uncertainty ranges are depicted in each column. It can be seen that the results of these independent methods correspond very well.

Determination of VOC mass flows

Since the VOCs could be sampled only at a few places, the distribution could not be determined in the same way as the NO_x distribution. Therefore, on the windward as well as on the leeward side VOC/NO_x concentration ratios were determined for the sampling points. By means of these ratios, the incoming and outgoing mass flows were calculated for several volatile organic compounds according to the NO_x mass flows. The results of this geometric plume assessment method are shown in Fig. 3.30 as first columns for several compounds.

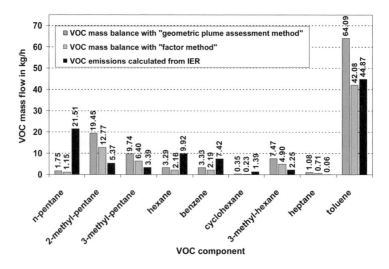

Fig. 3.30. Comparison of the VOC mass flows (emisson rates) of individual VOCs from the city of Augsburg, determined by two different mass balancing methods, based on the measurements, and compared with emission simulations by IER

By another method for extrapolation the VOC mass flows from the *small measurement window* to the whole plume the VOC/NO_x ratio is based on the NO_x mass flows determined by aircraft measurements by the IMK team (second column in Fig. 3.29) (Kalthoff 2001). The results of this VOC mass flow calculation are depicted in Figure Fig. 3.29 as the second column. The third column indicates the VOC mass flows resulting from emission modelling by IER (Kühlwein). Whereas the first balancing method of geometric plume assessment can be considered as independent, the second method is based on results of aircraft measurements and calculations of other groups. The calculated mass flows of different VOC components determined by the *geometric plume assessment method* are higher than these of the *factor method* based on the IMK NO_x mass flows. However, the problem is not the factor between these two extrapolation methods but the difference between the VOC mass flows determined experimentally in the real plume of the city and the mass flows won by emission modelling. The differences are not uniform. For some compounds the modelled values are much higher than those experimentally determined (n-pentane, hexane and benzene). In other cases the modelled values are much lower (2-methylpentane, 3-methylhexane and toluene).

Conclusions

On the one hand the results show a good correspondence between experimental and modelled NO_x mass flows. On the other hand the difficulties and uncertainties of VOC modelling and experimental determination become evident. More research has to be done in this field.

3.3.3.8 Comparison of modelled and measured CO and NOₓ emissions

F. Slemr

Modelled and measured CO and NO_x emissions from the investigated area for 5 intervals on working days (21.–23.10. and 27.3.1998) and for one interval on Saturday, 10.10.1998 are summarized in Table 3.16 and Table 3.17, respectively. To take account of emission variability, all measured emissions were normalized to the emission estimate 2 modelled for the time of the experiments. The relative emissions for working days were averaged and the average uncertainty calculated by propagation of individual uncertainties. Statistical significance of differences between average of modelled relative emissions and those measured by IMK or IFU or ASS was assessed using standard statistical procedures (Kaiser and Gottschalk1972). CO emission estimates at level 2 for working days are in agreement with the measured ones within the combined uncertainties of the emission model and of the measurements by IMK, IFU, and ASS. Although insignificant in statistical terms, the CO emission estimates tend to be smaller than the measured ones by all three techniques, and this tendency seems to be most pronounced on Saturday October 10, 1998. NO_x emission estimates at level 2 for the working days are in agreement with the measured ones by IMK and IFU within the combined uncertainties. Emission model tend to overestimate the NO_x emissions on the working days and especially on Saturday. The differences on Saturday for CO and NO_x emissions suggest problems with temporal allocation functions of emissions from industrial activities and with vehicle fleet compositions on weekends.

Table 3.16. Modelled and measured absolute CO emissions of the city of Augsburg in kg/h. The given uncertainties were estimated from the propagation of uncertainties of contributing measurements or estimates and correspond to 1 σ.

Day	IER estimate 2	IER Estimate 3	IMK mass balance	IFU mass balance	ASS tracer
October 10	1200 ± 570	1480 ± 550	1980 ± 470	1850 ± 330	1800 ± 400
October 21	1280 ± 610	1590 ± 590	1840 ± 610	2480 ± 280	-
October 22[a]	1100 ± 520	1370 ± 510	1600 ± 300	1540 ± 370	980 ± 300
October 22[b]	1110 ± 530	1380 ± 510	1220 ± 220	1010 ± 250	
October 23	1170 ± 560	1460 ± 540	-	1220 ± 730	900 ± 200
March 27	1570 ± 740	1940 ± 720	-	-	3000 ± 1000

[a] 12:21 – 13:56 local time (10:21 - 11:56 UTC)
[b] 13:45 – 15:00 local time (11:45 – 13:00 UTC)

The emission estimates at level 3 assume a 30 % higher light duty vehicle (LDV) activity and a 10 % lower heavy duty vehicles (HDV) activity in Augsburg than the estimates at level 2, which are based on the number of inhabitants of Augsburg and scaled by the nationwide activity per town inhabitant in Germany. The higher LDV driving performance is probably due to the large catchment area of Augsburg that causes more intensive commuter and shopping traffic in the town district. The more intensive LDV activity leads to CO emission estimates at level 3 which are larger by 24 % than the estimates at level 2. Because of decreasing HDV activity and other important NO_x emission sources besides traffic, NO_x emission estimates at level 3 are increased only by 5 % despite the more intensive LDV activity. The emission estimates at level 3 are in better agreement with the measured CO emissions on workdays but the difference on Saturday still remains notable. Slightly higher NO_x emission estimates at level 3 lead to a slightly, but in view of the uncertainties insignificantly worse agreement with the measured ones than estimates at level 2.

Table 3.17. Modelled and measured absolute NO_x emissions of the city of Augsburg in kg/h (as NO_2). The given uncertainties were estimated from uncertainty propagation and correspond to 1σ.

Day	IER estimate 2	IER estimate 3	IMK mass balance	IFU mass balance	ASS tracer
October 10	280 ± 120	300 ± 100	110 ± 50	150 ± 50	250 ± 60
October 21	440 ± 190	460 ± 160	430 ± 210	600 ± 180	-
October 22[a]	420 ± 180	450 ± 150	300 ± 120	380 ± 130	240 ± 70
October 22[b]	420 ± 180	450 ± 150	230 ± 100	240 ± 110	-
October 23	430 ± 190	460 ± 160	-	420 ± 170	220 ± 50
March 27	400 ± 170	420 ± 150	-	-	350 ± 120

[a]12:21 – 13:56 local time (10:21 - 11:56 UTC)
[b]13:45 – 15:00 local time (11:45 – 13:00 UTC)

On working days estimates at level 2 and 3 of CO and NO_x emissions for the city of Augsburg agreed with the measured ones within the combined model (CO: 48% estimate at level 2 and 37% estimate at level 3; NO_x: 43% estimate at level 2 and 35% estimate at level 3) and mean experimental uncertainties (CO: 24, 27, and 23%; NO_x: 43, 35, and 25% for IMK, IFU, and ASS measurements, respectively). Further reduction of these uncertainties and/or increase of number of measurements could thus transform the observed tendencies to differences of significance. Model simulations suggest that the uncertainty of CO measurements could be reduced by more detailed measurement of the vertical CO transport. Uncertainty of measured NO_x emissions could be substantially reduced by using better instruments. Two or more additional measurements on weekend could also allow a statistical treatment of the differences observed on Saturday October 10, 1998.

3.3.3.9 Comparison of modelled and measured emission ratios of pure hydrocarbons

K. Mannschreck, D. Klemp, J. Kühlwein

Introduction

In order to quantify the uncertainty of emission data the individual steps by which the emission data are calculated have to be evaluated (for vehicle exhaust emissions see e.g. chapter a). For this purpose it is necessary to investigate the emissions of the individual sources in detail by determining the emission factors as well as the variability in time. In reality a huge amount of different sources contribute to the emissions of a certain region. Therefore it is reasonable to evaluate the emission data of an urban region. The modelled emission data have to be compared to results of an independent method with known uncertainties.

The quality of highly resolved emission data was thus experimentally evaluated in the joint experiment EVA. In previous chapters the measurement techniques, the design of the experiment and the methodology for data evaluation together with first results were outlined. This study focuses on a comparison between modelled and experimentally determined ratios of pure C_2-C_{10} hydrocarbons (HC_i), the sum of HC_i ($:= HC_{sum}$) and NO_x.

Methodology

The comparison of experimentally determined and modelled emission data requires two data sets which are comparable to each other with respect to their resolution in space, time and species. The concentration measurements downwind of the city were shown to be representative for the urban area under investigation (for a discussion see chapter b and references therein). Since a higher spatial resolution is not possible due to mixing processes and changing wind directions, the modelled data can only be evaluated based on averages over the entire urban area.

Monthly mean values of experimental data reveal the highest statistical significance due to the large number of data points and are used for analysis presented here. Another approach would be to compare daily means which reveal a high temporal resolution required for the input data of air quality models. This approach is discussed elsewhere (Mannschreck et al. 2002a).

The resolution in species of modelled and experimental data is not in any case equal. In this study only pure hydrocarbons (HC_i) i.e. compounds containing only H- and C-atoms are considered since oxygenated and halogenated HC were not quantified experimentally. On the other hand modelled emission data are in some cases given as specific compounds, in other cases as classes of unspecified compounds or solvent mixtures. The comparison therefore requires a detailed analysis of data both measured and modelled which will be explained in the following.

Measurements

The measurements of C_2-C_{10} hydrocarbons and NO_x were performed downwind of the city of Augsburg. Since measured concentrations of pollutants downwind of a source are dependent on dilution processes with background air, they cannot be used for the comparison with modelled emission data. Ratios of trace gases, however, are not influenced by dilution since both compounds underlie the same processes during transport from the source to the measurement site. In this study ratios of trace gases were calculated via regression analysis by also considering background concentration levels. Hence the given ratios are ratios of two emitted trace gases (or groups of trace gases) and do not include background levels. The measurement techniques as well as the data evaluation method are described in the previous chapter and references therein.

Hydrocarbons with a molecular weight higher than n-decane cannot be measured quantitatively since they are partly lost in the water trap of the gas-chromatographic (GC) system. Loss rates of C_{11}-C_{14} hydrocarbons were determined as 28%, 49%, 86% and >95% for undecane, dodecane, tridecane and butadecane, respectively. With the used GC-system the order of elution of a compound is according to its molecular weight. Taking into account the losses, the sum of all peak areas with a retention time higher than of n-decane is only 10% of the total peak area of C_2-C_{10} HC. Thus only a small share of total C_2-C_{14} HC is not measured quantitatively and thus not included in the experimental data.

Model

The calculation of emissions of NO_x and volatile organic compounds (VOC = oxygenated, halogenated and pure HC) is described in chapter c. The contributions of the individual source types to total VOC calculated for the Augsburg area are as follows (monthly means for October 1998 based on mass units): 74% solvent evaporation, 13% traffic exhaust, 7% fuel evaporation, 5% industrial processes, 1% residential heating.

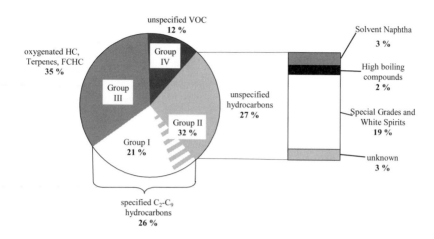

Fig. 3.31. Composition of VOC emissions for October 1998 at model default level (monthly values based on mass units).

Group I: pure hydrocarbons which 1) can be specified by the emission model as individual compounds, 2) the share in summarised compound classes (group II, see below) is smaller than 50 %, and 3) they can be measured quantitatively by the GC system. Emissions included in this group are dominated by traffic related emissions (72%) followed by solvent evaporation (12%) and other combustion processes (16%). Since all HC are specified they can be compared directly to the measured data.

Group II: pure hydrocarbons not fulfilling the criteria for group I HCs. Firstly, this group includes the specified HCs with high shares in unspecified summarised compound classes and two HCs which are not properly identified by the measurement system (striped area in Fig. 3.31). Secondly, this group contains HC which could not be further specified by the model, i.e. summarised compound classes of solvent naphtha, high boiling compounds, special grades and white spirits and hitherto unspecified HCs. Except the latter these classes are solvent mixtures which are defined according to their boiling points and consist of C_6- C_{14} hydrocarbons with different boiling limits in each of these mixtures. Although a comparison based on single compounds is not possible the share of these compounds identified and quantified by the GC-system has to be taken into account when comparing measured and modelled data.

Group III: oxygenated and halogenated HC and terpenes. These compounds were not identified in the chromatograms within this study and could therefore not be used for comparison.

Group IV: unknown VOC, no clear assignment to group I-III possible due to a lack of information on the chemical characteristics of these compounds. These emissions are mainly caused by solvents. A comparison with the measured data is not possible either.

Results and discussion

The relative contributions of specified HCs contained in group I are shown in Fig. 3.32 together with the experimental emission ratios.

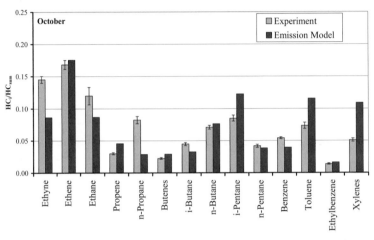

Fig. 3.32. Modelled and measured emission ratios of group I hydrocarbons (based on units of ppb). Error bars of measured data refer to the uncertainty (1 □) of the slope of the regression line.

Ethyne which predominantly originates from traffic exhaust is underestimated by the model by about a factor of two. Since ethyne is removed effectively by a properly working catalyst (Lies et al. 1998) vehicles without or with a defect catalyst and cold start emissions are important sources of ethyne (Sawyer et al. 2000). The underestimation of ethyne thus points to possibly incorrect assumptions about the share of vehicles without or with defect catalyst in the vehicle fleet and to inaccurate emission factors of cold start emissions. The emission ratio of propane is also underestimated, possibly due to incomplete emission data from individual point sources which are in some cases important propane emitters. Ratios of toluene and xylenes are overestimated by the model. According to the emission model solvent evaporation is an important source of these compounds which gives an indication of an overestimation of these emissions and will be discussed below. Also i-pentane is overestimated by the model. I-pentane emissions predicted by the model originated to a large percentage from fuel evaporation, hence the overestimation of its emissions may suggest a very high emission factor of this source or incorrect composition of fuel vapour or both. Despite these differences the agreement between modelled and experimental results is generally good.

However, for a complete comparison all modelled HC which were also measured, i.e. C_2-C_{10}-HC, have to be taken into account. Group II contains a wide range of hydrocarbons which are partly included in the measurements others are not. In order to find out the share of group II compounds identified and quantified by the GC-system the given compound classes have to be looked at in some detail. The solvent mixtures are pure HC and are defined according to their boiling limits. Since the order of elution from the chromatographic system is according to the boiling points of the respective compounds, the approximate retention time of any HC in the chromatogram can be estimated from its boiling point. Taking into account the boiling limits of the solvent mixtures given in group II and their predicted mass percentage results in a share of 50 – 70 % of group II emissions which are detectable by the used GC-system (for more details see Mannschreck et al. (2002a)).

A comparison which takes these additional HC emissions into account cannot be based on individual compounds since these are not known. However, the parameter HC_{sum}/NO_x can be used. Absolute emission rates of NO_x were experimentally determined by two methods and compared to the modelled NO_x emission rates (chapter d and e). As a result modelled and measured emission rates were in good agreement for October. During the March campaign only one experiment was carried out and the experimental uncertainty was high so that the absolute emission rates could not be assessed. In Table 3.18 experimentally determined HC_{sum}/NO_x ratios for October are compared first for 14 compounds of group $I(HC_{sum-14}/NO_x)$ and secondly for all detectable compounds ($HC_{sum-all}/NO_x$). In the latter case the total peak area in the C_2-C_{10} region was considered in the experimental data and the additional 50% (lower limit) of unspecified solvent emissions was considered in the modelled data. Since absolute NO_x emission rates agreed well, deviations can be attributed to HC-emission. In the case of group I HCs the two data sets agree well which was already concluded considering the HC_i/HC_{sum} ratios (Fig. 3.32). However, the modelled $HC_{sum-all}/NO_x$ ratios are substantially higher than the measured ones which can be attributed to an overestimation of group II emission.

Table 3.18. Monthly mean values of calculated and measured HC_{sum-14}/NO_x and $HC_{sum-all}/NO_x$ emission ratios for October.

Parameter [ppbC/ppb]	Model	Experiment
HC_{sum-14}/NO_x	1.9	2.0 ± 0.1
$HC_{sum-all}/NO_x$	3.9	3.2 ± 0.2

The predicted HC emissions from group II which originate mainly from solvent evaporation cannot be found in the measured data. This points to an overestimation of solvent emissions by the emission model. According to the emission model ca. 60% of all pure HC-emission originate from solvent evaporation. In the following approach the latter result can be assessed by regarding the HC-patterns, i.e. relative contribution of a HC to the total HC (HC_i/HC_{sum}). Every source type (e.g. traffic exhaust, fuel evaporation, solvents) has its characteristic HC-pattern. If the model prediction of 60% solvent emissions of pure HC was correct the measured HC pattern should exhibit large similarities or at least show the characteristic features of a solvent HC pattern. Fig. 3.33 shows the chromatograms of 5 types of grades and white spirits most frequently used in

Germany. Grades and white spirits account for approx. 65% of predicted HC-emissions from solvent evaporation and contain large shares of HC in the region of C_9-C_{12}. The relative contributions of measured HCs (dotted line) show a completely different pattern. Toluene and benzene, originating mainly from traffic exhaust, and some lighter HC have high shares whereas C_9-C_{12} HC have low or non-detectable contributions. In the case of n-decane, a compound which is exclusively emitted by solvent emissions (traffic related emissions of n-decane are below 1% (Schmitz et al. 2002)), this deviation is more than a factor of ten.

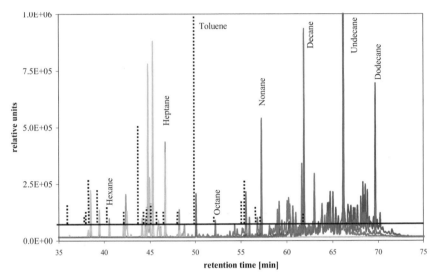

Fig. 3.33. Chromatogram of 5 grades and white spirits used most frequently in Germany and recorded with a similar GC-system as the ambient air measurements discussed here. The dotted lines show the relative contributions of HC_i to total HC measured in Augsburg and are normalised to toluene.

Conclusions

Comparisons between emission ratios (HC_i/HC_{sum} and HC_i/NO_x) derived from ground-based measurements downwind of a city and modelled emissions have been carried out to evaluate the emission model. The measurements show that urban HC-emissions are dominated by traffic related emissions. For HC which originate predominantly from traffic emissions the modelled and measured emission ratios agreed mostly within 30%.

According to the model 27% of all VOC emissions are pure hydrocarbons originating from solvent evaporation. These emissions cannot be resolved into single species but are rather given as summarised compound classes. About half of these HC-emissions are detectable by the GC system. If these additional HC-emissions are accounted for in the comparison, large deviations are found due to an overestimation of solvent emissions. In the case of n-decane this deviation is more than a factor of ten.

A further improvement of the emission model requires more detailed investigation. Oxygenated and halogenated hydrocarbons should also be included in further experiments since they are contained in many solvent mixtures. In order to determine the apportionment of a source type to a mixture of several source types (e.g. urban emissions) the Chemical Mass Balance (CMB) method was found to be useful. The use of this method requires the knowledge of the VOC composition (VOC_i/VOC_{sum}) of all important source types in a high resolution with respect to the number of species and a known data quality. The aim of further investigations should therefore be to characterise the VOC compositions of individual source types and to close the balance for total VOC.

Summary

The agreement between the measured and calculated CO and NO_x emissions was generally better than expected on the basis of previous reports, especially on the working days. For Saturdays, however, the temporal pattern of activities contributing to emissions should be improved.

Reasonably good agreement was also observed between most of the measured and calculated NMHC/CO and NMHC/sumNMHC emission ratios for traffic related hydrocarbons. Significant differences for several substances can be attributed to specific problems, such as incorrect emission factor e.g. for ethyne. The EVA experiment revealed also significant differences between average CO/NO_x and $sumHCi/NO_x$ emission ratios measured in March and October which were unpredicted by the modelled emission ratios. This suggests that seasonal profile of activities in emission models may also need improvement.

Solvent emission seems to be the most important problem revealed by the EVA project. Modelled emission ratios of nonane and decane relative to the total measured hydrocarbons were found to be substantially higher than the measured ones. Nonane and decane are assumed to originate almost exclusively from the use of white spirits as solvents. Significant differences were also observed for compounds which are not exclusively but to a substantial degree emitted by solvents such as toluene, xylenes and higher aromatic compounds. The difference of solvent emissions obtained by measurements and by model calculations cannot presently be explained. Further studies with improved models and extensive measurements of other solvent related compounds, especially of oxygenated hydrocarbons, are needed to resolve this problem.

The EVA project presents, to the best of our knowledge, the first comprehensive evaluation of the calculated highly resolved emissions by comparison with measurements. For this purpose the technique of mass balance has been refined and the tracer technique has been used for the first time on the scale reported here. As discussed by Kalthoff et al (2002), Panitz et al (2002), and Möllmann-Coers et al. (2002a), both techniques can be further refined to reduce the uncertainties of the measured emissions. Reduction of the uncertainties in the measured emissions would result in a more stringent test of the calculated emissions. Measurement of specific solvent compounds could also help to resolve the solvent problem.

The evaluation of calculated emissions for the city of Augsburg, though successful, still leaves at least one question open: Are the results obtained in Augsburg specific for this city or can they be generalized? This question can only be answered by further evaluations on other cities with different source composition and in other regions.

3.3.3.10 Evaluation of the Swiss emission inventory by statistical analysis of ambient air measurements from Zurich and Wallisellen

Johannes Staehelin

Introduction

The planning of the most cost effective strategy to reduce air pollution and greenhouse gases is based on inventories of anthropogenic emissions. The reliability of such emission inventories is therefore very important for any appropriate political action in this field. However, the evaluation of emission inventories, which are usually based on emission models of varying degree of sophistication by an independent method is not a simple task. Measurements of road tunnels can be used for comparison with road traffic emission models (see Staehelin and Sturm 3.3.2.1.). Another approach is the statistical analysis of ambient air measurements of primary air pollutants. Chemical Mass Balance models (CMB) have been often used for this purpose. Such studies are particularly valuable, if the respective emission sources emit simultaneously a large number of compounds and if the relative composition of the dominant sources is characteristically different. They are based on the following matrix equation:

$$X = S * Q + \varepsilon \tag{3.11}$$

Matrix of ambient air concentrations	=	Activities of sources	*	Source Profiles	+	error term
measured		to be determined		prescribed		to be minimized

In CMB models the observed measurements are interpreted as a superposition of emission sources with constant relative composition (source profiles (Q)). The fit of the measured air pollutant concentrations (X) allows the determination of the contributions of the sources to the ambient air quality (activities of the sources (S)). However, CMB-models require the knowledge of the composition of all relevant emission sources, which is often not adequately known. In the presented study we show another recently developed method which does not require the knowledge of the source profiles (Staehelin et al. 2001). The method is based on the mathematical method of the Principal Component Analysis (PCA) and a visual inspection of the data.

This method was applied to extensive ambient air pollutant measurements performed from 1993 to 1998 in the city Zürich (Switzerland) and at a suburban site close to Zürich (Wallisellen), which are summarized in the next section. After a short description of the method we show the results of the analysis for the two receptor sites, which are compared with the Swiss emission inventory. Solvent use makes the largest contribution in the Swiss emission inventory of anthropogenic Total Non-Methane Volatile Organic Compounds (t-NMVOC) (seeTable 3.19). The numbers are calculated for the reference year 1990 and estimated for 1995 based on emission changes predicted for the different sources, e.g. the increase of the proportion of gasoline driven vehicles equipped with catalytic converters.

Table 3.19. Swiss emission inventory of anthropogenic t-NMVOC (in tons per year, in brackets their relative contributions in percent), based on BUWAL (1995, 1996) (from Staehelin et al. 2001).

emission source	1990	1995
motor traffic [a]	116 550 (41 %)	76 850 (36.5 %)
industry	151 200 (51.8 %)	109 200 (51.9%)
agriculture and forestry	10 740 (3.7 %)	10 840 (5.1 %)
domestic use	13 110 (4.5 %)	13 610 (6.4 %)

[a] Motor traffic includes also gasoline vaporization and off road traffic emissions; gasoline vaporization and off-road traffic emissions are assumed to be constant from 1990 to 1995

Measurements

The site in Zürich (Kasernenhof) is located in the city centre of Zürich close to a parking lot surrounded by buildings, which border on busy streets. The used measurements of the city centre of Zürich were performed from Sept. 1993 to Sept. 1994 (except for the measurements of the passive samplers which were performed at both sites during the entire year of 1995). The site at Wallisellen is located in an agricultural area in open terrain at a distance of approximately 7 km to the city centre of Zürich. Frequently used high ways are at distances of 1.5 km and 2.5 km respectively, whereas the closer streets only carry a small number of vehicles. The measurements from Wallisellen were performed from Febr. 1996 to Febr. 1997. The very volatile organic compounds (VVOC) were measured by a completely automated instrument manufactured by Chrompack. The sampling included 30 min. of every hour, the quantified compounds are listed in Table 3.20. For the ambient air measurements of VOCs with lower volatility (see Table 3.20) passive samplers were used. However, these measurements have a much longer sampling time of 2 weeks which prevents their use in the statistical analysis. Total Non-Methane Volatile Organic Compounds (t-NMVOC) are monitored by the

Swiss Federal Laboratories for Material Testing and Research (EMPA) by a Flame Ionisation Detector (FID) at Zürich, Kasernenhof, 50 m away from the location of the Chrompack instrument (Nabel 1999). From the total amount of ppbC measured by FID the concentrations of the VOCs measured by VA and PS were subtracted. These readings mainly consist of polar VOCs including oxygenated species. The ppbC amount of this fraction of VOCs was estimated based on FID sensitivity of 0.5, because it is well known that oxygenated hydrocarbons exhibit a much lower sensitivity to FID than alkanes.

Table 3.20. Single VOC compounds quantified in this study. Compounds measured by the Chrompack analyser are marked by VA, those quantified by the passive samplers by PS (see text) (from Staehelin et al. 2001).

Alkanes	*Alkenes*	*Aromatics, cont.*
Ethane (VA)	Ethene (VA)	1,3,5-Trimethylbenz. (PS)
Propane (VA)	Propene (VA)	o-Ethyltoluene (PS)
Butane (VA)	1.3-Butadiene (VA)	m-Ethyltoluene (PS)
2-Methylpropane (VA)	*Alkines*	p-Ethyltoluene (PS)
n-Pentane (VA)	Ethine (VA)	n-Propylbenzene (PS)
2-Methylbutane (VA)	*Aromatics*	i-Propylbenzene (PS)
n-Hexane (VA)	Benzene (VA, PS)	*Chlorinated hydrocarbons*
2-Methylpentane (VA)	Toluene (VA, PS)	Heptane+Tetrachl.eth. (VA)
3-Methylpentane (VA)	o-Xylene (PS)	Trichloromethane (PS)
2 + 3-Methylhexane (VA)	m-Xylene (PS)	Tetrachloromethane (PS)
2,2,4-Trimethylpentane	p-Xylene (PS)	1,1,1-Trichloroethane (PS)
Ethylbenzene (PS)	1,1,2-Trichloroethane (PS)	n-Heptane (PS)
1,2,3-Trimeth.benz. (PS)	1,2,2,2-Tetrachloroeth. (PS)	
n-Octane (PS)	1,2.4-Trimeth.benz. (PS)	Trichloroethene (PS)
n-Nonane (PS)	Tetrachloroethene (PS)	n-Decane (PS)

The concentrations of O_3, NO, NO_2, SO_2 and CO are continuously measured at Zürich, Kasernenhof by EMPA and at Wallisellen by the Amt für Wasser, Energie, Abfall und Luft (AWEL), based on commonly used sensors. Half hourly mean values of the measurements of the inorganic compounds and of t-NMVOC were used in this study. For more details of the measurements see Locher, 2000 and Staehelin et al. 2001.

Emission source contributions to ambient air concentrations

The following method was developed at the Seminar for Statistics at ETHZ and does not require information on the number and relative composition of the sources but is solely based on the assumption of constant relative emissions of the relevant emission sources (source profile Q) and is applicable as long as the chemical degradation is negligible. (The use of one single source profile Q for industrial solvents is expected to be valid as long as the mixing time of the species from various industrial sources is short compared to the relatively long chemical lifetimes of the compounds in the polluted planetary boundary layer.) The method is based on a linear model, in which the measured pollutant concentrations (X) are the superposition of the contributions of individual sources with constant relative emissions (Q) multiplied by their activities (S), leading to the following matrix equation:

$$X = S * Q + \varepsilon \qquad (3.12)$$

Matrix of ambient =	Activities of *	Source Profiles +	error term
air concentrations	sources		
measured	*to be determined*	*to be determined*	*to be minimized*

The normalization of the data allows to eliminate most of the meteorological variability. Thereafter a Principal Component Analysis (PCA) is applied on the normalized measurements. Visual inspection, i.e. the projection of the data in the space of the eigenvectors, shows characteristic structures. The number of the calculated Principal Components represents the number of the most important sources. Corner points in these projections represent measurements, in which single emission sources are dominant. This allows the determination of the appropriate linear combination of principal components which describe the respective emission sources. Thereafter, negative elements in the source activity matrix (S) and in the source profile matrix (Q) are set to zero and the procedure is repeated again until convergence is obtained. This method therefore yields the number of dominant sources and the source profiles. Finally, the calculated source profiles, i.e. the relative chemical composition or its chemical fingerprints, can be associated to the profiles of the emission sources. The time pattern of the evolution of the sources from the activity matrix allows confirming the association. The method was recently refined by Wolbers (2002) leading to very similar results.

The method was applied to the large data sets of temporally highly resolved ambient air measurements of Zürich and Wallisellen including VVOCs, inorganic pollutants and t-NMVOC. The first source profile (Comb) was associated to road traffic emission because it is closely related to the results of the Gubrist tunnel study (Staehelin et al. 1998). Profile "Evap" contains large amounts of highly volatile organics and was associated to gasoline evaporation. Profile "SO_2" has a large contribution of sulphur dioxide and was attributed to room heating. The temporal evolution of the source activities and the seasonal variation of the proportion of the sources confirmed these associations. Another source containing large amounts of toluene and substantial NO_x concentrations was identified in the analysis of the measurements of both sites (see Table 3.21). At the centre of Zürich, sometimes large toluene concentrations of this source were observed during night. The contribution of motor traffic to this source is therefore rather speculative. The interpretation of the other sources is more difficult (Locher, 2000). The two first sources listed in Table 3.21 could be clearly attributed to motor traffic emissions, whereas the interpretation of the less dominant sources "Tol", "NM" and "Et1" and "Et2" was less certain.

Table 3.21. Relative contribution (in percent) of the different sources to ppbC and their association to motor traffic, obtained by PCA, see text (Staehelin et al. 2001).

Sources	Comb	Evap	SO$_2$	Tol	NM	Pr	Et1	Et2
Zürich: Proportion to VVOC	34	27	8	20	7	4		
Contrib. of motor traffic to the source	100	100	0	50-100	100	0		
Wallisellen: Proport-. tion to VVOC		36	29	2	11		15	7
Contrib. of motor traffic the source	100	100	0	100			0-50	0-50

Table 3.22. Summary of the results of motor traffic source contributions to t-NMVOC (in ppbC) and comparison with the Swiss emission inventory (including gasoline vaporization and off-road traffic, see Table 3.19) (from Staehelin et al. 2001).

VOC-class	prop. to t-NMVOC (FID measurements)	prop. of motor traffic in the respective class	Swiss t-NMVOC inventory 1990 1995
VVOC	63% (Chrompack anal)	76-88 % (PCA)	
Apolar VOC (volat. < toluene)	14 % (measurements of passive sampler)	0-80 % (assumption)	
Polar VOC	23 % FID measurements	0 (assumption)	
Total		48%-67%	34-48% 36.5%

In a further step we estimated the road traffic contribution to the t-NMVOC, which was measured by an FID instrument yielding a contribution of 63% to t-NMVOC from which 76-88% were found to be originated from motor traffic according to the statistical analysis (see Table 3.22). Additionally, 14% of t-NMVOC can be explained by the compounds with lower volatility quantified by the measurements of the passive sampler, which could not be directly included in the statistical analysis because of the poor time resolution of the measurements. The attribution of these VOC to emission sources is more difficult and we made a rather conservative estimate of a contribution of road traffic 0-80%. The rest of the compounds, probably mainly consisting of oxygenated and other polar species, were assumed to be entirely caused by non-traffic sources. By this analysis we found that 48-67% of the organic compounds have to be attributed to road traffic emissions (including gasoline evaporation), which is considerably more than expected from the Swiss emission inventory for the years 1990 and 1995.

The motor traffic contribution in the Swiss t-NMVOC emission inventory is much lower than the estimate of our study which is deduced from ambient air measurements (see Table 3.22). The emission inventory is based on the basis year 1990. The series of t-NMVOC measurements by FID from Zürich (NABEL, 1999) shows a large decrease from 1990 (annual mean of 0.18 mg/m^3) to 1995 (0.10 mg/m^3), which is substantially larger than the decrease in t-NMVOC emissions predicted by the Swiss inventory (comp. Table 3.19). An underestimate of the decrease of t-NMVOC by non traffic sources such as solvent use by industry and households between 1990 and 1995 may explain this difference for 1995 (comp. Table 3.19).

The close correspondence of the main results of the sites of Zürich and Wallisellen seems to preclude any major influence of local air pollution sources on the analysis of the measurements. There is no reason to assume large regional differences of the source contributions in the densely populated Swiss plateau. The road traffic contribution in the Swiss emission inventory is based on a road traffic model, thus a bottom up-approach using the knowledge of the emission factors. The results of the Gubrist tunnel study validated the road traffic emission model for organic substances (John et al. 1999), at least for high way driving conditions. Solvent use is the other main category of the t- NMVOC emissions which includes industrial and domestic uses. The emission estimate of these sources are first based on the import-export balance of VOC in Switzerland providing a reliable basis since all fossil fuels in Switzerland are imported. However, the proportion of solvents released to the atmosphere is more difficult to quantify. Therefore we suggest, that the emission from solvent use compared to motor traffic might be one major reason of the differences of Table 3.22. However, also the statistical method inherently includes several uncertainties. Small and irregular sources might be picked up by large sources such as motor traffic. These uncertainties are difficult to quantify.

3.4 Sensitivity analysis

B. Wickert, M. Memmesheimer and A. Ebel

3.4.1 Introduction

Anthropogenic emissions contribute to a large extent to the ambient air quality problems in Germany and Europe. In this context emissions of precursors for groundlevel ozone are of special interest, since they cause episodes of high ozone concentrations during summer months with suitable meteorological conditions. Scientific research can contribute to the understanding of the processes of ozone formation and to analyse possibilities for the decrease of ozone production by emission control. However, the uncertainties, which are inherent in model-supported air quality studies, are mostly unknown.

The objective of the work described here is the development of a model for the calculation of anthropogenic emissions in Germany with high spatial and temporal resolution. In a second step, sensitivity analyses with a coupled emission and chemistry transport model have been performed to examine the effects of uncertainties in the emission model on the results of atmospheric modelling.

In order to achieve these goals, first a complex emission model for Germany was developed to calculate data sets of highest possible accuracy. Therefore the emission model should take into account all relevant source groups and the most important ozone precursor substance classes NMVOC and NO_x, and, in addition, also other classical pollutants like CO, NH_3 and SO_2. In order to develop a state of the art model, it was necessary to transfer and modify existing methodologies for Germany, in particular for combustion plants and road traffic. Further improvements concern the methods for the calculation of spatial and temporal distributions which are applied here for the first time to the whole of Germany.

Variations of the emission model supply input data for a chemistry transport model, which simulates near-surface ozone concentrations for given land use, orographical and meteorological data. In this way sensitivity analyses can be carried out with respect to changes of emission input data. A large variety of cases is examined for different emission strengths of ozone precursor substances as well as modifications of the spatial distribution of the emissions, the temporal allocation and the VOC composition (Wickert 2001). Furthermore, the different effects of variations of the NMVOC emissions from biogenic and anthropogenic sources are analysed. The evaluation is performed regarding concentration totals, threshold exceedances, AOT60 values as well as a special parameter for the investigation of sensitivities, the 0.95 quantile of concentration differences.

The results of the sensitivity analysis show for the episode of investigation that uncertainties in emission data can have a strong influence on the results of the chemistry transport calculations. Of great importance is the correct estimation of NO_x emissions and the inclusion of large emitters as point sources. The allocation of the daily increase and decrease of emissions is obviously more important than the shape of the time curve. The use of many different VOC profiles does not improve the modelling quality significantly compared to the application of a single profile.

3.4.2 Scope

The impacts of air pollution to human beings – directly by affecting human health of indirectly by impairing his environment – have been subject of scientific research for a long time. Regarding the extension of the European community, international cooperation in air pollution control is of vital importance. Governments are intervening by setting up laws and directives, accommodating the increasing scientific and public ecological awareness.

Frequent threshold exceedances of near-surface ozone concentrations in Germany are to be regarded particularly critically (UBA 2000). Stimulated by meteorological conditions, ozone is formed in the troposphere due to anthropogenic emitted ozone precursors.

Ozone is, even in minimum concentrations, one of the strongest irritant gases since it has a strong oxidative capacity. It has negative impacts on pulmonary functions and can cause chronic tissue damages. Additionally, it causes airway and eyes irritation and raises susceptibility to infections. Apart these impacts on human health, ozone also causes damages on material goods and plants. (Baumbach 1990; Benk 1994)

With identifying and quantifying ozone-forming processes, science contributes to a better understanding of ozone-related issues and assists politics in developing ozone reduction instruments. Ozone concentrations can be simulated with computer-based models; measures can be determined that are most efficient in ozone reduction, both in ecological and economical terms. Such models consist of three major components, a meteorological model, an emission model and a chemistry transport model (Fig. 3.34). While the metrological model is simulating and forecasting metrological conditions, the emission model is to describe the emission situation most precisely. Both models provide the input data for the chemistry transport model which calculates the concentrations and depositions of substances within the modelled area.

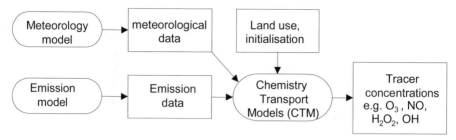

Fig. 3.34. Model system for computer based air quality analysis

Within the context of this analysis, two key objectives are pursued:

- The development and implementation of a model for the calculation of anthropogenic emissions in Germany, in high spatial, temporal and substantial resolution. (Wickert 2001)
- The performance of sensitivity studies with new instruments in interconnexion with an atmospheric dispersion model for the examination of the impacts of uncertainties within the emission model on air quality studies.

The following procedure is implemented:

Firstly, state of the art instruments for the calculation of emission data in high resolution for Germany are developed for use in chemistry transport models. These instruments are then executed exemplarily for the year 1994. Model development focuses on ozone precursors NO_x, NMVOC (VOC without Methane) and CO.

In respect of reactions of the wet phase chemistry, emission sources of SO_2 and NH_3 are also considered. Future German emission development is calculated by a 2010 trend scenario.

Secondly, sensitivity calculations are conducted with an attached model system from the newly developed emission model and a chemistry transport model.

In doing so, algorithms of individual modules of the emission model are variegated in order to obtain different emission data.

This data is then implemented in the chemistry transport model to simulate ozone concentrations for a two day summer smog period. The ozone concentrations are then compared with a reference case, resulting from the most accurate emission calculation. With this analysis it is possible to identify the parts of the emission model in which uncertainties have more or less impact on simulating ozone concentrations. Thereby, it is possible to identify both uncertainty sources and further need for research and development.

3.4.3 Sensitivity calculations using the EURAD CTM

Sensitivity calculations with a chemistry transport model

The underlying objective is to include preferably all emission processes in highest detail for the regional scale – as far as it is possible with limited resources. The calculated emission registers are to serve now as input data for chemistry transport models.

Information on uncertainties of emission registers does not allow estimates on the applicability of the data for use in chemistry transport models (CTM), since the impacts of uncertainties in the CTM have to be determined first. This determination is just spirit and purpose of the sensitivity calculations described in this chapter. These experiments shall allow statements on which parameters affect CTM the most. The following describes the approach used and implemented. A detailed listing of all sensitivity results can be found in the annexe.

Method

Initial point is the question where to find the major causes for uncertainties in the results in a complex emission model embedded in a system of models of emission simulation, dispersion, transformation, concentration and deposition. Since chemistry transport models issue a large amount of data, it is necessary to determine one or more target values for which uncertainties shall be investigated. Within the frame of this analysis, ozone is employed as a target value since it is a core element both in prevailing political discussion on air pollution and in throposheric research. Furthermore, the emission model presented in this analysis suits ozone concentration modelling perfectly.

The experimental set-up is illustrated in Fig. 3.35. The lines drawn through indicate the part of the model usually realised in model-based air quality studies, with evaluation of concentrations of other air pollutants and meteorological fields being natural in this studies.

The emission calculation model is variegated in a most diverse manner for the sensitivity calculation. The modified ozone concentrations are compared to the reference case, resulting from an execution of the model with most accurate emissions (dashed lines). This comparison allows a categorisation of the sub-models, according to their relevance on emission data quality.

The lines drawn through indicate basic ozone modelling with emission model and CTM, the dashed lines indicate additional steps for the sensitivity calculations.

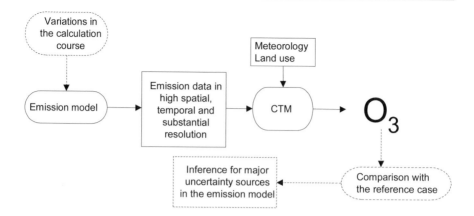

Fig. 3.35. Schematic illustration of the sensitivity calculations

Ancillary conditions of sensitivity experiments

For a comprehensive investigation of the emission model developed for Germany, it is doubtless desirable to implement as many different chemistry transport models with long episodes in different resolutions as possible. As of today's state of research, a series of problems goes along:

It is not possible to scale the spatial resolution in chemistry transport models arbitrarily. Models are commonly only implemented within certain resolution levels. Small-scale models require ancillary data from large-scale models and are thus dependent from the quality of their results.

A comparison of different chemistry transport models often provides different results, even when input data is about the same. As a consequence, it makes only sense to conduct sensitivity calculations for models producing results close to reality.

The quality of chemistry transport models can also be evaluated with a comparison of the simulation results with measured data, but measurements are time-consuming and expensive. Thus, most chemistry transport models are usually only evaluated with short episodes (several days up to weeks).

For most chemistry transport models, computing capacity needs to be enormous. Even when computing with high efficiency computers, models need several CPU hours for computing a few-day episode. The same applies for the preceding simulation of the necessary meteorological fields with a meteorological model.

In most cases, chemistry transport models are developed in research units and demand a high level of both scientific and technological knowledge from the users. It can take weeks up to months, before a model can be operated after the installation.

For this reasons, the following ancillary conditions for the sensitivity calculations are anticipated:

Selection of an appropriate episode: In order to limit the necessary computing efforts for the calculation of as many sensitivity cases as possible, all calculations are performed for a two-day episode (July 25 – July 26, 1994). Focussing on these two days means a substantial reduction of both utilisable output data and necessary computing time compared to a several-day simulation episodes commonly computed.

Furthermore, this two-day episode is part of the FLUMOB campaign (Abraham et al. 1995), performed from 23 July 1994 to 17 July 1994 in the Berlin-Brandenburg region with an area of investigation of 100 x 100 m². Within this campaign, near-surface ozone concentrations of more than $200\mu g/m^3$, thus exceeding both the WHO 1999 threshold of 120 $\mu g/m^3$ and the BImSchV1994[6] threshold for the information of the population ($180\mu g/m^3$), were measured. This makes this episode for an air-hygienic analysis of particular interest.

Chemistry transport model: The EURAD-CTM model is implemented for the sensitivity calculations as described here. The EURAD-CTM model, initiated 1987 and continuously improved, and the necessary input data were provided by the Institute of Geophysics and Meteorology, Cologne.

The model is in particular applicable due to several reasons:

Nesting capability: It is possible to zoom into the areas of investigation with adequate initialisation. This allows analysing for different spatial resolutions.

Spatial spectrum: Standard analysis in Europe starts with a European-wide grid with a resolution of 50 km. With four nesting steps, it is possible to focus on areas of several 1000 km², with a spatial resolution of 2 km. This makes the model independent from boundary values from other models for small-scale applications.

Computing time for processing the model is about three hours (nest at 56 x 58 grid points), on a NEC-SX5- array processor (16CPU, 4 GFLOPS peak performance, 32 GByte central memory, world ranking list number 167 in November 2000), which allows the processing of many individual calculations.

The standard EURAD-CTM is implemented in FORTRAN and operated with shell scripts. This makes it easily portable.

Additionally to CTM, the EURAD working group simplified preparatory work by providing all necessary programmes (calculation of plume superelevation, biogenous emissions, conversion of VOC categories) and input data (metrological fields, land use, J-values) in three spatial resolutions.

The EURAD-CTM was evaluated for four different episodes within the framework of the troposphere research programme (GSF 2000). For the FLUMOB-episode, there is also a comparison of ozone concentrations calculated by EURAD-CTM with plane measurements available (Schaller and Wenzel 2000). The EURAD CTM made second place out of six different model systems. Speaking in a European context, the EURAD-CTM, together with local model (meteorological model) of the German Weather Service, is leading in terms of ozone concentration accuracy (Tilmes et al. 2000). Thus, the EURAD-CTM can be considered to be state of the art for meso-scale ozone simulations.

---

[6] Administrative by-law to the German Federal Emission Control Act (*Bundes Immissions-Schutz Verordnung*)

Area of investigation for small-scale calculations: Since calculation time increases with increasing number of grid points, it is necessary to confine oneself to one area for performing sensitivity calculations in high spatial resolution. For this purpose, the FLUMOB-area is especially appropriate, since high ozone concentrations were measured and, additionally, the CTM was already evaluated for this area. As shown in Fig. 3.36, three nested areas are resulting:

Nest1:	56 x 68 cells,	Dx = 18 km
Nest2:	59 x 59 cells,	Dx = 6 km
Nest3:	77 x 77 cells,	Dx = 2 km
Germany	*Eastern Germany*	*Berlin-Brandenburg*

Sensitivity calculation areas of investigation

Each of the large-scale areas determines the boundary values for the areas in high spatial resolution. The largest area (*Nest1*) contains Germany, Switzerland, Austria, The Netherlands, Denmark and parts of Poland, the Czech Republic, Slovakia, Hungary, Slovenia, Italy, France and Sweden with a grid width of 18 km and a total area of about 1.2 million km². The *Nest2* area (approximately 125 000 km²) comprises mostly the German *New Laender*, parts of Poland and the Czech Republic with a grid width of 6km. The grid with the highest spatial resolution of 2 km x 2 km focuses on an area of approximately 24 000km² covering Greater Berlin-Brandenburg. Boundary values for the *Nest1* model are derived from a large-scale European simulation and are implemented alike for all sensitivity cases.

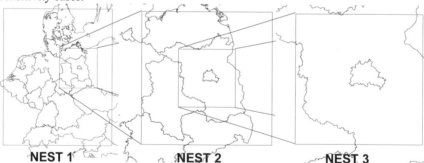

NEST 1   **NEST 2**   **NEST 3**

Fig. 3.36. Areas of investigation for the sensitivity calculations. The large-scale areas provide boundary values for the areas with high spatial resolution.

Target values

Chemistry transport models are usually able to provide time variable 3D concentration fields for all substances processed in transport and reaction equations. Ozone concentrations calculated with the EURAD-CTM are provided for all altitudes analysed by the model. But the interesting layer for emission control politics is the near-surface layer since here impairing impacts to humans and environment occur. For this reason, only changes in ozone concentrations in the lowest layer (up to 75 m above ground) are accounted, although, from a purely scientific point of view, concentration changes of all substances and in all layers are of interest.

Selected results

„Switching off" of the anthropogenic (OANT) resp. biogenic (OBIO) emissions can have different impacts on near-surface ozone concentrations, as illustrated in Fig. 3.37. During the night, ozone is depleted by NO mostly emitted from traffic. As anthropogenic emissions in Germany are not included into the simulation, there are higher ozone concentrations in urban areas, at individual point sources or along highways in the OANT case than there are in the reference case. During daylight, in the photo-chemically active phase, anthropogenic emissions contribute to the formation of ozone. Thus, „switching off" these emissions leads to a reduction of ozone of $50\mu g/m^3$ compared to the reference case. In the graph below, the origin of a part of the ozone plume above the North Sea (Fig. 3.37) becomes clear: Among other things, ozone precursors contribute to the formation of ozone. These precursors, that do not exist in the OANT case, were transported north-western due to corresponding meteorological conditions. In contrast, high ozone concentrations above the North Sea, as of 26 July, result from the transportation of a plume from the Netherlands due to western winds.

The lack of biogenous emissions (OBIO case) does not lead to such a dramatic concentration decrease as in the OANT case, but has similar tendency. During the night time, especially in areas without increasing ozone concentrations in the OANT case, exiguously increasing concentrations can be found. During daytime, there is less ozone produced, but the greatest difference is only minus $33\mu g/m^3$.

In temporal course, the impacts of OANT and OBIO are different: In case of lacking of biogenous emissions, the greatest concentration decreases compared to the reference case can be found at dawn, while in the case without anthropogenic emissions, minima of the differences (= maxima of ozone reduction) can be found in the early afternoon and are much broader. Ozone increase at night, as from 25 July to 26 July, is in the OBIO case about twice as much as in the OANT case, presumingly due to missing ozone lysis of the biogenous emitted substances.

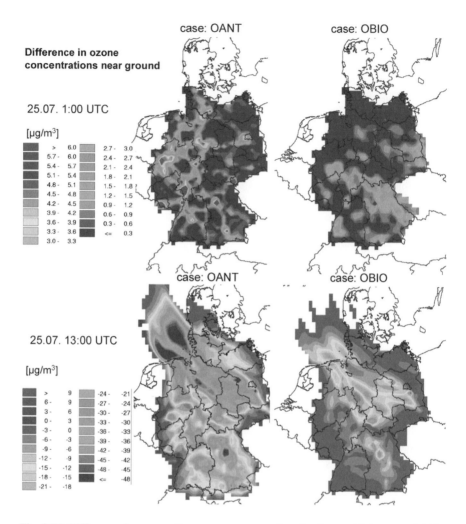

Fig. 3.37. Differences in near-surface ozone concentrations for the sensitivity cases OANT and OBIO for two different points in time on 25 July 1994. At each case, the difference between concentrations in the sensitivity case and the reference case is shown (e.g. difference = *conc*OANT – *conc*REF)

A summary of impacts in the OANT and the OBIO case on ozone concentrations in Germany is given in Table 3.23. For the entire grid and the whole episode considered, average ozone concentration decreases about 10% in the OANT case and about 3-4% in the OBIO case.

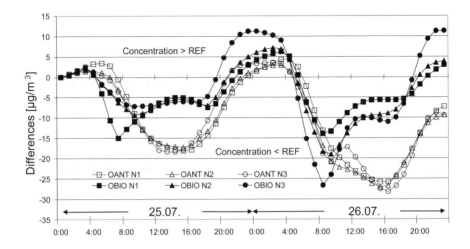

Fig. 3.38. Differences in mean O_3-concentrations in the three nesting areas over time, comparing the reference case to OANT and OBIO – time stamps in UTC

Only in the urban area (tilled area of the grid cells < 25 %), the Nest1 grid in the OANT case stands out against both others. Here, average ozone concentrations increase by 2% due to lacking large-scale NO emissions for ozone depletion, while Nest2 and Nest3 grids show decreases of 7%.

Table 3.23. Summary of ozone concentrations for the OANT and OBIO cases

		Nest1		Nest2		Nest3	
		Total	Urban	Total	Urban	Total	Urban
REF	absolute [µg/m3]	111.0	106.3	101.8	102.7	103.3	102.6
OANT	absolute [µg/m3]	101.3	108.2	91.4	95.2	93.4	95.3
	Deviation to REF [µg/m3]	-9.6	1.9	-10.4	-7.5	-9.9	-7.4
	Deviation to REF [%]	-9%	2%	-10%	-7%	-10%	-7%
OBIO	absolute [µg/m3]	106.9	101.3	97.7	97.6	100.2	99.4
	Deviation to REF [µg/m3]	-4.0	-5.0	-4.1	-5.0	-3.1	-3.2
	Deviation to REF [%]	-4%	-5%	-4%	-5%	-3%	-3%

Conclusions

Large-scale analysis for the modelling of air quality rely on the computer based compilation of anthropogenic emission registers since area-wide measurements of all relevant emissions processes are not possible. With the instruments presented in the framework of this analysis, it is possible to calculate emissions in Germany in high spatial, temporal and substance resolution, serving as input data for chemistry transport models for the determination of chemical and physical processes in the troposphere. For this purpose, already existing approaches were enhanced and implemented exemplarily for the year 1994. Improvements in calculation methods and output data are listed in the following:

The results of the emission calculation in the year 1994 show that traffic is one of the major sources for anthropogenic air pollution in Germany. Apart from road traffic, solvent use is the second main source for NMVOC emissions, while almost all NH_3 emissions come from the agricultural sector. Despite technical progress in flue gas treatment, large point sources are still responsible for almost all SO_2 emissions and for about a third of NO_x emissions within the last three decades. The spatial resolution of NO_x, CO and NMVOC emissions mirrors the level of urbanisation and industrial activity: In North Rhine-Westphalia, with it's high population density and it's large number of industrial firms, most emissions are generated.

In terms of temporal course of NMVOC emissions, the hourly course of the day is of importance, since it shows considerable higher fluctuations than the course of the day or the course of the year does. The higher the share of emissions of a pollutant from room heating installations, the higher seasonal fluctuations, as for example for CO, where combustion processes in households contribute considerably to the total emissions.

Based upon the emission data for the year 1994, a trend scenario for 2010 was developed taking into account all relevant air pollution control policies in place and in pipeline. In spite of significant emission reductions for NO_x and NMVOC, it is to be expected that the emission ceilings as set in the EC National Emission Ceilings Directive (NECD) cannot be achieved by Germany. And even by switching off all anthropogenic emissions of ozone precursors in Germany the health threshold for ozone at 120 $\mu g/m^3$ is frequently exceeded during episodes, strongly advocating the continuation of transboundary efforts for air pollution control to achieve compliance with the health threshold AOT60 as it is set by the WHO.

In addition to the results displayed here, a large number of sensitivity calculations have been conducted to assess the uncertainties involved when running CTMs on the basis of high resolution emission data. Based on the comparison of concentration changes due to variations in emission inputs, far reaching conclusions could be achieved as to the relevant importance of influencing factors within the whole modelling system and regarding input data quantity and quality.

Anthropogenic as well as biogenic emissions have to be carefully modelled for the simulation of tropospheric ozone. In summer weather conditions during the investigated period, a high share of NMVOC emissions stems from biogenic sources, thus leading to a NO_x limitation of groundlevel ozone concentrations, i.e. ozone concentration changes occur typically due to changes in NO_x emissions.

However, this cannot be averaged over the whole year, as only a quarter of annual NMVOC emissions are biogenic (Steinbrecher 1999).

Regression analyses to determine correlations of concentration changes of groundlevel ozone with the uncertainty bands of anthropogenic resp. biogenic emissions lead to the conclusion, that errors (relative and absolute) in the modelling of biogenic emissions induce far larger deviations in ozone simulations, than errors in anthropogenic emissions. This is quite important as uncertainties of biogenic emissions are typically much higher than those related to anthropogenic emissions (up to a factor of 10 according to Guenther et al. (2000)). On the other hand, errors in the emissions of anthropogenic NO_x emissions have a higher influence than errors in anthropogenic NMVOC emissions, which results in the recommendation to aim towards a as high as possible temporal and spatial resolution accuracy for the calculation of NO_x sources.

Not taking into account the plume superelevation of point sources can lead to a significantly greater error in the results than for instance a general over- or underestimation of NO_x emissions. Based upon this finding, the consideration of all major sources is modelled as point sources, in particular ozone simulations on small regional scale. In a similar way, taking into account the exact location of road transport emissions based on road networks improves the emission modelling results significantly.

The incorporation of land-use data to improve the spatial resolution of area source emissions seems to only increase the quality of emission date to a minor extent, as long as large point sources are accurately covered as individual sources.

Outlook

To increase the representativeness of the aforementioned conclusions, further investigations should be conducted in a similar way. These should include, among other aspects, evaluations with other CTMs and evaluations of other regions and episodes with the same model. As the investigations providing the basis for this analysis have been conducted in an area with strong NO_x limitation due to dominating biogenic NMVOC contributions to ozone formation, further experiments should include NMVOC limited areas, as well as episodes with lower temperatures and thus less biogenic emissions.

In addition to that, it would be favourable to conduct similar exercises on a regional scale for other regions in Europe, provided a sufficient level of detail. This could be accompanied by scenario calculations to assess the importance of individual components of emission calculation models on future emission levels. The relevance of land-use data for the improvement of the spatial resolution should definitely be investigated more thoroughly.

As indicated above, specific attention should be devoted to the modelling of biogenic emissions, as they have a very high impact on ozone formation in particular during summer smog episodes, when they can exceed anthropogenic emissions of NMVOC by far. This has an impact not only on the absolute concentrations, but as well on the inherent uncertainty of the overall modelling system, especially with anthropogenic NMVOC emissions decreasing due to emission control activities.

The improvement of the methodology of ozone modelling is essential to provide robust and reliable policy decision support, as more reliable results on cause and characteristics of high ozone episodes are needed. Research is not only necessary with regard to the emission calculation models, but to the further improve whole system of emission, meteorology, dispersion, chemical transformation and deposition models. Promising trends currently are emerging on variable grids, which allow an adaptation of the spatial resolution by taking into account the dynamics of specific modelling regions. Furthermore, integrated concepts taking into account all relevant air pollutants as well as greenhouse gases, as well as different environmental media (air, water, soil) need to be developed, as most often pollutants and effects are interrelated and thus bear potential synergies when trying to solve the problems they cause.

References

Abraham HJ, Lenschow P, Lutz M., Reichenbächer W, Reimer E, Scherer B, Stark B (1995) Flugzeug- und Bodenmessung von Ozon und Vorläuferstoffen zur Abschätzung der Wirksamkeit von emissionsmindernden Maßnahmen im Großraum Berlin-Brandenburg (FLUMOB-Projekt) – Abschlußbericht, Berlin.

Baumbach G (1990) Luftreinhaltung. Springer-Verlag, Heidelberg.

Baumbach G and Vogt U (1995) A Tethered-Balloon Measurement System for the Determination of the Spatial and Temporal Distribution of Air Pollutants such as O_3, NO_2, VOC, Particles and Meteorological Parameters. Eurotrac Newsletter 16, 23 – 29

Bayerisches Landesamt für Umweltschutz (1994).

Benk M (1994) Basiswissen Umwelttechnik: Wasser, Luft, Abfall, Lärm, Umweltrecht. Vogel Buchverlag, Würzburg.

Bishop GA, Stedman DH, Ashbaugh L (1996) Motor vehicle emissions. variability *J. Air Waste Manage. Assoc.* 46, 667–675.

Bishop GA, Stedman DH, Peterson JE, Hosick TJ, Guenther PL (1993) A cost-effectiveness study of carbon monoxide emissions reduction utilizing remote sensing *J. Air Waste Manage. Assoc.* 43, 978–988.

Bishop GA, Stackey JR, Ihlenfeldt A, Williams WJ, Stedman DH (1989) IR long-path photometry: A remote sensing tool for automobile emissions. Anal. Chem. 61:671A.

Bruckmann P, Kesten W, Funke W, Balfanz E, König J, Theisen J, Ball M, Päpke G (1988) The occurence of chlorinated and other organic trace components in urban air. Chemosphere, 17: 2363–2380 .

Bruckmann P, Beier R, Krautscheid S (1983) Immissionsmessungen von Kohlenwasserstoffen in den Belastungsgebieten Rhein-Ruhr. Staub-Reinhaltung der Luft, 43:404–410.

BUWAL (1996) Schadstoffemissionen und Treibstoffverbrauch des Off-Road-Sektors. Umweltmaterialien 49, Swiss Agency for the Environment, Forests and Landscape (BUWAL), Bern, Switzerland.

BUWAL (1995) Vom Menschen verursachte Luftschadstoffemissionen in der Schweiz von 1900 bis 2010. Schriftenreihe Umwelt 256, Swiss Agency for the Environment, Forests and Landscape (BUWAL), Bern, Switzerland.

Catenacci G, Riva M, Volta M, Finzi G (1999) A Model for Emission Scenario Processing in Northern Italy. In: Borrell PM, Borrell P (eds) Proceedings of the EUROTRAC Symposium 1998 Vol. 2, WIT Press, Southampton. pp720-724.

Cirillo MC, de Lauretis R, del Ciello R (ENEA) (1996) Review study on European urban emission inventories(final draft).European Environment Agency. European topic center on air emission, Topic report 1996/ Document EEA/96.

Dasgupta P, Dong S, Hwang H, Yang HC, Genfa Z (1988) Continuous liquid-phase fluorimetric coupled to a diffusion scrubber for the real-time determination of atmospheric formaldehyde, hydrogen peroxide and sulfur dioxide. Atmos Environ 22: 949–963.

Derwent RG, Jenkins MG, Sauders SM, and Pilling MJ (1998) Photochemical ozone creation potential for organic compounds in northwest europe calculated with a master chemical mechanism. Atmos Environ 32:2429-2441.

Ebel A, Friedrich R, Rodhe H (eds1997) Tropospheric Modelling and Emission Estimation [P. Borrell et al., Series Eds., Transport and Chemical Transformation of Pollutants in the Troposphere, Vol. 7] Springer, Berlin-Heidelberg.

Ellermann K, Herkelmann K, Gladtke D, Pfeffer HU (1995) LIMES-Jahresbericht, Landesumweltamt Nordrhein-Westfahlen.

EVA-Database, IER, prepared in connection with the EVA intercomparison in March (2000) Stuttgart.

Fahey DW, Eubank CS, Hübler G, and Fehsenfeld FC (1985) Evaluation of a catalytic reduction technique for the measurement of total reactive odd-nitrogen NO_y in the atmosphere. J Atmos Chem 3: 435-468.

Fujita EM, Watson JG, Chow JC Magliano K (1995) Receptor model and emission inventory source apportionments of nonmethane organic species in California's San Joaquin Valley and San Francisco Bay Area, Atmos Environ 29:3019-3035.

Gerbig C, Schmitgen S, Kley D, Volz-Thomas A, Dewey K, Haaks D (1999) An improved fast-response vacuum-UV fluorescence CO instrument. J Geophys Res 104:1699-1704.

Gerbig C, Kley D, Volz-Thomas A, Kent J, Dewey K, McKenna DS (1996) Fast-Response Resonance Fluorescence CO Measurements Aboard the C-130: Instrument Characterization and Measurements Made during NARE '93. J Geophys Res 101:29229-29238.

Glaser K (2001) Methoden der Qualitätssicherung bei der Messung von Luftverunreinigungen bei Feldexperimenten. Dissertation, submitted to the University of Stuttgart, 2000

GSF (2000) Förderschwerpunkt Troposphärenforschung GSF – Forschungszentrum für Umwelt und Gesundheit, GmbH, Neuherberg.

Guenther A, Geron C, Pierce T, Lamb B, Harley P, Fall R (2000) Natural emissions of non-methane volatile organic compounds, carbon monoxide, and oxides of nitrogen from North America. Atm. Env. 34: 2205 – 2230.

Guenther PL, Stedman DH, Bishop GA, Bean JH, Quine RW (1995) A hydrocarbon detector for remote sensing of vehicle emissions. *Rev. Sci. Instr.* 66, 3024–3029.

Harris GW, Mackay IG, Iguchi T, Mayne K, and Schiff HI (1989) Measurements of formaldehyde in the troposphere by tunable diode laser absorption spectroscopy. J Atmos Chem 8:119–137.

Hassel D, Jost P, Weber FJ, Dursbeck F, Sonnborn S, Plettau D (1995) Abgasemissionen von Nutzfahrzeugen in der Bundesrepublik Deutschland für das Bezugsjahr 1990. Abschlußbericht zum Vorhaben: UFO Plan-Nr. 104 05 151/02, Berlin.

Hassel D, Jost P, Weber FJ, Dursbeck F, Sonnborn KS, Plettau D (1994) Das Abgas-Emissionsverhalten von Personenkraftwagen in der Bundesrepublik Deutschalnd im Bezzugsjahr 1990. Technischer Überwachungsverein Rheinland. Berichte Umweltbundesamt 8/94. Eruch Schmidt Verlag, Bonn.

Hausberger S, Rodler J, Sturm P (2002) Emission factors for HDV and validation by tunnel measurements 11th International Conference "Transport and Air Pollution" 2002, Graz, VKM-THD Mitteilungen 81:93-100.

Hayman G, Bartzis J, Dore C, Ekstrand S, Goodwin J, Licotti C, Olsson B, Rabasco C, Sjodin Å, Steinnocher K, Tamponi M, Vlachogiannis D, Winiwarter W (2001) IMPRESAREO - Improving the Spatial Resolution of Air Emission Inventories Using Earth Observation Data. Final report. AEAT/ENV/R/0693, Culham, UK.

John C (1999) Emissionen von Luftverunreinigungen aus dem Straßenverkehr in hoher zeitlicher und räumlicher Auflösung. Untersuchung am Beispiel Baden-Württembergs. Dissertation, Institute for Energy Economics and the Rational Use of Energy (IER), University of Stuttgart, Forschungsberichte Band 58.

John C, Friedrich R, Staehelin J, Schläpfer K, Stahel W (1999) Comparison of emission factors for road traffic from a tunnel study (Gubrist tunnel, Switzerland) and from emission modelling. Atmospheric Environment 33:3367-3376

Kalthoff N, Corsmeier U, Schmidt K, Kottmeier CH, Fiedler F, Habram M, Slemr F (2002) Emissions of the city of Augsburg determined using the mass balance method, Atmos. Environ. 36: S19-S32.

Kanter HJ, Mohnen VA, Volz-Thomas A, Junkermann W, Glaser K, Weitkamp C, Slemr F (2002) Quality assurance in TFS for inorganic compounds. J. Atmos. Chem. 42: 235-253.

Keller M, Hausberger S et al. (1998) Handbuch Emissionsfaktoren für den Strassenverkehr in Oesterreich (Handbook on Emissionfactors for the street traffic in Austria) im Auiftrag des BMUJF und UBA Oesterreich, Wien.

Klemm O, Schaller E (1994) Aircraft measurement of pollutant fluxes across the borders of eastern Germany. Atmos. Environ. 28: 2847-2860.

Klemp D, Mannschreck K, Pätz HW, Habram M, Matuska P, Slemr F (2002) Determination of anthropogenic emissions ratios in the Augsburg area from concentration ratios: results from long-term measurements. Atmos. Environ. 36: S61-S80.

Klemp D, Kern T, Beck H, Mihelcic D (1994) Design and development of a one-channel tunable diode laser spectrometer for the measurement of atmospheric formaldehyde mixing ratios – A contribution to EUROTRAC subproject JETDLAG. in: EUROTRAC Annual REPORT 1993, pp. 17 – 25, Int. Sci. Secr. Fraunhofer Inst. (IFU), Garmisch-Partenkirchen, Germany.

Kraftfahrt-Bundesamt (1994) Statistische Mitteilungen. Reihe 2: Kraftfahrzeuge, Sonderheft 2: Bestand an Kraftfahrzeugen u. Kraftfahrzeuganhängern am 1. Juli 1994 nach Zulassungsbezirken in Deutschland. Metzler-Poeschel-Verlag, Stuttgart.

Kühlwein J (2003) Unsicherheiten bei der rechnerischen Ermittlung von Schadstoffemissionen des Straßenverkehrs und Anforderungen an zukünftige Modelle. Ph.D. thesis, University of Stuttgart, in preparation.

Kühlwein J and Friedrich R (2000) Uncertainties of modelling emissions from road transport, Atmospheric Environment 34 (2002) 4603-4610.

Kühlwein J, Friedrich R, Kalthoff N, Corsmeier U, Slemr F, Comparison of total modelled and measured CO and NO$_x$ emission rates. *Submitted to atmospheric environment.*

Kühlwein J, Wickert B, Trukenmüller A, Theloke J, Friedrich R (2002a) Emission modelling in high spatial and temporal resolution and calculation of pollutant concentrations for comparisons with measured concentrations. Atmos. Environ. 36: S7-S18.

Kühlwein J, Friedrich R, Kalthoff N, Corsmeier U, Slemr F, Habram M, Möllmann-Coers M (2002b) Comparison of modelled and measured total CO and NOx emission rates. Atmos. Environ. 36: S53-S60.

Kühlwein J, Friedrich R, Obermeier A, Theloke J (1999) Estimation and Assessment of uncertainties for high resolution NO$_x$ and NMVOC emission data, PEF - Projekt Europäisches Forschungszentrum für Maßnahmen zur Luftreinhaltung, research report PEF 2 96 002, Karlsruhe.

Lawson DR (1995) The costs of "M" in I/M - Reflections on Inspection/Maintenance programs. *J. Air Waste Manage. Assoc.* 45, 465–476.

Lawson DR (1993) "Passing the test" - Human behavior and the California's Smog Check program. *J. Air Waste Manage. Assoc.* 43, 1567–1575.

Lies KH, Schulze J, Winneke H, Kuhler M, Kraft J, Hartung A, Postulka A, Gring H, Schröder D (1998) Nicht-limitierte Automobil-Abgaskomponenten. Studie der Volkswagen AG, Wolfsburg .

Locher R (2000) VOC-Immissionsmessungen in den Kantonen Zürich, Schaffhausen und Luzern, 1993-1998. Umweltmaterialien 118, Swiss Agency for the Environment, Forests and Landscape (BUWAL), Bern, Switzerland.

Maffeis G, Longoni M, De Martini A, Tamponi M (1999) Emissions estimate of ozone precursors during PIPAPO campaign, in Photochemical Oxidants and Aerosols in Lombardy Region, Milan, Italy .

Mannschreck K (2001) Experimentelle Bestimmung von städtischen Emissionen anhand von Konzentrationsmessungen im Lee einer Stadt – Untersuchungen zum Beitrag verschiedener Quelltypen und Vergleich mit einem Emissionsberechnungsmodell. Berichte des FZ Jülich, JÜL-3846.

Mannschreck K (2000) Experimentelle Bestimmung von städtischen Emissionen anhand von Konzentrationsmessungen im Lee einer Stadt – Untersuchungen zum Beitrag verschiedener Quelltypen und Vergleich mit einem Emissionsberechnungsmodell. Berichte des Forschungszentrums Jülich, JÜL-3846, Jülich, Germany.

Mannschreck K, Klemp D, Kley D, Friedrich R, Kühlwein J, Wickert B, Matuska P, Habram M, Slemr F (2002a) Evaluation of an emission inventory by comparisons of modelled and measured emission ratios of individual HCs, CO and NO$_x$. Atmospheric Environment 36:81–94.

Mannschreck K, Bächmann K, Barnes I, Becker KH, Heil T, Kurtenbach R, Memmesheimer M, Mohnen VA, Schmitz T, Steinbrecher R, Obermeier A, Poppe D, Volz-Thomas A, Zabel F (2002b) A Database for Volatile Organic Compounds, J Atmos Chem, 42:281-286.

MEET (1999) Methodology for calculating transport emissions and energy consumption. European Commission. Transport Research. Forth framework programme. Strategic research, DG VII – 99.

MEET Deliverable No. 14, Samaras Z, Ntziachristos L, Kylindris K (1998) Average Hot Emission Factors for Passenger Cars and Light Duty Trucks; European Commission DG VII.

MEET Deliverable No. 17, Jorgensen M, Sorensen SC (1997) Estimating Emissions from Railway Traffic, European Commission DG VII.

MEET Deliverable No. 18, Kallivoda M, Kudnra M (1997) Methodologies for estimating emissions from air traffic, European Commission DG VII.

MEET Deliverable No. 19, Trozzi C, Vaccaro R (1997) Methodologies for Estimating Air Pollutant Emissions from Ships; European Commission DG VII.

Möllmann-Coers M, Klemp D, Mannschreck K, Slemr F (2002a) Determination of anthropogenic emissions in the Augsburg area by the source-tracer-ratio method. Atmos. Environ. 36: S95-S108.

Möllmann-Coers M, Klemp D, Mannschreck K, Slemr F (2002b) Statistical study of the diurnal variation of modeled and measured NMHC contributions. Atmos. Environ. 36: S109-S122.

Möllmann-Coers M, Klemp D, Mannschreck M, Slemr F (2002c) Statistical Investigation of the Consequences of Predicted HC-Source Contributions. Atmos Environ 36, Suppl. 1:109-122.

NABEL (1999) Luftbelastung 1998. Schriftenreihe Umwelt 311, Swiss Agency for the Environment, Forests and Landscape (BUWAL), Bern, Switzerland Orthofer R., Winiwarter W.; Spatial and Temporal Disaggregation of Emission Inventories. In. Power H, Baldasano J.M.(ed.) Air Pollution Emission Inventory, ISBN: 1853125172 pp 250, 1998.

Neftel A et al. (2002) LOOP final report.

Ntziachristos L, Tourlou PM, Samaras Z, Geivanidis S, Andrias A (2001) National and central estimates for air emissions from road transport. Technical report No 74, European Topic Centre on Air and Climate Change. EEA, Copenhagen.

Panitz HJ, Nester K, Fiedler F (2002) Mass budget simulation of NOx and CO for the evaluation of calculated emissions for the city of Augsburg (Germany), Atmos. Environ. 36: S33-S52.

Pasquill F (1974) Atmospheric Diffusion, 2nd Edition. John Wiley & Sons, New York.

Pätz HW, Corsmeier U, Glaser K, Vogt U, Kalthoff N, Klemp D, Kolahgar B, Lerner A, Neininger B, Schmitz T, Schultz M, Slemr J, Volz-Thomas A (2000) Measurements of trace gases and photolysis frequencies during SLOPE96 and a coarse estimate of the local OH concentration from HNO$_3$ formation. J Geophys Res. 105:1563–1580.

Placet M, Mann CO, Gilbert RO, Niefer MJ (2000) Emissions of ozone precursors from stationary sources: a critical review. Atmos. Environ. 34: 2183-2204.

Popp PJ, Bishop GA, Stedman DH (1998) Development of a high-speed ultraviolet spectrometer for remote sensing of mobile source nitric oxide emissions. . J. Air Waste Manage. Assoc. 49, 1463–1468.

Pulles T, Mareckova K, Kozakovic L, Esser P, Berdowski J (1996) Verification of CORINAIR 90 Emission Inventory by Comparison with Ambient Air Measurements. TNO-MEP - R96/001, Delft.

Sawyer RF, Harley RA, Cadle SH, Norbeck JM, Slott R, Bravo HA (2000) Mobile sources critical review: 1988 NARSTO assessment. Atmos. Environ 34: 2161-2181.

Schaller E and Wenzel A (2000) Evaluierung regionaler atmosphärischer Chemie-Transport-Modelle, Fall3: FLUMOB, 22.7.94 bis 27.7.94, Lehrstuhl für

Umweltmeteorologie – Fakultät 4: Umweltwissenschaften und Verfahrenstechnik, Brandenburgische Technische Universität, Cottbus.

Schmid H, Pucher E, Ellinger R, Biebl P, Puxbaum H (2001) Decadal reductions of road traffic emissions on a transit route in Austria - results of the Taurentunnel experiment 1997. Atmospheric Environment 35:3585-3593.

Schmitz S, Haserich D, Oppermann F, Otto I, Pütz T, Siedhoff M, Viktorin D (1997) Entwicklung eines planungsrelevanten Emissionskatasters Straßenverkehr. Bundesforschungsanstalt für Landeskunde, Bonn. Materialien zur Raumentwicklung, Heft 80.

Schmitz T, Hassel D, Weber F (2000) Zusammensetzung der Kohlenwasserstoffe im Abgas unterschiedlicher Fahrzeugkonzepte. Berichte des Forschungszentrums Jülich Jül-3646.

Schmitz T, Klemp D, Kley D (1997) Messungen der Immissionskonzentrationen verschiedener Ozonvorläufersubstanzen in Ballungsgebieten und an Autobahnen. Berichte des Forschungszentrums Jülich, JÜL-3457.

Schwarz U, Wickert B, Obermeier A, Friedrich R (1999) Generation of Atmospheric Emission Inventories in Europe with High Spatial and Temporal Resolution. In: Borrell PM, Borrell P (eds) Proceedings of EUROTRAC Symp. 1998, Vol. 2, WITPress, Southampton, pp 261-265.

Sjödin Å, Sjöberg K, Svanberg PA, Backström H (1996) Verification of Expected Trends in Urban Traffic NOx Emissions from Long-Term Measurements of Ambient NO_2 Concentrations in Urban Air. Sc. Total Environ. 189/190, 213.

Sjödin Å (1994) On-road emission performance of late-model TWC-cars as measured by remote sensing. J. Air Waste Manage. Assoc. 43, 397–404.

Slemr F, Friedrich R, Seiler W (2002a) The research project EVA – general objectives and main results, Atmos. Environ. 36: S1-S6.

Slemr F, Baumbach G, Blank P, Corsmeier U, Fiedler F, Friedrich R, Habram M, Kalthoff N, Klemp D, Kühlwein J, Mannschreck K, Möllmann-Coers M, Nester K, Panitz HJ, Rabl P, Slemr J, Vogt U, Wickert B (2002b) Evaluation of modeled spatially and temporarily highly resolved emission inventories of photosmog precursors for the city of Augsburg: the experiment EVA and its major results. J. Atmos. Chem. 42: 207-233.

Slemr F, Friedrich R, Seiler W (2001) The research project EVA – general objectives and main results. Submitted to atmospheric environment.

Slemr J (1999) Qualitätssicherung von Kohlenwasserstoffen für das TFS-Subprojekt EVA. Projektträger Umwelt. GSF Forschungszentrum für Umwelt und Gesundheit, München.

Slemr J, Kanter HJ, Mohnen V Measurements made during the EVA campaigns: quality control and assessment. Submitted to atmospheric environment.

Staehelin J, Locher R, Mönckeberg S, Stahel WA (2001) Contribution of road traffic emissions to ambient air concentrations of hydrocarbons: The interpretation of monitoring measurements of Switzerland by Principal Component analyis and road tunnel measurements. Intern. J. Veh. Design 27:161-172.

Staehelin J, Keller C, Stahel W, Schläpfer K, Wunderli S (1998) Emission factors from raod traffic from a tunnel study (Gubrist tunnel, Switzerland). Part III: Results of organic compounds, SO_2, and speciation of organic exhaust emission. Atmospheric Environment 32:999-1009.

Staehelin J, Keller C, Stahel W, Schläpfer K, Steinemann U, Bürgin T Schneider S (1997) Modelling emission factors of road traffic from a tunnel study, Environmetrics 8: 219-239.

Staehelin J, Schläpfer K, Bürgin T, Steinemann U, Schneider S, Brunner D, Bäumle M, Meier M, Zahner C, Keiser S, Stahel W, Keller C (1995) Emission factors from road traffic from a tunnel study (Gubrist tunnel, Switzerland). Part I: Concept and first results. Sci. Tot. Env. 169:141-147.

Stedman DH, Bishop GA, Aldrete P, Slott RS. (1997) On-road evaluation of an automobile emission test program. *Environ. Sci. Technol.* 31, 927–931.

Steinemann U and Zumsteg F (2001) Verkehrs- und Schadstoffmessungen 2000 im Gubristtunnel, Bericht US 89-16 für das Amt für Abfall, Wasser, Energie und Luft des Kantons Zürich (AWEL), Schweiz.

Steisslinger B, Baumbach G (1990) Vertikalprofilmessungen von O_3, NO_2 und SO_2 bis 500 m über Grund mit einem Fesselballonsystem. -- VDI-Berichte Nr. 838, S. 527 -- 542, VDI-Verlag, Düsseldorf.

Steven H (1995) Emissionsfaktoren für verschieden Fahrzeugschichten, Straßenkategorien und Verkehrszustände (1). 2. Zwischenbericht zum Forschungsvorhaben 105 06 044: Erarbeitung von Grundlagen für die Umsetzung von §40.2 BImSchG. Commissioned by Umweltbundsamt, Berlin (2nd edition).

Sturm P, Rodler J, Lechhner B. Almbauer A. (2001) Validation of emission factors for road vehicles based on street tunnel measurements. Internat. Journal of Vehicle Design 27:65-75.

Thijsse T, Oss van RF (1997) Hydrocarbons in the Berlin air. TNO report R97/057, Apeldoorn.

Tilmes S, Brandt J, Flaty F, Langner J, Bergström R, Christensen JH, Ebel A, Flemming J, Friedrich R, Frohn LM, Heidegger A, Hov O, Jacobsen J, Jakobs H, Reimer E, Wickert B, Zimmermann J (2000) Intercomparison of Eulerian ozone prediction systems within GLOREAM for summer 1999 using the German monitoring data. GLOREAM Workshop 20 – 22 September 2000. S. Tilmes, Cottbus.

UBA (2000) Ozonsituation 2000 in der Bundesrepublik Deutschland, Umweltbundesamt, Berlin.

UBA (1999) Umweltbundesamt, Berlin; Bundesamt für Umwelt, Wald und Landschaft, Bern; INFRAS AG, Bern: Handbuch Emissionsfaktoren des Straßenverkehrs, Version 1.2. Berlin, 1999 (published as software on CD-ROM).

UBA (1995) Umweltbundesamt Berlin und INFRAS AG Bern, Handbuch für Emissionsfaktoren des Strassenverkehrs Version 1.1. Berlin. Software on CD-ROM.

Vautard R, Martin D, Beekmann M, Drobinski P, Friedrich R, Jaubertie A, Kley D, Lattuati M, Moral, Neininger B, Theloke J (2003) Paris emission inventory diagnostics from the ESQUIF airborne measurements and a chemistry-transport model, J. Geophys. Res.-atmospheres, accepted 2002, in press.

Volz-Thomas A, Slemr J, Konrad S, Schmitz T, Apel EC, Mohnen VA (2002) Quality assurance of hydrocarbon measurements for the German Tropospheric Research focus (TFS). J. Atmos. Chem. 42: 255-279.

Volz-Thomas A, Gerbig C (1998) Abschlußbericht: PC-gestützte Auswertung von CO-Meßdaten aus Luftproben des EVA-Experiments, Forschungszentrum Jülich .

Winiwarter W (2002) Performance of emission inventories – what can modellers expect ? In: Neftel A et al. (ed) LOOP final report.

Winiwarter W (1993) Temporal Disaggregation of Emission Inventories. In: P. Anttila (ed.), Proceedings of the EMEP Workshop on the Control of Photochemical Oxidants in Europe, 20-22 April 1993, pp. 145-149. Finnish Meteorological Institute, Helsinki, Finland (1993).

Winiwarter W, Dore C, Hayman G, Vlachogiannis D, Gounaris N, Bartzis J, Ekstrand S, Tamponi M, Maffeis G (2002) Methods for comparing gridded inventories of atmospheric emissions – application for Milan province, Italy, and the Greater Athens Area, Greece. The Science of the Total Environment, in press.

Winiwarter W, Loibl W, Maffeis G, Longoni MG, Tamponi M, Schwarz U, Friedrich R (1999) Spatial statistics for the intercomparison of independent emission inventories of the same area. Paper prepared for the joint BERLIOZ/LOOP workshop, Varese, Italy, Dec. 1999.

Winiwarter W, Musalek G, Züger J, Baumann R, Neininger B (1997) Validierung der POP-Emissionsinventuren mit Hilfe von Meßdaten. OeFZS-A—4125, Seibersdorf.

Wolbers M (2002) Linear Unmixing of Multivariate Observations. Ph.D. thesis, Swiss Federal Institute of Technology Zurich, Nr. 14723.

York D (1966) Least-Squares-Fitting of a Straight Line. Can J Phys 44:1079-1086.

Zeuner G, Möllmann-Coers M (1994) Dispersion Experiments in Complex Terrain. Meteorol Zeitschrift N.F. 3:107-110.

Zhang Y, Stedman DH, Bishop GA, Beaton SP, Peterson JE, McVey IF (1996) Enhancement of remote sensing for mobiles source nitric oxide. J. Air Waste Manage. Assoc. 46, 25–29.

Zongmin W (1986) Die Krigingmethode zur Lösung mehrdimensionaler Interpolations-probleme. Göttingen, Univ., Diss., (Diss. 1986/4452), Identifikationsnummer Titel: 2230500

ANNEX:

The following tables represent the format required to report the background of the emission data

How to use these tables: All fields with headers in *italics* need to be filled with a short verbal description plus a citation of the source of information

Annex 1: Emissions

Element	Parameters						
Emissions	*choice of indicator*	*value of indicator (activity)*	*emission factor*	*model structure*	*surrogate for spatial disaggregation*	*surrogate for temporal disaggregation*	*choice of VOC-split*
Public power plants							
District heating plants							
Petroleum refining plants							
Solid fuel transform. plants							
......							

Annex 2: Completeness analysis

Element	Parameters		
Completeness analysis		emission covered ? (Y/N)	(If emission not covered: Comment why considered irrelevant)
Public power plants			
District heating plants			
Petroleum refining plants			
Solid fuel transform. plants			
......			

Annex 3: Uncertainty

Element	Parameters						
Uncertainty		Applicability of indicator	uncertainty of indicator (activity)	uncertainty of emission factor	representativity of emission factor	model structure	uncertainty of spatial disaggregation	uncertainty of temporal disaggregation
Public power plants							
District heating plants							
Petroleum refining plants							
Solid fuel transform. plants							
......							

Annex 4/5: Retrospective Assessment and Validation (Results to be presented at separate sheets)

Element	Approaches				
Retrospective analysis	relative contribution of source groups	sensitivity analysis (optional)	overall uncertainties		
Validation (at least one approach is required)	Alternative emission assessment	Emission trends vs. ambient air concentratio n trends	Flux estimations from downwind measure-ments	Atmo-spheric models	Finger-prints and receptor modelling

4 Emission Models and Tools

4.1 Introduction

S. Reis, P. Blank, R. Friedrich

4.1.1 Overview

Air quality monitoring and modelling is a key element of foresighted action under conditions of global change. Good quality emission data are a prerequisite for a number of EU directives and other international agreements. In particular, they are needed for the Kyoto Protocol to the UNFCCC, the European Climate Change Programme ECCP, and for the UNECE Convention on Long Range Transboundary Air Pollution, CLRTAP, and its protocols. In order to assess data ranges and uncertainties, data quality has to be known. This is not only important because of the use of emission data as input to atmospheric dispersion modelling, but as well as compliance with emission ceilings (NEC Directive, Gothenburg Protocol, Kyoto targets) has to be assessed.

4.1.2 Methodology

Emission models and tools are thus needed to generate emission data as input into CTMs, for compliance monitoring, and various other applications, providing data with sufficient speed, in good quality, and with known uncertainties. In general, these models apply methods that combine emission factors (EF's) with activity data (A) to calculate emissions (E) in the following way:

$$E = EF \times A \qquad (4.1)$$

However, even though this formula suggests a simple and straightforward approach, both the emission factors and the activity rates comprise a vast number of individual parameters which have an impact on the resulting emissions.

4.1.2.1 Emission factors

Emission factors typically depend on a number of factors, which are influenced by technology, environmental conditions and other source-specific characteristics. The following list shall serve as an example for the influence factors on the specific emission factor in g per km for a passenger car:

- Engine type, size, age, technology, load factor
- Vehicle weight
- Fuel type (gasoline, diesel, LPG, CNG, etc.)
- Exhaust aftertreatment equipment (e.g. three-way catalyst, particle trap, etc.)
- Road gradient
- Driving pattern (e.g. urban, rural, highway, congested roads, etc.)
- Driving speed
- Ambient temperature (e.g. for gasoline evaporation, hot soak losses etc.)
- …

This coarse list of factors indicates, how complex the task of deriving emission factors for all relevant emission sources is, and even more, how detailed the information requirements for activity rates are (see below). Furthermore, it becomes obvious how important a comprehensive methodology, along with the assessment of uncertainties, is to achieve a high quality of calculated emission data.

4.1.2.2 Activity rates

In addition to emission factors, activity rates the second indispensable input for emission calculation. To continue the example set above, activity rates for the calculation of emissions from passenger cars would comprise the number of vehicles per time on a specific road segment, the distribution of driving speeds, the number of gasoline and diesel fuelled vehicles and suchlike. For other source categories, typical activity rates would, for instance, be the number of livestock on farms (agriculture), the amount of electricity produced and the annual utilisation of a power plant (energy), or the volume of solvent containing products applied (solvent use).

Both activity rates and emission factors are used on different levels of aggregation, depending on the availability of information and the requirements for the resolution of emission data from the side of the data user.

4.1.3 Structure

While the previous chapters have focussed on individual pollutants and the state-of-the-art of methodologies for the generation of emission data in high temporal and spatial resolution, the following sections will address a selection of models that have been developed and applied in the course of the GENEMIS project. To begin with, Sect. 4.2 elaborates on the importance of land use data as an essential input to emission models and as the basis for any emission inventorying work, in particular for the spatial distribution of emissions. Sect. 4.3 contains the description of three individual emission models which have been developed and applied within GENEMIS, and which present the current knowledge in this field. Finally, a set of emission inventorying activities in Turkey is described in Sect. 4.3.4, giving an in-depth view on the procedures and data requirements, as well as an analysis of the latest results.

4.2 Land cover and topography mapping

G. Smiatek

4.2.1 Introduction

Meteorology models (MM), climate models (CM), chemistry transport models (CTM) as well as anthropogenic and biogenic emissions models are widely used in the development of clean air policy and air pollution abatement strategies. As in many environmental models the quality of their outcome is strongly influenced by the spatial and temporal accuracy of land cover and topography data used in their development and operation.

To be used with models or their preprocessors land cover and topography data have to meet some model-related specifications. Models usually accept only certain formats, nomenclatures and map projections. And finally, the extent of the data has to cover the entire modelling area. Thus, the available geographical data must be pre-processed according to the specific needs of a particular model. The processing steps include transformations of nomenclatures, scales and map projections as well as merging of data from different sources. Here Geographical Information Systems (GIS) and Relational Database Management Systems (RDBMS) provide suitable data processing tools.

This article describes the Land Cover data for Europe (LCE) compiled at IMK-IFU as well as major land cover and topography data suitable for use with environmental models especially with the Meteorology Model (MM5, Grell et al. 1993) and the Multiscale-Climate-Chemistry Model (MCCM, Grell et al. 2000). The data compilation is a part of a GIS/RDBMS-based system of input data processing for environmental models.

4.2.2 Data models and data processing

Land use and land cover definitions are often used as synonyms. In environmental modelling, however, a clear distinction should be made. Land use refers to the way and purpose how land resources are used. An example is agriculture or mining. On the other hand land cover describes the physical state of the land surface, such as deciduous or coniferous forests, grass, surface waters or soil properties. MM5 and MCCM derive physical parameters like albedo, moisture, emissitivity, roughness length and thermal inertia from land cover data. Therefore in the model application land cover and appropriate data with a detailed nomenclature are of primary interest. Unfortunately available land cover data utilize different nomenclature systems. They differ in both number of land cover categories and the category content. One of the major steps in data processing is the creation of appropriate transfer look-up tables relating land cover categories of one nomenclature to another. It must be stated here that in many cases the classification schemes are not hierarchical. Also there are some incompatibilities between the nomenclatures.

Within a GIS digital geographic feature data can be represented in vector and raster data form. Topography is usually stored in raster. Both raster and vector are suitable to land cover data. In the vector data model the geographic land cover features are characterized as areas. In a coordinate system referencing the real world locations polygons represent homogenous area features. In the raster data model, each location is represented as a cell. The matrix of cells, organized into rows and columns, is called a grid. Each row contains a group of cells with values describing a geographic phenomenon. Cell values are numbers, which represent nominal data, such as land cover classes.

The accuracy of raster data is associated with the geometrical resolution of the real-world area represented by each grid cell. The larger the area covered by a grid cell, the lower the resolution and the land cover features are represented less accurately. The accuracy of vector data is related to the scale of the map used in the survey and digitalisation process. Land cover is often derived from air-borne and satellite images and converted to vector format. Their accuracy depends on the geometrical and spectral resolution of the imagery and the image classification procedure. A GIS provides various tools for vector to raster and raster to vector conversion. However, the data conversion might significantly influence the data accuracy.

The compilation of the LCE land cover data as well as the data provision to the models is entirely based on available existing digital data sources. Detailed information on the land cover nomenclature, map extent, map projection parameters, type of the feature data representation and geometrical resolution of each data set must be known and be available to the processing system. Therefore appropriate meta-information has been created and stored in a relational database. The data compilation as well as the data provision to models has been facilitated in a GIS/RDBMS – based class system written in object-oriented Perl (Smiatek 2002). GIS administrates all geo-data, such as land cover, administrative boundaries, plant species distribution and topography in raster or in vector form. In order to ensure the most possible flexibility and accuracy all available data sets used in the compilation processes are stored in their original data models.

The procedure itself includes the following steps:

- Resampling of the input data into geographical projection and vector-raster conversion of the vector data with a geographical resolution of 30 arc seconds by 30 arc seconds
- Reclassification of the land cover categories into the chosen land cover nomenclature. For each nomenclature a transfer scheme look-up table is retrieved from the relational database.
- Merging of the raster data sets into the final land cover map
- Integration of high resolution topography data

- Creation of specific products for land cover and topography suitable as input for the MM5/MCCM model

The system is able to provide land cover and topography data in general data format required by the TERRAIN preprocessor of the MM5/MCCM model as well as data tailored to the defined model domain.

4.2.3 Land Cover data for Europe (LCE)

For regional environmental research and modelling applications within Europe the "Land Cover for Europe" (LCE) data set with a geometrical resolution of 1 arc minute by 1 arc minute has been compiled (Smiatek 2000). Recently the LCE data set has entirely been updated by incorporating newly available data. Its current version number is 2.5. The database covers an area between 36° West and 56° East and 29° North and 75° North. While maintaining the geographic projection the geometrical resolution of the raster data has been increased to 30 arc seconds by 30 arc seconds. For each grid cell of the LCE data the predominant land use category is given. The land use categories depicted in the data set were chosen according to USGS nomenclature used in the MM5-Model (see Table 4.1.)

The described extent of LCE can only be reached by merging data from different sources. Corner stones of the LCE data set are the land cover data collected within the CORINE (Coordination of Information on the Environment) and Phare (Poland, Hungary Aid for the reconstruction of the Economy) Programmes of the European Union. One of the core projects of both is the land cover mapping (EEA 1992). LCE has been compiled from 10 national CORINE and 5 national Phare datasets. For areas where CORINE/Phare data are not yet available, land cover data from the PELCOM (Mücher et al. 2000) database and the Global Land Cover Characteristics (GLCC) (USGS 2002b) have been used.

CORINE Land Cover project has been started in 1985 with a pilot study in Portugal. At this time the majority of the EU countries have completed the work. Also, the mapping program has been extended to the Eastern European countries (within the Phare Programme) and to some other non-EU members, such as Morocco and Tunisia. Major advantage of the CORINE land cover program is the use of a standard mapping procedure. It allows achieving a very high data quality by all involved mapping agencies and their subcontractors. The nomenclature distinguishes 44 different categories, which are grouped in a 3-level hierarchy. The land cover survey methodology is based on computer aided photo-interpretation of Earth observation satellite imagery and simultaneous consultation of ancillary data. The mapping scale is 1:100000.

For regional environmental research and modelling applications within Europe the PELCOM land cover characterization (Mücher et al. 2000) is available since 1999. The land cover monitoring project was aiming at the provision of high spatial accuracy land cover information which can periodically be updated. The approach is based on use of multi-spectral and multi-temporal satellite imagery

Table 4.1. Land cover categories of the LCE 2.5 data set. (USGS nomenclature)

No.	Code	Category	Area [1000 km^2]	Proportion [%]
1	1	Urban and build-up areas	333.6	0.55
2	2	Dryland cropland and pasture	5952.9	9.77
3	3	Irrigated cropland and pasture	174.6	0.29
4	4	Mixed dryland/ irrig. Cropland and past.	0	0.00
5	5	Cropland/grassland mosaic	986.9	1.62
6	6	Cropland/woodland mosaic	672.6	1.10
7	7	Grassland	3742.0	6.14
8	8	Shrubland	2317.9	3.80
9	9	Mixed shrubland/grassland	181.0	0.30
10	10	Savannah	79.4	0.13
11	11	Deciduous broadleaf forests	1654.7	2.72
12	12	Deciduous needle leaf forests	6.4	0.01
13	13	Evergreen broadleaf forests	26.2	0.04
14	14	Evergreen needle leaf forests	3470.9	5.70
15	15	Mixed forest	1478.1	2.43
16	16	Water bodies	33465.3	54.92
17	17	Herbaceous wetland	832.8	1.37
18	18	Wooded wetland	3.1	0.01
19	19	Barren or sparsely vegetated	3600.6	5.91
20	20	Herbaceous tundra	9.9	0.02
21	21	Wooded tundra	340.9	0.56
22	22	Mixed tundra	70.0	0.11
23	23	Bare ground tundra	6.2	0.01
24	24	Snow and ice	1534.0	2.52
25	-999	Missing data	0	0.00
		Total area	60940.8	100.00

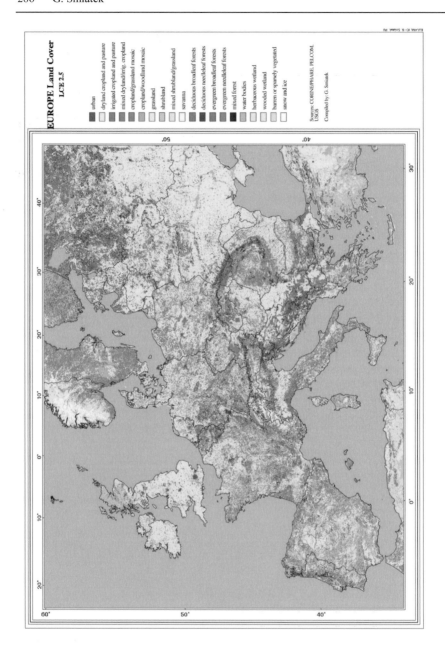

Fig. 4.1. Land Cover map of Europe (Version 2.5)

from the NOAA-AVHRR system and ancillary data. Both unsupervised and supervised image classification schemas were applied together with vegetation index (NDVI – normalized difference over sum vegetation index) monthly maximum value composites. The data set depicts 12 different land cover categories and its geometrical resolution is 1.1 km x 1.1 km.

The Global Land Cover Characteristics database (GLCC) is the only available global land use data set suitable for use with MMs and CTMs at global scale. The data is now available in the updated version 2.0. It is part of the National Aeronautics and Space Administration (NASA) Earth Observing System Pathfinder Program and the IGBP -Data and Information System focus 1 activity. The data set has been derived from 1km x 1-km Advanced Very High Resolution Radiometer (AVHRR) data from a 12-month period ranging from April 1992 until March 1993 in a joint effort of the U.S. Geological Survey's (USGS) Earth Resources Observation System (EROS) Data Center, the University of Nebraska-Lincoln (UNL) and the Joint Research Centre of the European Commission (USGS 2002b). The applied mapping procedure is based on seasonal land cover regions composed of homogeneous land cover associations showing distinctive phenology and common levels of primary production.

4.2.4 Digital Elevation Models

Digital Elevation Models (DEM) are equally spaced lattices of discrete values of elevation over a surface, i.e. mean sea level. DEMs can result from a wide range of mapping procedures utilizing data from topographic surveys, photogrammetry, air-borne altimeters, lidars and the newly available technique of radar interferometry. In the years 1993 - 1996 a global digital elevation model GTOPO30 was compiled by the U.S. Geological Survey's EROS Data Center in Sioux Falls, South Dakota (USGS 2002a). GTOPO30 was developed to cover the needs of the geographical data users for regional and continental scale topographic data. The data set covers the latitude from 90° South to 90° North and the longitude from 180° West to 180° East. Its geometrical resolution is 30 arc seconds by 30 arc seconds. The horizontal coordinate of the data is decimal degrees of latitude and longitude. The vertical units represent elevation in meters above mean sea level.

Data with higher spatial resolution up to 30 meters by 30 meters exist at national level. However, legal restriction and especially high charges often prevent the data from being used in the scientific community. The X-SAR/SRTM - Shuttle Mission (http://www.dlr.de/srtm), conducted in the year 2000, is supposed to provide state of the art digital elevation data for large parts of the world at a geometrical resolution of 3 arc seconds by 3 arc seconds and 1 arc second by 1 arc second (Schmulius et al. 2000). However, it is not known yet, when these data become available to the entire scientific community. The GIS/RDBMS–based data processing system provides tools suitable for merging available high resolution regional topography data with the GTOPO30 data set.

4.2.5 Data accuracy

Quality of land cover data substantially determinates the quality of a model outcome. The magnitude of daily VOC emission from biogenic sources can vary up to 70% depending of the land cover data used to simulate the meteorological parameters (Smiatek 2000). Area totals calculated for a specified area can give a first indication of the data quality but in modelling the spatial accuracy is much more important. Table 2 shows the area of major land cover categories in Germany derived from different data sources. Compared to the statistical information available from the Federal Statistical Office, Germany, (StaBuA) CORINE land cover figures show only a marginal difference. GLCC data deviate significantly.

Assessment of the spatial accuracy of land cover data is very complex. For that purpose an independent sample of ground information is necessary. This information is not available. Due to the mapping scale of 1:100000 and the precise mapping procedure CORINE/Phare data show the highest spatial accuracy. Thus, the GLCC and PELCOM data sets can be verified against the spatial information of the CORINE land cover data. The results of that comparison are depicted in Table 4.3 and Table 4.4. For the area of between 5° East and 16° East and 47° North and 55° North a random sample of 2000 grid cells has been drawn. At each sample point the land cover category of the CORINE, GLCC and PELCOM data has been retrieved and contingency tables for the land cover categories "agriculture" (see Table 4.3) and "forests" (see Table 4.4) have been calculated. GLCC data overestimates agriculture area and underestimates the forest area significantly. The goodness of fit described by the κ coefficient is with 0.278 (agriculture) and 0.37 (forest) quite

Table 4.2. Area of the major land use and land cover categories in Germany derived from different data sources

Category	GLCC	PELCOM	ELC2.5	CORINE	StaBuA[1]
Urban	8.4	15.8	25.9	26.9	42.1
Forest	53.3	114.7	103.8	104.2	104.9
-Deciduous	50.8	32.9	23.3	23.6	-
-Coniferous	2.3	48.0	57.1	57.1	-
-Mixed	0.2	33.8	23.4	23.4	-
Waters	4.3	2.1	3.4	5.0	7.9
Agriculture	287.7	216.7	217.5	216.5	193.1
-Arable land	287.1	185.6	173.0	173.2	-
-Grassland	0.1	31.1	44.5	43.26	-
Other	0	0	0	4.5	4.7
Totals	354.3	354.3	354.3	354.3	257

[1]Statistical data, Statistische Ämter des Bundes und der Länder 2001, www.brandenburg.de

Table 4.3. Contingency table for the land cover category "Agriculture"

a. CORINE vs. GLCC		CORINE Land Cover		
		Yes	No	Σ
GLCC data	Yes	530	718	1248
	No	75	677	752
	Σ	605	1395	2000

Proportion CORINE = 0.3025, Proportion GLCC = 0.624, κ = 0.278

		CORINE Land Cover		
		Yes	No	Σ
PELCOM data	Yes	478	368	846
	No	127	1027	1154
	Σ	605	1395	2000

Proportion CORINE = 0.3025, Proportion PELCOM = 0.423, κ = 0.473

Table 4.4. Contingency table for the land cover category "Forest"

		CORINE Land Cover		
		Yes	No	Σ
GLCC data	Yes	210	112	322
	No	314	1364	1678
	Σ	524	1476	2000

Proportion CORINE = 0.262, Proportion GLCC = 0.161, κ = 0.371

		CORINE Land Cover		
		Yes	No	Σ
PELCOM data	Yes	359	270	629
	No	165	1206	1371
	Σ	524	1476	2000

Proportion CORINE = 0.262, Proportion PELCOM = 0.315, κ = 0.472

low. The possible range of the κ value is from 0 indicating independent data to 1 indicating full agreement. PELCOM data quality is much higher. Proportion estimate for agriculture is overestimated by 12% and the forest proportion by 5.3%. The κ coefficient is with 0.47 for the agriculture and forest categories significantly larger.

Due to the fact that the geometrical resolution of the GTOPO30 elevation data is 30 arc second by 30 arc seconds it is difficult to compare the depicted elevations with elevations found in analogue maps especially in complex terrain. On the

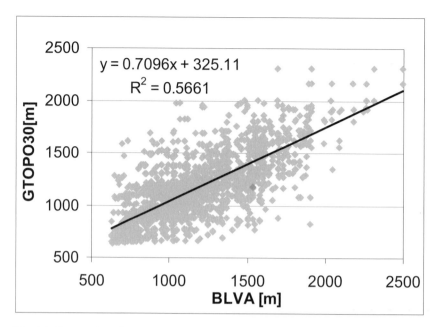

Fig. 4.2. Scatter plot of a BLVA DEM derived from 30 m x 30 m data and resampled to 30 arc sec. x 30 arc second geometrical resolution versus GTOPO30 DEM. Loisach valley, Garmisch-Partenkirchen, Germany.

other hand if the data is used in numerical models the correct representation of complex terrain is extremely important. In order to get some indication of the data quality, the GTOPO30 data covering the Loisach valley close to Garmisch-Partenkirchen, Germany has been compared with a DEM provided by the Bayerisches Landesvermessungsamt (BLVA). The DEM data has been resampled from the original geometrical resolution of 30 m by 30 m in Gauß-Krüger map projection to a 30 arc seconds by 30 arc seconds grid in geographic projection. Elevation contours created from both data sets show quite similar patterns. GTOPO 30 underestimates the higher elevation by 100 – 300 meters. There is however a systematic shift in the GTOPO30 data to northwest in a range of one or two grid cells. Figure 4.2 shows a scatter plot of the grid values. The overall correlation coefficient is 0.75. Taking into account the rather coarse geometrical resolution of the GTOPO30 data it documents a high quality of the global topography data suitable for mesoscale applications.

4.2.6 Conclusions

With the compiled European Land Cover data an important input data set has been provided to the modelling community. It is compatible in the resolution to the ETOPO30 topography. LCE is being used in several modelling activities within the EUROTRAC subprojects GENEMIS and GLOREAM as well as in national programmes, such as AFO2000 in Germany. Within areas where CORINE/Phare data were available the data set has reached its final status. These are the countries Ireland, Spain, Italy, France, Luxembourg, Belgium, The Netherlands, Germany, Austria, Rumania, Hungary, Slovakia, Czech Republic, Poland and parts of Tunisia and Morocco covering approximately 25% of the LCE2.5 land area. For all other countries within Europe (additional 45% of the data set area) the PELCOM land cover data is available with a comparable quality status. The quality of the data in the remaining area (approximately 30%) mostly in Asia and Africa is still not satisfactory.

Together with the LCE data a set of GIS-based tools has been developed. It allows provision of various data based on LCE land cover GTOPO30 topography tailored to the specific needs of the environmental models.

4.3 EMIMO: An emission model

R. San José, J.I. Peña, J.L. Pérez and R.M. González

4.3.1 Introduction

In this contribution we describe the EMIMO model which is an emission software tool to generate anthropogenic and biogenic emission data for being used in many different applications and particularly on Air Quality Modelling Systems (AQMS). Emission inventories are essential information for simulating dispersion of pollutants in the atmosphere. Modern AQMS's require detailed and accurate information on emission sources in order to have the best possible input datasets on these complex AQMS's. The AQMS's are complex mathematical tools which simulate atmospheric flow over a geographical domain (mesoscale, microscale, etc.) and over a period of time (day, weeks, years, etc.). Short term air quality modelling systems require a high spatial and temporal resolution emission data sets. Emission datasets are part of a group of input data sets which should be provided to the mathematical tools (Dispersion models) to perform the simulations in a satisfactory way. The emission inventory is perhaps the input data set which contains a higher level of uncertainty due to the inherent chaotic and uncertain nature of the sources which provides the emission data.

Traditionally, emission inventories have been divided into two main sections: A) *anthropogenic* and B) *biogenic*. Importance of biogenic emissions is growing in the last years due to the high impact on secondary pollutant formation (ozone, PAN, etc.). Great diversity of sources are involved in emission of pollutants, traffic, industry, etc. In this contribution we describe the EMIMO emission model which is using a top-down approach. The EMIMO model was developed based on the experience gained by UPM from developing the EMIMA model (a bottom-up approach emission modelling tool, see San José et al. 1995, 1996). The EMIMO model is developed by using global and continental yearly emission inventory as initial data sets and also land use, population files and road vector files (global information). In addition to this, training areas and specific emission data files such as UK Environmental Ministry emission data sets and IER emission data for Madrid and Bilbao (hourly emission information) are also used to generate EMIMO emission data sets. EMIMO incorporates a module called BIOEMI developed by UPM in 1998 to calculate biogenic emissions over any area in the world based on the algorithms of Guenther et al. (1995) and Steinbrecher et al. (1997).

The EMIMO model is mainly developed for providing emission data (high spatial and temporal resolution) for AQMS's models. The 3rd generation of AQMS is represented by MM5-CMAQ (MM5 is a non-hydrostatic mesoscale meteorological model developed by NCAR/PSU and CMAQ is a Community Multiscale Mesoscale Air Quality Model in Models-3/EPA U.S.). The MM5-CMAQ has a nesting capability which allows the user to simulate the atmospheric flow over several days in the forecasting (in forecasting mode) horizon. The mother and nesting domains cover from continental or national levels to city levels (nesting level 3, 4 or even 5). In Figs. 4.3 and 4.4 we show an example of how these model domains for MM5-CMAQ look-like for an application centred over the Iberian Peninsula (Madrid). The emission information should be provided as total amount of pollutant (SO_2, NO_x, CO, CO_2, VOC's, PM) per hour, per cell for the specific required period of time.

THE MM5-CMAQ MODELLING SYSTEM

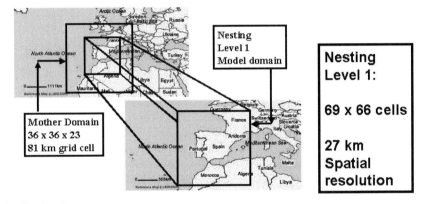

Fig. 4.3. Mother and nesting level 1 in the design of an MM5-CMAQ modelling exercise.

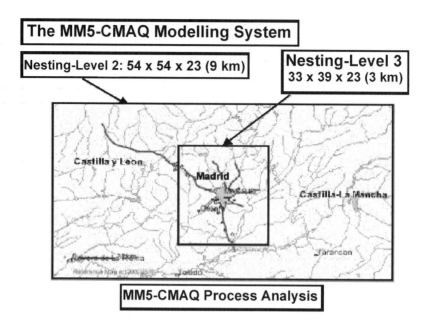

Fig. 4.4. Nesting levels 2 and 3 in the design of an MM5-CMAQ modelling exercise.

4.3.2 Methodology

EMIMO model uses the following initial data sets which are distributed spatially and temporally over the required model domain:

4.3.2.1 GEIA

The Global Emissions Inventory Activity (GEIA) was created in 1990 to develop and distribute global emission inventories of gases and aerosols emitted into the atmosphere from natural and anthropogenic (human-caused) sources. The long-term goal is to develop inventories of all trace species that are involved in global atmospheric chemistry. GEIA's mission is to provide high quality data for timely, relevant assessments. The requirements for a data set to be accepted as a GEIA database include substantial peer review as reflected by acceptance for journal publication and agreement among the individual GEIA project teams. To ensure that the best available data can be easily accessed by users, GEIA provides links to "pre-GEIA" data for projects that do not yet have a formal GEIA database and projects where the formal GEIA data need to be updated. These interim pre-GEIA data and data requiring updates have not yet undergone the formal GEIA review requirements. To enhance development and distribution of the highest quality information, GEIA also facilitates communication with other related database and assessment activities by providing links to these activities.

GEIA provides global data with 1 degree resolution under yearly basis.

4.3.2.2 EMEP

EMEP is a scientifically based and policy driven program under the Convection on Long-Range Transboundary Air Pollution for international co-operation to solve transboundary air pollution problems. EMEP provides European data with 50 km spatial resolution under yearly basis.

4.3.2.3 EDGAR

The EDGAR database is a joint project of RIVM and TNO and stores global inventories of direct and indirect greenhouse gas emissions from anthropogenic sources including halocarbons both on a per country basis as well as on $1° \times 1°$ grid. The database has been developed with financial support from the Dutch Ministry of the Environment and the Dutch National Research Programme on Global Air Pollution and Climate Change (NRP), in close cooperation with the Global Emissions Inventory Activity (GEIA), a component of the International Atmospheric Chemistry Programme (IGAC) of the International Geosphere-Biosphere Program (IGBP). EDGAR Version 3.2 comprises an update of the 1990 emissions and new emissions for 1995, both on 1x1 degree and summary tables per region/source for different years. In addition, EDGAR 3.2 emissions will be made available as reference dataset on a per country basis. For the direct greenhouse gases CO2, CH4 and N2O also emissions estimates for 1970 and 1980 will be made available. Data for the precursor gases CO, NO_x, NMVOC and SO2 will be limited to 1990 and 1995.

4.3.2.4.DCW

The Digital Chart of the World (DCW) is an Environmental Systems Research Institute, Inc. (ESRI) product originally developed for the US Defense Mapping Agency (DMA) using DMA data. We used the DCW 1993 version at 1:1,000,000 scale. The DMA data sources are aeronautical charts, which emphasize landmarks important from flying altitudes.

4.3.2.5 Global Land Cover

The U.S. Geological Survey (USGS), the University of Nebraska-Lincoln (UNL), and the European Commission's Joint Research Centre (JRC) have generated a 1 km resolution global land cover characteristics database for use in a wide range of environmental research and modelling applications. The global land cover characteristics database was developed on a continent-by-continent basis. All continental databases share the same map projections (Interrupted Goode Homolosine and Lambert Azimuthal Equal Area), have 1-km nominal spatial resolution, and are based on 1 km Advanced Very High Resolution Radiometer (AVHRR) data spanning April 1992 through March 1993. Each database contains unique elements based on the geographic aspects of the specific continent. In order to provide flexibility for a variety of applications, a core set of derived thematic maps produced through the aggregation of seasonal land cover regions are included in each continental database. The continental databases are combined to make seven global data sets, each representing a different landscape based on a particular classification legend. The global data sets are also available in two map projections (Interrupted Goode Homolosine and Geographic).

4.3.2.6 CIESIN

CIESIN's mission is to provide access to and enhance the use of information worldwide, advancing understanding of human interactions in the environment and serving the needs of science and public and private decision making. The Center for International Earth Science Information Network (CIESIN) was established in 1989 as an independent non-governmental organization to provide information that would help scientists, decision-makers, and the public better understand the changing relationship between human beings and the environment. In 1998, CIESIN became a centre within Columbia University's Earth Institute. From its offices at Columbia's Lamont-Doherty Earth Observatory campus in Palisades, New York, CIESIN continues to focus on applying state-of-the-art information technology to pressing interdisciplinary data, information, and research problems related to human interactions in the environment.

4.3.2.7 GLOED

The EPA Office of Research and Development has developed a powerful software package called the Global Emissions Database (GloED). GloED is a user-friendly, menu-driven tool for storage and retrieval of emissions factors and activity data on a country-specific basis.

The meted uses global emission data sets and detailed information from the land cover (land use, population and road km.) to perform a multiple linear regression between these data sets by using "reference" maps and two options: 1) Option EMEP and 2) Option GEIA. Software uses GRADS software tool for visualizing and projection assistance. The Grid Analysis and Display System (GrADS) is an interactive desktop tool that is used for easy access, manipulation, and visualization of earth science data. The format of the data may be either binary, GRIB, NetCDF, or HDF-SDS (Scientific Data Sets). GrADS has been implemented worldwide on a variety of commonly used operating systems and is freely distributed over the Internet. GrADS uses a 4-Dimensional data environment: longitude, latitude, vertical level, and time. Data sets are placed within the 4-D space by use of a data descriptor file. GrADS interprets station data as well as gridded data, and the grids may be regular, non-linearly spaced, gaussian, or of variable resolution. Data from different data sets may be graphically overlaid, with correct spatial and time registration. Operations are executed interactively by entering FORTRAN-like expressions at the command line. A rich set of built-in functions is provided, but users may also add their own functions as external routines written in any programming language.

EMIMO has also a Tcl/ Tk visual interface to manage all above information in a visual way.

4.3.3 Results

EMIMO has to be configured to define the model domain we are interested on. The model has in the configuration file the capability to define the spatial resolution and the period of time we need with a time resolution of 1 hour. The spatial resolution is defined always in geographic coordinates and the highest spatial resolution is set at 0,01 degree. Fig. 4.5 shows an example of the combination of GRADS and the visual Tcl/ Tk tool to visualise the emission data before exporting it to a data file.

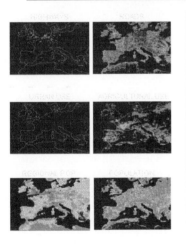

multiple regression process

9 distribution variables :

♦ **3 Roads:**

   Digital Chart of the Word (Pennsylvania State University)

♦ **4 Land uses:**

   USGS (*U.S. Geological Survey*)

♦ **2 Population:**

   • CIESIN (*Centre of International Earth Science Information Network*)

   • CGEIC (*Canadian Global Emission Interpretation Centre*)

Fig. 4.5. EMIMO friendly user visual interface in Tcl/Tk and where GRADS mapping tool is incorporated for assistance on projection and map visualization.

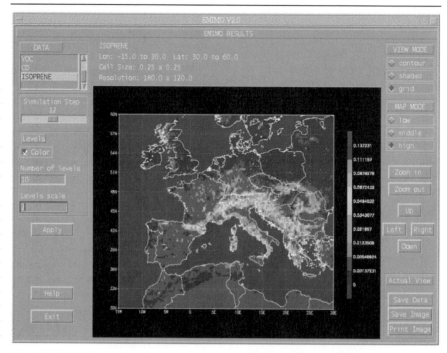

Fig. 4.6. EMIMO uses three data sets to calculate the appropriate parameters of the multiple linear regression approach.

4.3.4 Summary

We have developed an emission model, EMIMO to calculate spatial and temporal emission pollution data. The output of the model is particularly appropriate for being used into state-of-the-art Air Quality Modelling Systems such as MM5-CMAQ. The model has a general top-down approach which allows the user to calculate emission data sets over any area in the world with a temporal resolution of 1 hour and spatial resolution up to 0,01 degree. Due to the poor information on industrial emission global data sets and particularly poor information on DCW of industrial locations in the world, it is recommended to add industrial emissions manually by the user for a particular model domain application by collecting local data.

4.4 MIMOSA: a road emission model using average speeds from a multi-modal traffic flow model.

N. Lewyckyj, A. Colles, L. Janssen, C. Mensink

4.4.1 Context and objectives

A road emission model (Antigoon) was earlier developed and implemented for the city of Antwerp (Mensink 2000; Mensink et al. 1999, 2000a and 2000b). The model was based on the 'static' approach proposed in Copert-II (Ahlvik et al. 1997) (i.e. using hourly average speeds) and was limited to six pollutants: CO, NO_x, SO_2, NMVOC, PM and Pb. According to the EMEP/CORINAIR methodology (EMEP 1996) thirteen vehicle types (regrouped in three main categories PC, LDV and HDV) were identified, together with three fuel types (diesel, gasoline en LPG). Six classes of road types were implemented and average speeds for each vehicle category were assigned to each type of road. Hot emissions as well as cold emissions were simulated, including the evaporation emissions for the NMVOC. Total NMVOC emissions were split according to Veldt (1992). Distribution of the vehicle amount on the road network was based on electronic counting provided by the police of the Antwerp city.

   With the publication of the Copert-III (Ntziachristos and Samaras 2000) and the MEET (1999) documents it was decided to improve (*state of the art*) the former version of the emission model and to extend its application area to all Flanders. The new MIMOSA model was therefore developed.

4.4.1.1 Main features of the MIMOSA model

The MIMOSA software is a combination of Fortran-77 and Fortran-90. The program is structured in a modular way allowing easy changes in both input parameters and subroutines dealing with calculation of the emission factors.

4.4.2 Methodology and considered pollutants

The MIMOSA road traffic emission model also uses the 'static' approach, i.e. hourly average speeds of the different vehicles instead of short-term fluctuations in speed (accelerations and decelerations) as proposed in the INFRAS-UBA/ BUWAL methodology (BUWAL 1998, 1999). However, option in the MIMOSA model offers the user the possibility to perform simulation using a 'dynamic' version, provided that the required data are available (see section' Static *versus* dynamic versions').

The emission factors used were partially extracted from experimental data collected by Vito ('on-road' measurements) as well as from the Copert-III document. For missing data (some specific pollutants) emission factors from MEET were implemented. Distinction was also made between hot emissions, cold emissions and emissions due to evaporation (diurnal losses, running losses).

The number of pollutants initially considered in Antigoon was extended to sixteen: CO, NO_x, VOC, PM, SO_2, CO_2, N_2O, CH_4, NH_3 and the heavy metals Pb, Cu, Cr, Ni, Se, Zn and Cd.

4.4.2.1 Definition of the vehicle fleet composition

Five classes of vehicles are distinguished (Passenger Cars, Light Duty Vehicles, Heavy Duty Vehicles, Motorcycles and Mopeds) with a further subdivision depending on the age of the vehicle and its cylinder capacity for the passenger cars and for the two-wheels vehicles, while age and weight criteria are used to distinguish LDV and HDV (including busses and coaches). The operated classification is analogue to the sub-classification proposed in the Copert-III methodology, however limited to 105 classes in total. Distinction is made between four fuel types (gasoline, diesel, LPG and 2-stroke gasoline), with lead and sulphur contents depending on the year of the simulation. The road network is divided in six road classes (highways, national roads, main roads outside the city, main roads inside the city, secondary roads and harbour area). A specific year-dependent fleet composition is associated to each road class.

The relative fleet composition has been updated for the years 1996 to 2000 using the most representative and actual data. Estimated vehicle park compositions for the years 2005 and 2010 were also implemented.

4.4.2.2 Coupling the MIMOSA model with the TRIPS/32 traffic flow model

The input for the MIMOSA model consists of the output of the TRIPS/32 road traffic flow model distributed in Belgium by TRITEL N.V. The TRIPS/32 model is a peak-hour model (17h00-18h00) which uses more than 8000 segments for the definition of the main roads in Flanders. TRIPS/32 offers also the possibility to take into account all the roads of one province at once. A network covering the totality of the Flemish road network (five provinces) is actually under construction.

A road network data-file is constituted of the Lambert coordinates of each road segment, together with the number of vehicles on this segment and their hourly average speed. Realistic speed limitations were however implemented for the HDV as only one average speed is used for all vehicles on a particular road section. The average trip length used for the calculation of the cold emissions is automatically determined from the data-file. The major difference between the MIMOSA model and the Copert-III software lies in the fact that MIMOSA uses different speeds for the different road segments instead of using generic average values per road type as proposed in Copert-III.

The MIMOSA model can work with any network data-file exhibiting the same structure and is thus totally independent of the network definition. Moreover, the MIMOSA model can directly be linked to multi-modal models as the TRIPS/32 model, allowing the user to test the effect of mobility scenarios on the emissions.

Finally, the time-dependency of the emissions is simulated using normalised distribution factors expressing the fluctuations of the traffic flow as a function of the hour of the day, the day of the week and the month of the year. These distribution factors are directly derived from vehicle counting carried out in Belgium. Emission calculations are performed for each road segment separately, hour per hour. The model can thus be used from the street level to the country level, for time periods fluctuating between one hour and several years.

4.4.2.3 Meteorological data

Beside the different time-dependent parameters already mentioned here above, an important input parameter for the calculation of the cold emissions concerns the meteorological data. Hourly temperature (averaged for all Flanders) is provided for the years 1996 to 2000 and can also be used for simulations performed in the future (years 2005 and 2010).

4.4.2.4 The scenario manager

A scenario manager was implemented to allow the user to change some parameters affecting the simulation results. As for now, default values are always provided but the user is allowed to change:
- the simulated period and the units for expressing the results,
- the definition of the road network,
- the type of emissions to be calculated (with or without cold emissions),
- the fleet composition (relative contribution of the different vehicles classes),
- the percentage of PC as a function of the cylinder capacity,
- the relative vehicle distribution per fuel type, and
- the relative vehicle distribution per road type

Furthermore, thanks to the coupling of the TRIPS/32 traffic flow model to the MIMOSA model, all options originally foreseen in the TRIPS/32 model are still possible. It is therefore possible to change both the maximum occupancy capacity of the different road segments as well as the maximum allowed speed of the different vehicles on these road sections. Specific roads can be supposed to be closed and the future construction of new roads can also be tested.

4.4.2.5 Output files

The output produced by the MIMOSA program consists of five different files, the largest one providing emission data (hot, cold, evaporation and total emissions) integrated over the whole simulation period for each road segment for the 16 pollutants. However, if desired, a supplementary file containing the same information can easily be created on an hourly base

4.4.2.6 The graphical user-interface

A user-friendly interface was developed under ARCVIEW 3.2 offering to the user different options to visualise (and print) the results using GIS. These options concern:

- the choice of the pollutant(s),
- the type of emissions (hot, cold, evaporation or total),
- the road category to take into account (one or more), and
- the choice between expressing the results using road segment(s) or per area units (emissions integrated per km^2)

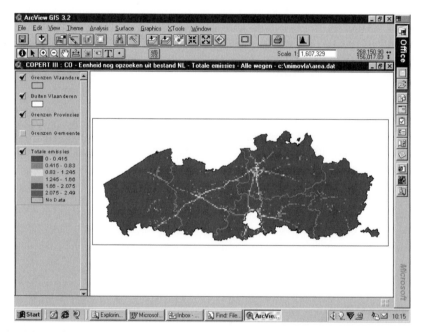

Fig. 4.7. Total CO emissions per km^2 at the level of Flanders for a specific simulation period

All options offered by the ARCVIEW software remain valid. As for now, visualisation of the emissions for each separated vehicle category is not yet implemented. As an example, Fig. 4.7 represents the CO total emissions per km^2 at the level of Flanders for the considered period.

4.4.3 Static *versus* dynamic versions

An optional approach offered by the MIMOSA software deals with a more dynamic simulation of the emissions. This approach is based on a simplified version of the INFRAS-UBA/BUWAL methodology. The emission factors are not expressed through speed-dependent functions but via arrays containing for each vehicle type and each considered traffic situation specific emission factors. Despite the simplified version implemented in MIMOSA (only 9 different traffic situations) this approach requires much more memory space for the program. However it has the advantage to run much faster then the 'static' version. Moreover, the temporal accuracy is much higher allowing the user to simulate acceleration and deceleration processes (provided that the required speed information is available).

4.4.4 Advantages of the MIMOSA software and further improvements

As compared to the Copert-III software, the MIMOSA model presents four main advantages:
- the calculations are realised on hourly basis. This fact allows to make simulations for any kind of time interval (from one hour up to several years),
- the calculations are realised separately for each segment constituting the road network, allowing the visualization of the results with maps,
- the obtained results are more realistic since the emissions are based on vehicle speeds provided by a traffic flow model instead of using generic values,
- the emission model is coupled to a traffic flow model that allows quantifying (in terms of emissions) the effect of mobility scenarios.

The major disadvantages of the MIMOSA model lies in the required runtime (about 8 hours on a Pentium-4^{TM} for one year simulation at the level of Flanders) and memory space (several Mbytes per run). Moreover, at this stage, only one single speed is used for the different vehicle categories on one specific road section. Further improvement consists thus in using different speeds for the different vehicles. A visualization of the results per vehicle category would also be an asset.

4.4.5 Comparison exercise and sources of discrepancies

The emissions for the main roads of Flanders for the year 2000 were calculated with both the MIMOSA model and the Copert-III software. Using identical input data (i.e. generic values for the speeds) for both models leads to very similar output results. However, when simulations were performed with MIMOSA using the speeds provided by the TRIPS/32 traffic flow model as input, discrepancies up to 50% of the total annual emissions were obtained for pollutants which are highly speed-dependent (like NO_x and PM). Oppositely, similar results were obtained for pollutants which are relatively speed independent like NH_3.

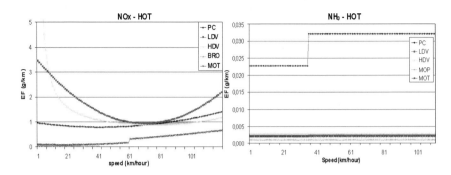

Fig. 4.8. Speed dependency of the integrated (over the whole vehicle park) hot emission factors for NO_x and NH_3 (year 2000) for the five vehicle categories.

Knowing that MIMOSA uses the full range of speeds while Copert-III is limited to one generic speed per road type (e.g. 25 km/h in urban area and 105 km/h on highway for PC), the explanation for the obtained differences is straightforward. Indeed, Fig. 4.8 shows the speed dependency of the integrated hot emission factors (over the whole vehicle park for the year 2000) for NO_x and NH_3 for the five vehicle categories. Looking to the major contributors to the total emissions (PC, LDV and HDV), big differences in NO_X emission factors are observed for small variations at both high speeds (between 105 and 120 km/h) and low speeds (between 20 and 40 km/h). The use of correct speeds for PC, LDV and HDV on both highway and in urban area is thus critical for correct calculations of the NO_x emissions. This is not the case for NH_3 as the integrated emission factors are nearly speed independent for the different vehicle categories.

4.4.6 Summary

The emissions factors used in MIMOSA are based on the most recent available data and the model takes into account changes of motor technology. The number of pollutants has been extended to 16, including now 7 heavy metals. The motor and mopeds categories have been added and the 2-stroke fuel was implemented, bringing the number of different vehicle types to 105. The definition of the vehicle park has been updated and expected distributions for the years 2005 and 2010 have been implemented. However, the major improvement realised concerns the coupling of the model with the multi-modal TRIPS/32 road traffic flow model. This last option allows simulating the effect of mobility scenarios on traffic emission patterns. Both temporal (from one hour up to several years) and spatial (from the street level to the country level) variations can now be visualised using the user-friendly interface. A comparison exercise performed between the MIMOSA model and the Copert-III software points out the importance of the speed used for the different vehicles on the different segments constituting the road network.

4.5 Improvement and application of methodology and models to calculate multiscale high resolution emission data for Germany and Europe

S. Reis, P. Blank, B. Wickert and R. Friedrich

4.5.1 Introduction

For the development of air quality improvement strategies emission data in high temporal and spatial resolution, covering all emission sources and all relevant pollutant species, is needed. Usually computer aided models are used to provide this emission data, because it is not possible to measure all. A large amount of data has to be handled. For this purpose the new emission-model CAREAIR has been developed. Modelling technique and data structures of this emission model are described in the following. The most important part of the model is a database containing data and specifications for possible data manipulations. The information in the database is specified using sets of references to attributes of different categories. Data manipulation is performed using flexible programs which are mostly controlled by their actual input data. The emission calculation is carried out combining several executions of these data manipulation operators. The emission model allows the determination of emissions of different pollutants in a high spatial and temporal resolution, e.g. hourly values for a 1km x 1km-grid for any investigation project. CAREAIR is in fact more a flexible modelling toolbox than just a single emission model. CAREAIR was and is further used for preparing emission data for various investigation areas including parts in Germany in detail, Germany, Europe and other regions.

4.5.2 Development of CAREAIR/ECM

Air pollution problems are existing on global, regional and local levels. Despite successful clean-air policy during the last years, further emission reduction measures have to be implemented. This means that complex planning tasks have to be carried out under various social, political, economical and technical conditions, which require scientifically based decision support (see Obermeier et al. 1994; Müller et al. 1990; Laing 1991).

As measures to reduce air pollution become more and more expensive, it is important for the acceptance to show, that the envisaged environmental objectives can be achieved by these measures and that they can be fulfilled with the least costs possible. Scenario analyses should be performed to assess the potential and the costs of possible emission reduction measures. Final objective is to determine air-quality improvement strategies which are cost-efficient.

Detailed information about emissions and source structure is needed to provide background knowledge for planners and decision makers. Concerning regional and local air-quality in particular, data on emissions of various species in a high spatial and temporal resolution is necessary, to provide input for models that analyse atmospheric transport and photochemical transformation processes. Atmospheric models for the calculation of ozone concentrations need hourly values of NO_x, CO and VOC (Volatile Organic Compounds). The VOC-emissions have to be disaggregated into more than 30 substances and substance-classes. A spatial resolution in grid cells is necessary, e.g. for total Europe as investigation area typically cell sizes of 30 km x 30 km are used. For smaller regions cells of 1km x 1km are typical, e.g. the amount of cells in the grid used to cover Baden-Württemberg is over 70.000. For an episode of one week (168 hours) up to 500 millions of single emission values have to be provided.

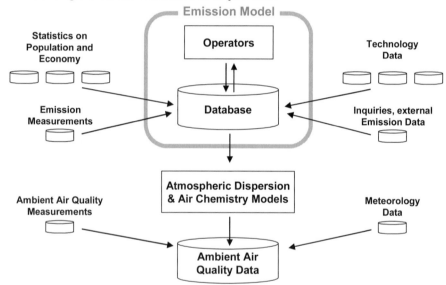

Fig. 4.9. Data structure of CAREAIR

Evaluation of existing emission models e.g. GLOED (1992), MINERGG (1995) revealed limitations in particular concerning their capabilities for analyses in high temporal resolution and their flexibility with regard to spatial resolution and the structure of source-categories. Such limitations of existing emission models have been the motivation to develop the new emission model CAREAIR.

CAREAIR is the acronym for "Computer aided Analysis of Reduction strategies for Emissions and ambient AIR pollution". In the further development stage, CAREAIR was transferred into a modular architecture and termed ECM (Emission Calculation Module).

An overview is given in Fig. 4.9. CAREAIR/ECM is an integrative information system for all air pollution analysis and clean-air planning tasks. Major components are a database with data base management system including scenario management and a modelling system, which first part is a flexible emission model, developed at the IER. Secondly several atmospheric dispersion and transformation models are coupled to the emission model. This enables to estimate changes of concentrations and deposition rates of pollutants resulting from emission reduction measures.

In the following the emission model is described in detail.

4.5.3 Model structure

4.5.3.1 Modelling technique and data structures

It is the aim of the emission model to calculate emission data as described in Chap. 2. This is usually done by first calculating activity data (e. g. energy consumption, solvent use, traffic densities per road segment a. s. o.) from available e.g. statistical data and next multiplying these activities with appropriate emission coefficients. Following steps concern the enhancement of the spatial and temporal resolution of the emission data. Certain requirements concerning the efficiency of the computing are due to the fact, that very large amounts of data have to be handled.

4.5.3.2 Basic concepts of the emission model

An integrating data model and several programs to manipulate the data have been developed. The CAREAIR data model gives flexibility to incorporate any data from external data sources and to meet the necessities for different calculation methods within several projects. There are two basic concepts.

1) A shared database serves as information system and for modelling purposes. All data are described using a combination (set) of several attributes of different categories. Algorithms specifying possible data manipulation are considered being "data" as well and they are part of the database. The description of algorithms is implemented using two or more attribute-sets, possibly including parameters of equations. In most cases such an 'algorithm table' defines, which input data to combine to which output (the variables) while the principle algorithm (the equation) is defined implicitly.

2) A modular pool of data manipulation tools is provided in form of executable programs and a function library. Flexible control of these elementary operators is primarily given by the combined data- and algorithm tables ('table' in the sense of logical combined part of the data base, physical representation could be different). Secondly various execution control parameters influence the behaviour of the operators. Modelling calculations are performed combining these operators.

4.5.3.3 The attributes

Sets of attributes are used to specify and characterize data and algorithms. A set consists of the actual instance for at least one or more among the attributes. Hereby numerical or character-based 'keycodes' are used to refer to attributes in a data table. The keycodes are defined in attribute catalogue tables. The most important attribute categories are shown in Table 4.5.

Table 4.5. Categories of attributes in CAREAIR

Category	Attribute	Description
GEO (spatial specification)	ARE, REG, CLZ, LIN, PNT, CEL (GRD)	Area, Region, Climate Zone, Line, Point, Cell (defined within a Grid)
TMP (temporal specification)	DTE, PER, TMZ	Date, Period, Time Zone
PRO (emission-relevant process specification)	SEC, SIZ, PUR, TEC, SUB, POL, DIM	Sector, Size-Class, Purpose, Technique, Substance, Pollutant, Dimension (incl. Unit)
VAL, DOC (value(s) and documentation)	VAL/ VEC, QLT, ORI, OWN, VER, CMT	Value/Series of Values, Quality-Classification, Data Origin, Owner, Version, Last Update, Textual Comment

4.5.3.4 Data structures

Data is divided into several classes. Activity data forms the first class. Activities are described using one attribute set. Examples are population, employees, amount of goods produced, agricultural area, amount of registered cars, energy consumption as well as the calculated emissions. Coefficients build the second class. They are described using two attribute sets (cross-references), because they are defined as quotient of two activities. According to this coefficients are data manipulation specifications. Examples are energy consumption per capita, solvent contents, and in particular all emission coefficients. The third class of data consists also of cross-references of two attribute sets, however without a value. Such cross-references are e.g. used to specify hierarchies.

4.5.3.5 Elementary operations

An analysis of data operations that are necessary to generate emissions data in high temporal and spatial resolution revealed, that all data manipulation can be comfortably carried out by combining only four elementary operations. These are 1) *select*, 2) *sort*, 3) *disaggregate* and 4) *aggregate*, which are to be performed in several variations on 1) *spatial*, 2) *temporal*) and 3) *process-sub-categories* (see Table 5). Selecting and sorting options are provided by commercial relational data base management systems, however their possibilities for disaggregating and aggregating are not sufficient to meet the needs for the emission calculation in high resolution. Therefore CAREAIR operators emphasize these two functionalities. The majority of dependencies is linear.

The basic form of the disaggregation is the linear transformation as shown in the following equation:

$$z_{ij} = c_{ij} * x_i \qquad (4.2)$$

The activities $x(j)$ are transformed to activities $y(i,j)$ according to the coefficients $c(i,j)$. The result from one value $x(j)$ consists of multiple values $y(i,j)$, one according to each of the transformation coefficients $c(i,j)$. Coefficient cross-references include this equation implicitly, they are algorithm tables.

The basic form of the aggregation is the summation: $\quad z_l = \sum_k (y_{k,l}) \qquad$ (4.3)

The selection of the values $y(k,l)$ that have to be summed up and the characterization of the result $z(l)$ is specified using an algorithm table, whose records consist of two attribute sets each (a cross-reference without coefficient). If several values $y(i,j)$ after a data manipulation (e.g. disaggregation) turn out to have identical attribute sets, linear super-position is assumed, so no explicit specification in an algorithm table is necessary in such cases. This combination of disaggregation and linear super-position is the standard-case for a complete disaggregation.

4.5.3.6 The operator coefficient-disagg

The operator coefficient-disagg (*cdisagg*) is applied for several purposes during the emission model. One example is the basic equation "emission is equal to activity multiplied with emission coefficient". Other applications are e.g. the calculation of solvent-use proportional to the number of employees, the split of vehicle-kilometres into several vehicle-classes or the disaggregation of total-VOC-emissions into various VOC-substance-classes (VOC-Split-calculation, see *Chap. 3*). Fig. 4.10 illustrates the logical data model and data access of the operator *cdisagg*. An input activity-table is scanned producing the result activity table using the coefficients from the coefficient cross-reference table.

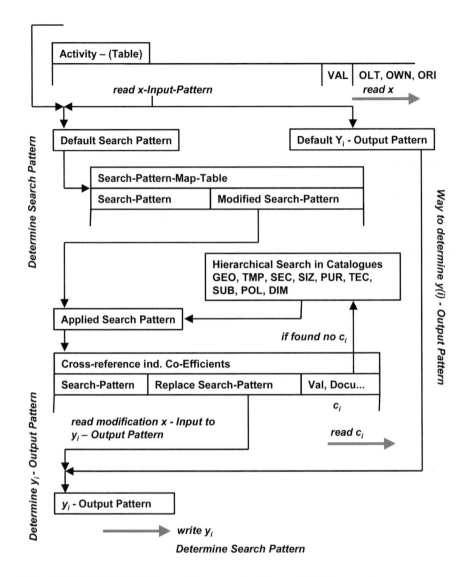

Fig. 4.10. The CAREAIR data model – a search-pattern controlled approach

The graph sketches the approach for one activity record. The attribute set (x-pattern) of each input record determines the search-pattern used to determine which coefficients c_i to apply. However before searching the search-pattern can be modified, either through command-line options, or through a search-pattern-map-table or according to hierarchies within each attribute category. The x-pattern also determines the default output attribute set (y-pattern), which is further modified according to the information found with the coefficient applied (replace-pattern).

4.5.3.7 Aggregation operators

For several purposes operators for aggregations are provided. Aggregation can be performed based on geographical-, temporal- or elements of the process-attributes. The standard form is summation of values, however other mathematical operations are useful too, e.g. to determine related data like per-capita-figures.

4.5.3.8 Pre-processing for temporal and spatial resolving

Emission sources are usually analysed on base of yearly data for all emission relevant processes and pollutants, geographically referring to territorial units like communities (*area*), road segments (*line*) and point-sources (all is spatially vector-oriented data). The temporal resolution is enhanced combining real and typical time-curves of different resolutions. Afterwards a transformation to spatially grid-oriented data is necessary e.g. to serve as input for atmospheric models, which need emission values for each pollutant, each cell of a grid and each hour of the period considered.

Most operations for spatial or temporal disaggregations are following linear approaches similar to cdisagg. However, in particular the temporal disaggregation needs various pre-processing to determine "year-in-hours" time-curves holding "hour-shares" which are finally used to resolve yearly emissions. Partly non-linear dependencies are to be applied, e.g. temperature dependencies, using operators, together called *tdisagg* (time-disagg). Geographical information is pre-processed with the help of a Geo-Information-System to produce grid vectors holding so-called "geo-to-cel-shares", e.g. area of cell in - per area of a community.

4.5.3.9 The operator matrix-disagg

After the pre-processing, as well the transformation of vector-oriented data to grid-oriented data as the temporal disaggregation of yearly data to e. g. hourly data are performed following linear transformations with super-position. Due to memory restrictions it was found to be most efficient to combine the two operations. For this purpose the matrix-disagg operator (*mdisagg*) was developed. Its functionality is sketched in Fig. 4.11.

The operator mdisagg can handle activity-data (y) (e.g. yearly emissions) in various spatial resolutions and of all process-sub-attributes in detail. "Year-in-hours" time-curves ("hour-shares" a) and grid coefficients ("geo-to-cell-shares" b) as much as necessary for all activities specified via any attribute-references are read in and applied. Which data to combine is selected using a similar search-pattern controlled approach as explained for cdisagg. The summation of all contributions to the emissions of each pollutant (m), for each hour (h) and each cell (g) of the grid is performed for all spatial-(j) and process-(l) attributes of the input data. The resulting values from mdisagg are computed according to the following equation:

$$z_{h,g,k,m} = \sum_{j,l} \left(a_{h,j,k,l} * b_{g,j,k,l} * y_{j,k,l,m} \right) \qquad \text{(4.4)}$$

Mdisagg performs the last step of the emission calculation in high resolution and provides the adequate interface to couple atmospheric dispersion and transformation models.

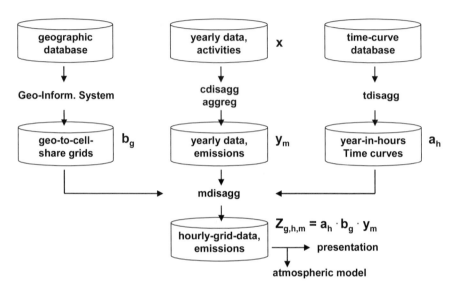

Fig. 4.11. CAREAIR Matrix-disagg illustrated

4.5.3.10 Software-technical aspects

The system has been implemented on RISC-Workstations based on the RDBMS (Relational Data Base Management System) Oracle[TM] and the GIS (Geographical-Information System) Arc/Info[TM]. The database follows a Client/ Server architecture and access is possible from Unix- and Standard PC-Clients. The operators have been implemented using the programming language C/C++. Memory is dynamically allocated only for minimum data necessary. Additionally SQL, various Unix tools, shell-scripts and pipes are used. Results of the models can be prepared for multimedia representations and are accessible for project partners via Internet.

4.5.4 Examples of applications

Within the project EUROTRAC-TRACT (Transport of pollutants over Complex Terrain) hourly emissions in a 1 km * 1 km grid have been determined for Baden-Württemberg and surrounding areas (parts of Hessen, Rheinland-Pfalz, Alsace,

Switzerland, see Friedrich et al. 1994). Within the EUROTRAC subproject GENEMIS the calculation of emission data for all European countries and in particular the enhancement of the temporal resolution of the data was performed /9/. Within the project SANA the area of the former GDR is investigated /10/. The figure below gives an example of a graphical representation, here the NOx-emissions on September 4th 1991, 16:00-17:00 hours. Major contributions originate from power plants and traffic. The example shows that CAREAIR is currently being applied successfully assessing air pollution problems.

4.5.4.1 Calculation of the FLUMOP and BERLIOZ cases

For the evaluation of 8 atmospheric dispersion models, spatial, temporal and substance high resolution emission data was generated on 12 different grid types in different projections and provided to modelling groups. Both evaluation cases, FLUMOP (for July 1994) and BERLIOZ (July 1998) focussed on the Greater Berlin area. (see Fig. 4.12)

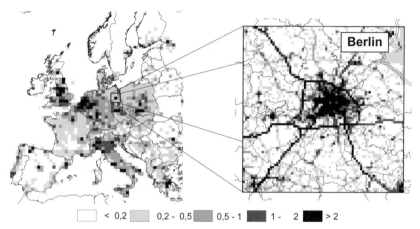

Fig. 4.12. Emission calculation for the TFS Evaluation Group: NOx-Emissions on July 20, 1998 at 6^{00} UTC in kg/km^2

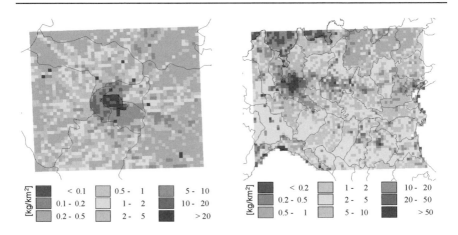

Fig. 4.13. NO_x Emissions on Sept. 1st 1998 around Paris 11^{oo} - 12^{oo} UTC (left) and CO Emissions on Sept. 1st 1998 in North Italy 11^{oo} - 12^{oo} UTC (right)

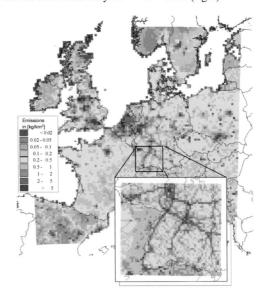

Fig. 4.14. CO Emissions on July 21st 1998 in Central Europe 6^{oo} - 7^{oo} UTC and details for the German Federal State of Baden-Wuerttemberg

4.5.4.2 Calculation of European emissions for different spatial scales

The above Figures give a variety of examples of emissions calculated with CAREAIR/ECM as input for CTMs, e.g. for AIRPARIF (Fig. 13) and other air quality related studies.

4.5.5 Conclusions

Emission data sets and their fast and flexible calculation remain to be vital for all modelling and evaluation issues. By improving the software and approaches for this calculation, it is now possible to provide data sets with up to several Giga-bytes of data within an acceptable time frame. CAREAIR/ECM is currently applied to provide emission data for the EURAD[1] forecasting system and is being further developed and improved.

---

[1] http://www.eurad.uni-koeln.de/index_e.html, 2003/04/02

4.6 Developing emission inventories for Turkey

A. Müezzinoğlu, T. Elbir, F. Dinçer, A. Bayram, M. Odabasi, E. Cetin, R. Seyfioglu

4.6.1 Overview

Within the scope of the EUROTRAC Subproject GENEMIS, activities to develop emission inventories for Turkey have been conducted. In this, comprehensive methods have been developed for specific regions of Turkey, for instance to cover the surroundings of the city of Izmir. These methods were developed by the partial sponsorship of TUBITAK, the EUROTRAC member representing Turkey. While a first national inventory for the whole Turkey was made available during the first part of GENEMIS, a more detailed inventory of emission sources with source characteristics based on regionally applied GIS was to be generated as a next step. For this approach, specific regions were selected, e.g. the city of Izmir, the Aegean and the western part of the Mediterranean regions on the basis of 1:25000 maps. A GIS based information recording was started and inventories for these regions included the largest point sources (such as large industries and power plants), domestic heating and traffic sectors.

Here, the most prominent results of these activities are described.

4.6.2 Introduction

The national emission inventory of Turkey was under study beginning with 1990's. However, the first results were published for the base year 1996 (Muezzinoglu et al. 1997a) followed by another update for the base year of 2000 (Elbir et al. 2000). But a detailed inventory base was started in the second phase of GENEMIS to show the local details in a computer-based retrieval system. These efforts started in the metropolitan city of Izmir, the first settlement for which the detailed emission inventory for 1996 was prepared. Izmir is the third largest city of Turkey with sizeable industrial and tourism activities. Similar studies in the Mediterranean region have started since 1998.

Throughout the study, regions have been divided into individual provincial areas to be able to collect basic information. Provinces in Turkey are the individual units for which many detailed statistics are recorded and made available. Fig. 4.15 shows the regions of Turkey. Fig. 4.16 shows the same regions with the names of major provincial centres indicated.

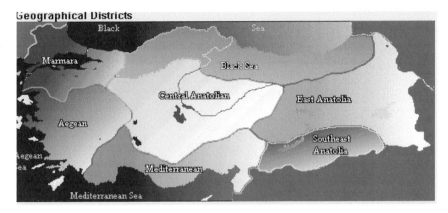

Fig. 4.15. Regions of Turkey

Fig. 4.16. Location of major provincial centres of Turkey

4.6.3 First steps towards an improved emission inventory

In all provincial areas the emissions which were calculated with respect to four pollutant types (SO_x, NO_x, NMVOC and CO) were evaluated in the point, area and line source categories.

4.6.3.1 Point sources

In point source category, almost all major industries were covered. Emissions from industries were calculated for individual provinces that were studied. It was assumed that no domestic heating source is large enough to be classified as a point source. Industrial emissions were estimated using fuel consumption data and approximate emission factors. For the calculation of emissions, Corinair Emission Factors were mostly used.

Throughout the study a questionnaire was distributed to collect the production records and other important data belonging to the industry. This questionnaire included questions concerning working hours, shifts, number of employees, types, quantities and utilization techniques of fuels in the plant, numbers and capacities of boilers, process description, production rates, power demands, stack properties, quantity of liquid and solid wastes removed from the plant etc. (Müezzinoğlu et al. 1997a).

4.6.3.2 Area sources

In this category, domestic heating sources as well as small sized industries and workplaces gathered in what is known as "industrial zones" were evaluated. Their apportionment on the grids applied to the GIS maps is based on population data belonging to the provinces. Emissions were calculated using Corinair emission factors and fuel consumption data taken from their municipalities.

4.6.3.3 Line sources

Emissions from urban traffic were evaluated as area sources in the study. This assumption is based on the urban driving characteristics and very low speeds due to generally congested traffic.

Gasoline and diesel fuels used in the traffic sector were taken from the refinery sales statistics. Then, gasoline and diesel fuel consumption data in all provinces were obtained from fuel sale statistics of individual petroleum companies inside the cities. The differences between the two were attributed to the mobile source emissions at the intercity roads. Corinair emission factors based on amount of traffic fuel use were used to calculate the total emissions. Then, these figures were distributed among major intercity roads on the basis of traffic counts made by the State Highways Department.

Urban traffic fuel consumptions on the other hand were distributed into cities with respect to population density, assuming that higher population per square grid generates more travel requirement and higher fuel consumption Then, emissions per grid were calculated using urban traffic Corinair emission factors on the basis of unit quantity of fuel combustion.

Traffic emission sector contains both the intercity source emissions distributed over the delineation of major highways and urban area source emissions distributed over the grids.

Calculated total sectoral and seasonal (both winter and summer) emissions of Izmir are summarized in Tables 4.6 and 4.7 (Müezzinoğlu et al. 1998).

Table 4.6. Winter emissions by sectors in Izmir, tons per day

Sector	SO$_x$	NO$_x$	VOC	CO
Industry	370.90	28.91	34.27	35.47
Domestic heating	49.50	6.07	8.95	11.04
Traffic	6.60	61.27	18.92	165.79
Total	427.00	96.25	62.14	212.30

Table 4.7. Summer emissions by sectors in Izmir, tons per day

Sector	SO$_x$	NO$_x$	VOC	CO
Industry	370.90	28.91	34.27	35.47
Domestic heating	-	-	-	-
Traffic	9.63	86.18	31.62	263.92
Total	380.53	115.09	65.89	299.39

Within the scope of GENEMIS, emission inventories of provinces have been also prepared in western part of the Mediterranean region. Table 4.8 shows the sectoral emission obtained for the cities of Burdur, Isparta and Antalya.

Table 4.8. Sectoral emissions in the provinces of *Burdur*, *Isparta* and *Antalya*, tons per day

	ISPARTA			
Sectors	SO$_x$	NO$_x$	NMVOC	CO
Industry	6.70	1.30	0.23	0.11
Domestic heating	5.72	0.70	1.03	1.28
Traffic	0.14	0.92	0.96	11.70
	BURDUR			
Sectors	SO$_x$	NO$_x$	NMVOC	CO
Industry	4.13	0.84	0.17	0.09
Domestic heating	5.87	0.67	0.84	1.06
Traffic	0.09	0.50	0.45	6.33
	ANTALYA			
Sectors	SO$_x$	NO$_x$	NMVOC	CO
Industry	0.93	0.19	0.04	0.02
Domestic heating	0.98	0.10	0.11	0.14
Traffic	0.45	0.86	4.39	57.42

For other major cities such as Istanbul and Ankara, different project groups in Turkey have prepared emission inventories. Following are taken from their publications. For example, Tayanç (2000) generated annual SO$_2$ emissions with respect to fuel types and population change in Istanbul for 1980-1990 period. Table 4.9 shows these annual emissions. Atımtay et al. (1995), studied emissions of Ankara for domestic heating and traffic sectors in 1992, as indicated in Table 4.10.

Table 4.9. Sulphur dioxide emissions generated by different fuel use and population change in Istanbul, tons per year (Tayanç 2000).

Years	Coal [tons SO$_2$]	Fuel-oil [tons SO$_2$]	Diesel [tons SO$_2$]	Gasoline [tons SO$_2$]	Total [tons SO$_2$]	Istanbul's Total Population
1980	33 516	54 154	2 070	486	90 226	4 741 890
1981	40 918	57 356	2 060	473	100 807	4 962 109
1982	46 278	66 484	2 308	511	115 581	5 182 329
1983	55 635	61 050	2 597	550	119 832	5 402 547
1984	77 560	60 331	3 031	544	141 466	5 622 766
1985	87 037	67 094	3 450	533	158 114	5 842 985
1986	117 521	70 062	4 691	592	192 866	6 136 226
1987	123 626	86 879	5 591	692	216 788	6 429 467
1988	119 159	92 350	5 981	786	218 276	6 772 708
1989	163 963	86 753	5 608	835	257 159	7 015 949
1990	246 641	79 958	5 882	880	333 361	7 309 190

Table 4.10. Sectoral emissions of Ankara in 1992, tons per year, (Atımtay et al. 1999).

	Industry	Domestic heating
CO [ton / year]	10 965	24 465
NO$_x$ [ton / year]	365	2 531
C$_n$H$_m$ [ton / year]	3 492	27 028
SO$_2$ [ton / year]	4 139	7 006

4.6.4 First results

When calculated emissions for the Aegean and Mediterranean regions already submitted were added up and compared with Turkish national emissions (Müezzinoğlu et al. 1998), it could be seen that the regional detailed emission inventories completed so far have covered only 4% of all national SO$_2$ emissions calculated earlier. Table 4.11 shows emission estimates of Turkey for 1985 – 2005.

Hence, the portion of SO$_2$ emissions that could be detailed in this study was small compared to the overall national SO$_2$ emissions. But if Istanbul and Ankara inventories were added to this figure, altogether inventoried provinces account for 20% of the national emissions. When one considers that provincial emissions excluded power station emissions accounting to more than 55% of the national SO$_2$ release, and add this 20% to this figure, only about 25% of the emissions are left unaccounted for. This 25% portion should be emitted from all other provinces and industrial zones in Turkey. Major problem areas making up these SO$_2$ emissions are Kocaeli and Adana-Iskenderun industrial regions, some ore smelter zones at the Black Sea coastline. Also it is thought that some harsh climate urban centres with high winter emissions at the middle and eastern Anatolia must be taken into consideration.

Table 4.11. Air pollutant emission estimates of Turkey calculated at five year intervals, tons per year (Müezzinoğlu et al. 1998b)

Years	PM	SO_x	NO_x	VOC	CO
1985	4 374 314	2 024 038	452 928	387 948	881 697
1990	4 932 347	2 560 088	640 244	432 950	1 378 372
1995	5 957 556	2 978 326	727 417	428 719	1 567 469
2000	6 974 196	3 501 711	844 887	448 495	1 821 720
2005	7 963 148	4 002 879	974 094	471 106	2 091 610

4.6.5 Emission inventory of the Aegean region in Turkey

In order to be able to solve significant air pollution problems in the country, dependable figures to show the main categories of pollution are necessary. Thus local administrators would have the chance to see which source categories need to be controlled in a priority order to achieve a better air quality. This could save them from making errors usually based upon widespread "beliefs" for the significance and nature of air polluting sources. To achieve this, a detailed study of the Aegean Region was conducted by generating detailed inventories for smaller provincial areas to be able to collect basic information which is usually provided on a provincial basis. Provinces are the individual units for which many detailed statistics are recorded and made available. Fig. 4.17 shows Aegean Region with the major provinces.

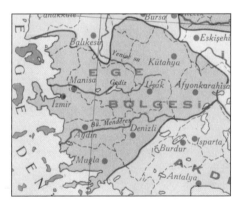

Fig. 4.17. Map of Aegean Region

In all provincial areas calculated emissions of five types of pollutants (PM, SO_x, NO_x, NMVOC and CO) were evaluated in three source categories of point, area and line sources.

4.6.5.1 Point sources

In point source category almost all major industries were included. Emissions from industries were calculated for individual provinces. Industrial emissions were estimated using geographically distributed fuel consumptions data and emission factors. For calculation of emissions, CORINAIR Emission Factors were mostly used. Otherwise EPA emission factors were used. For this part of the study the same methods used for Izmir as described previously were used.

4.6.5.2 Area sources

Domestic heating sources were included in this category. They were apportioned based on population data of the provinces. Emissions were calculated using CORINAIR emission factors and fuel consumption data taken from respective municipalities. Urban and highway emissions were separately evaluated as described for Izmir. For slow urban traffic sources, area source assumption was made.

4.6.5.3 Line sources

Intercity transport at the highways and expressways were inventoried as line sources, emissions were calculated using urban traffic CORINAIR emission factors.

4.6.6 Results

Calculated total sectoral emissions of all provinces in the Aegean Region are summarized in Tables 4.12, 4.13 and 4.14. Table 4.12 shows industrial emissions in all provinces in the Aegean Region. Domestic heating emissions are given in Table 4.13 and traffic emissions are presented in Table 4.14 (Müezzinoglu et al. 2000).

Table 4.12. Industrial emissions of all provinces in the Aegean Region, tons/year

Provinces	PM	SO_x	NO_x	NMVOC	CO
Afyon	735	156	25	4	3
Aydın	22	254	31	2	3
Denizli	309	2 075	266	15	22
İzmir	45 905	113 110	10 080	10 980	11 370
Manisa	14 110	172 805	20 870	2 400	1 340
Muğla	6 621	581 675	25 430	2 915	1 590
Uşak	8 835	15 485	2 754	169	239
Total	**76 537**	**885 560**	**59 456**	**16 485**	**14 567**

Table 4.13. Domestic heating emissions in the region, tons/year

Provinces	PM	SO$_x$	NO$_x$	NMVOC	CO
Afyon	2 418	989	68	82	111
Aydın	491	427	39	19	30
Denizli	4 018	1 289	129	132	173
İzmir	38 433	820	887	1 216	1 517
Manisa	4 805	1 035	112	152	190
Muğla	1 600	300	34	50	62
Uşak	3603	776	84	114	143
Total	**55 368**	**5 636**	**1 353**	**1 765**	**2226**

Table 4.14. Traffic emissions in the region, tons/year

Provinces	SO$_x$	NO$_x$	NMVOC	CO
Afyon	85	281	30	196
Aydın	157	239	26	167
Denizli	134	278	30	193
İzmir	533	991	103	682
Manisa	185	403	44	284
Muğla	128	230	25	161
Uşak	37	95	10	66
Total	**1 259**	**2 517**	**268**	**1 749**

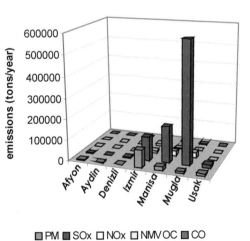

Fig. 4.18. Total emissions of the Agean region, tons/year

When calculated emissions for Aegean Region are compared with previously estimated national emissions (Müezzinoglu et al. 1998), it was seen that the detailed emission inventories of provinces completed so far have accounted only

25% of all national SO$_x$ emissions. Similar numbers were 2% for PM, 8% for NO$_x$, 4% for NMVOC and 1% for CO. Table 4.15 reminds emission estimates of Turkey for the year 2000. These percentages are shown in Fig. 4.19.

Table 4.15. Emission estimates of Turkey for 2000, tons/year

PM	SO$_x$	NO$_x$	NMVOC	CO
6 974 196	3 501 711	844 887	448 495	1 821 720

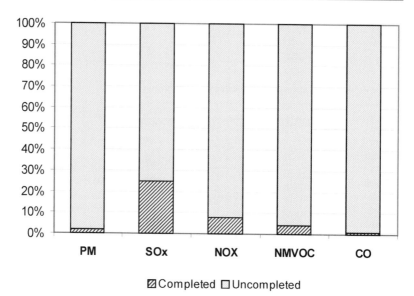

Fig. 4.19. The percentages of the national emissions calculated up to this date in the completed studies.

4.6.7 Air quality management in Izmir

In order to enable the city authorities in the city of Izmir to develop air quality plans and manage the urban activities so that the ambient air quality is improved, a decision support system for urban air quality management for the Izmir metropolitan area was developed at the scale of 50km x 50km. This system included mainly the calculation of a comprehensive emission inventory, dispersion modelling to predict air quality, analysis of predictions and scenario evaluation for air pollution abatement in the decision making process. Preparation of a comprehensive emission inventory for studied air pollutants of sulphur dioxide (SO$_2$), particulate matter (PM), nitrogen oxides (NO$_x$), volatile organic compounds (VOC) and carbon monoxide (CO) using appropriate emission factors was a beginning step. Dispersion modelling using ISCST 3 and California Puff (CALPUFF) model to predict urban air quality was applied to evaluate the ambient air quality in the area.

The results showed that industry was the most polluting sector in the study area contributing about 91% of total SO_2 emissions, 40% of total PM emissions, 90% of total NO_x emissions, 40% of total VOC emissions and 70% of total CO emissions per year. Especially, contributions from industries located at the outside of the metropolitan area were much higher than emissions from within the city centre. Ninety three percent of total industrial SO_2 emissions, 59% of total industrial PM emissions, 80% of total industrial NO_x emissions, 73% of total industrial VOC emissions and 80% of total industrial CO emissions were coming from outside of the Izmir metropolitan area.

Air quality predictions showed that most polluted areas were *Konak*, *Alsancak*, *Buca* and partly *Cigli* with respect to annual average ambient pollutant concentrations in the Izmir metropolitan area. *Buca*, *Konak*, *Alsancak* and *Karsiyaka* were also found as the most polluted residential areas. However, the effects of industries to urban air quality were found as the lowest in *Buca*, *Guzelbahce*, *Konak* and *Alsancak*.

4.6.7.1 Approach

A similar approach used in earlier emission inventories of Izmir 1996, the Aegean region and the Western Mediterranean region were repeated for the base year 2000. In the inventory of emissions contributions of industries, domestic sources and traffic are separately estimated and summed up. Time resolution of the study is annual, seasonal, monthly, daily and hourly. Geographical resolution is 1x1 km grids applied to the area. These resolutions were also applied to the dispersion studies.

4.6.7.2 Overview

To maintain air quality that does not harm human health and welfare in a city, air quality management is necessary. The goal of air quality management recognizes that air quality must be maintained at levels that protect human health, but must also provide protection of animals, plants (crops, forests and natural vegetation), ecosystems, historical places (buildings, statues, monuments etc.). To achieve this goal, it is necessary to develop policies and strategies by global, regional or local authorities.

In the last two or three decades, air quality management practices in the cities of many countries have been broadened in scope. However, the emphasis and success of management activities vary. On the other hand, although considerable progress to improve air quality has been achieved in some large cities of developing countries, many of them still face very significant challenges to achieve notable success. Also, it is now recognized that urban air pollution can travel long distances, affecting areas outside the local and national boundaries away from the areas where polluting event occurs. Polluted air crosses regional and national boundaries, affecting human health and environmental balances in rural areas and in neighbouring countries.

At present, computer based decision support systems for urban air quality management are applied in major cities of the world. This is an opportunity for improving air quality via better planning in the large cities.

The study focuses on the determination of air quality in Izmir and usage of a decision making system for urban air quality management. Preparation of a comprehensive emission inventory and air quality modelling for the year 2000 are the main themes of this research.

4.6.7.3 Objectives

The main objective of this study was to develop a decision support system for urban air quality management and apply it to the city of İzmir. This application intended to be an example for other major Turkish cities. Specific objectives of the study for Izmir were:

- to determine the present air quality in the city of İzmir,
- to prepare a comprehensive emission inventory indicating the sources of different air pollutants in the city,
- to map air quality levels over the city due to different source categories for different time periods,
- to produce example scenarios for air pollution abatement in decision making process,
- to provide a dynamic system for future decisions of local authorities.

4.6.7.4 Methodology

Air pollutant emissions were calculated from source categories of industrial and domestic heating sources as well as traffic. Inventory of emissions was assessed for five pollutant parameters; PM, SO_2, NO_x, VOC and CO for the year 2000. Industrial emissions were calculated at the metropolitan level (Fig. 4.20). Provincial sources outside the urban area were visited to determine the emissions, questionnaires were given out and collected. Specific information on production capacities, raw materials, manufacturing processes, fuel consumptions, stack characteristics (i.e. stack height, outlet diameter, flue gas temperature and exit gas velocity) were collected through these questionnaires for the year 2000. In addition to that, some important pollution sources outside the provincial Izmir borders such as Manisa Organized Industrial Zone were visited and similar data was collected in order to see their impact on the air quality in the Izmir metropolitan area.

Domestic heating sources were evaluated and allocated on the grid system with respect to the density of population in the study area. For the calculation of domestic heating emissions, the information collected included mainly number of inhabitants, type of fuels in use, fuel consumption figures and their emission characteristics.

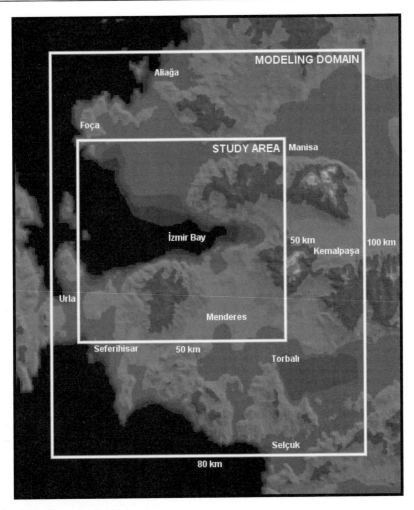

Fig. 4.20. The map of study area

Emissions from all sources were estimated using fuel data and appropriate emission factors. Selection of emission factors had to take into account the fuel types, types of pollutants in accordance with sectoral technology, production range and availability of pollution control devices. In this study, the emission factors were taken from Corinair (CITEPA 1992) and whenever they were not available in Corinair, they were taken from the United States Environmental Protection Agency's source emission factors catalogues (US EPA 1995).

In this study, two well known dispersion models, EPA approved ISCST-3 and *California Puff* (CALPUFF) Models both using Gaussian approach were used for air quality predictions.

For mapping of air quality predictions obtained by the dispersion models and the calculated emissions, GIS softwares were used in this study.

4.6.8 Results

4.6.8.1 Industrial emissions

374 industrial facilities contributing to air pollution in the city of İzmir and its surroundings were included in the study. 153 of them are located in the boundaries of
Izmir Municipality. Others are located around the city. The main manufacturing
sectors are food, textile, paper, metals, printing, chemical, mineral and ceramics,
petrochemical, petroleum refinery, electric and electronic products, cement, iron
and non-iron metallurgy although in fact a much wider pattern is available.

Industrial emissions were estimated using fuel consumption data and appropriate emission factors. The main fuel types used in industries are fuel oil, lignite,
LPG and others (wood, biomass, petroleum coke, biogas, naphtha). Lignite is the
primary fuel type in the metropolitan area while fuel oil is the mostly used fuel
type at outside of the city. Fig. 4.21 shows the main fuels used in the urban and
surrounding areas.

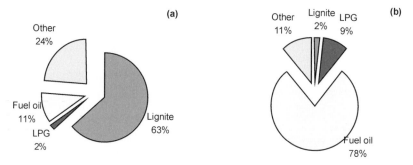

Fig. 4.21. The contribution of each fuel type to total daily consumptions in the city (a) and
the contribution of each fuel type to total daily consumptions in its surroundings (b)

Total annual industrial emissions calculated using fuel consumptions and appropriate emission factors for five pollutants (SO_2, PM, NO_X, VOC and CO) are
given in Table 4.16 over the area. Due to different working hours per day in each
industry, total daily emissions are different that the total annual emissions. The total daily emissions are also given in Table 4.17.

Table 4.16. Industrial emissions from the study area (tons/year)

	PM	SO_2	NO_x	VOC	CO
Metropolitan Area	3 941	5 539	2 631	208	230
Surroundings	5 674	68 904	10 313	564	904
TOTAL	9 614	74 443	12 944	772	1 134

The results show that for all pollutant emissions from the industries located outside of the city are higher than emissions within the city. Metropolitan area has the lowest portion of 7% for SO_2 and the highest portion of 41% for PM. This is due to the high fuel oil consumption in industries located outside the urban centre. For high PM emissions in metropolitan area two cement plants have the highest contribution to PM emissions in the study area. Stack gas measurements were mainly used for the PM emissions from these cement plants in the study.

4.6.8.2 Emissions from domestic heating

Lignite coal and fuel oil were only studied as the major fuels for domestic heating sector due to the insignificance of other fuel uses such as LPG, diesel and kerosene. The fuel use pattern in the city is generally controlled by population density and purchasing capacity of the inhabitants. Population projection data for 2000 was obtained from State Institute of Statistics of Turkey based on the census year 1997.

Table 4.17. Domestic source emissions from the study area (tons/year)

	PM	SO_2	NO_x	VOC	CO
Metropolitan Area	11 159	5 693	1 124	880	369
Surroundings	3 538	1 616	340	277	113
TOTAL	14 697	7 309	1 465	1 156	482

4.6.8.3 Traffic emissions

In Izmir and its surroundings, traffic sector was evaluated in two sub-sectors: inner city and outer city. Because of the differences in the type of data, the emissions from both sub-sectors were separately calculated.

For the city centre there was no data to perform detailed emission calculations. Data from a study involved in the construction of urban underground was used to estimate the traffic emissions from the metropolitan area. Vehicles with respect to their types (car, bus, truck, etc.) were counted at seven major streets for six hours on August 20, 1997. For modelling of air quality for traffic sector, this data was used. Although the model results from these seven streets do not fully represent the air pollution caused by inner city traffic, model runs were made with this data. These emissions are added to the figures given in Table 4.18 for estimated emissions from main intercity roads and submitted in Table 4.19.

Table 4.18. Emissions from major mighways, tons/day

PM	SO_2	NO_x	VOC	CO
3.72	5.14	53.24	14.06	142.33

4.6.8.4 Total emissions

In this study, total emissions were considered as the sum of emissions from industries and domestic heating sources due to the lack of traffic emissions. Table 4.19 summarizes the total emissions for the year 2000 in and around Izmir.

Table 4.19. Total emissions in the study area, tons/year

		PM	SO_2	NO_x	VOC	CO
Domestic	Metropolitan Area	11 159	5 693	1 124	880	369
	Surroundings	3 538	1 616	340	277	113
	Subtotal	14 697	7 309	1 465	1 156	482
Industrial	Metropolitan Area	3 941	5 539	2 631	208	230
	Surroundings	5 674	68 904	10 313	564	904
	Subtotal	9 614	74 443	12 944	772	1 134
Total	Metropolitan Area	15 100	11 232	3755	1 088	599
	Surroundings	9 212	70 520	10 653	841	1 017
	Total	24 311	81 752	14 409	1 928	1 616

Based on the emission calculations given above air pollutant dispersion model calculations were carried out. The map of the total annual average SO_2 concentrations calculated by CALPUFF model is given in the Fig. 4.22 as an example. Such maps were made available for all pollutants, all seasons and all categories at the end of the study. Results of any decision on urban development, industrial permitting, traffic management, fuel apportioning and air quality assurance can easily be seen by running the developed Izmir model. Fig. 4.22 is an example of the several pollution maps drawn for winter and summer halves of the year for all of the five pollutants studied.

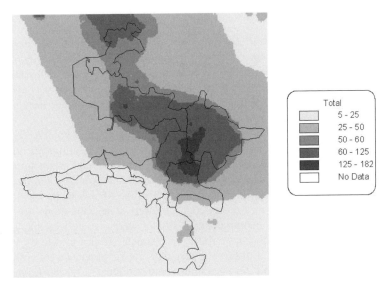

Fig. 4.22. Annual average SO$_2$ concentrations (all sources), μg/m^3

4.6.8.5 VOC studies in Izmir

During the Genemis II period several studies covering certain VOC fractions have been started. Some of them were completed and some are still under field investigation stage.

The first completed and published study which was based on a Thesis research by L. Onat supervised by Muezzinoglu in 1999. This work was published in Atmospheric Environment (Muezzinoglu et al. 2000). In this study a sampling program was conducted to determine the ambient VOCs in the urban area of Izmir, during daytime and overnight hours in summer of 1998. Samples were analysed for benzene, toluene, m, p, -xylene, o-xylene, alkylbenzenes, (ethylbenzene, 1,3,5-trimethylbenzene, 1,2,4-trimethylbenzene), n-hexane, and n-heptane. A good relationship between traffic emissions and ambient VOC levels were established for urban and sub-urban sites by examining the toluene-to-benzene ratios in the city.

Another project on VOCs was completed around a major petrochemical complex and a petroleum refinery air near Izmir. This project resulted in a submitted paper presently in peer review (Cetin et al. 2002).

In this project air samples were collected to measure ambient VOC concentrations between September 2000 and September 2001 in Izmir, Turkey at three sampling sites located around a petrochemical complex and an oil refinery. VOC concentrations were found 4-20 times higher than those measured at a suburban site in Izmir, Turkey. Ethylene dichloride, a leaded gasoline additive used in petroleum refinery and in the ethylene process of the petro-chemical complex was the most abundant volatile organic compound followed by ethyl alcohol and acetone (Cetin 2002).

VOC concentrations showed seasonal variations at sampling sites. Concentrations were the highest in summer followed by autumn, probably due to increased evaporation of VOCs from fugitive sources as a result of higher ambient temperatures. The relationship between wind speed and VOC concentrations was also investigated. When wind speed was ≥5 m/s, concentrations were higher for all sites compared to those measured at lower wind speeds.

In another study atmospheric mono-aromatic and oxygenated volatile organic compound (VOC) concentrations in Izmir were investigated (Cetin et al. 2002). Air samples were collected in Izmir, Turkey at two suburban and urban sampling sites to determine ambient concentrations of selected mono-aromatic and oxygenated volatile organic compounds (VOCs). The objective of this work was to determine the diurnal variation of ambient concentrations of selected mono-aromatic and oxygenated VOCs at the suburban site. Samples were also collected at an urban site close to a street with heavy traffic to characterize the profile of VOCs emitted from motor vehicles. For this eighteen ambient air samples were collected between June 11 and 20, 2001 on a sampling platform located on the Kaynaklar campus of the Dokuz Eylul University, Izmir, Turkey. The sampling site is located approximately 10 km southeast of Izmir's centre and is outside the city limits. Concentrations were compared with the ones measured previously inside Izmir and other cities around the world. Additional samples were collected at an urban site close to a street with heavy traffic on October 5-6, 2002. The urban sampling site is located on the roof of a seven story- building 50 m away from the street. Nine consecutive samples were collected at this sampling site.

A new Ph.D. research study is in progress related to airborne levels of formaldehyde (by Remzi Seyfioğlu) and its dry deposition under the supervision of Mustafa Odabasi.

Recently the sulphur containing organic air pollutants in the malodorous deltas of the creeks discharging into Izmir Bay were studied and published (Muezzinoglu 2003). In this study, the H_2S levels as well as the organic derivatives like methane thiol, ethane thiol, propane thiol, butane thiol, thiophene and dimethyl sulphide in the ambient air as well as emission sources were measured during summer 2001. Results were correlated with local SO_2 concentrations which increased up during strong insolation on hot and dry summer days.

4.6.9 Conclusions and outlook

Field investigations and emission studies aiming at VOC identification are becoming more popular among the young scientists in our group. We hope to be able to contribute more in this respect in the near future. Detailed inventory and modelling work is also going on as long as there is sponsorship and research time.

4.7 References

Ahlvik P, Eggleston E, Gorissen N, Hassel D, Hickman A-J, Joumard R, Ntziachristos L, Rijkeboer R, Samaras Z and Zierock K-H (1997) COPERT-II Computer Programme to Calculate Emissions from Road Transport, Methodology and emission factors, 2nd edition, European Environmental Agency, European Topic Center on Air Emissions.

Atımtay A, Güllü E and Yetiş Ü (1995) Ankara Hava Kirliliği Envanter Çalışması, Yanma ve Hava Kirliliği Kontrolu III. Ulusal Sempozyumu, Ankara (1995) 376-387 (in Turkish).

BUWAL (1998) NOREM Database for non-regulated emissions from motor vehicles, SAEFL, BUWAL, OFEFP, CD-ROM.

BUWAL (1999) Handbuch Emissionsfaktoren des Strassenverkehrs, INFRAS, BUWAL, OFEFP, CD-ROM

Cetin E (2002) Ambient volatile organic compound (VOC) concentrations around a petrochemical plant. MS thesis, Graduate School of Natural and Applied Sciences. Izmir, Turkey: Dokuz Eylul University, 2002.

Cetin E, Odabasi M, Seyfioglu R (2002) Ambient volatile organic compound (VOC) concentrations around a petrochemical complex and a petroleum refinery. (in review) The Science of the Total Environment.

CITEPA (1992) CORINAIR Software Instructions for use, Provisional Document. Centre Interprofessionel technique d'études de la pollution atmosphérique (CITEPA). CORINAIR Inventory, Commission of the European Community. Paris

EEA, Corine Land Cover. 1992, 22

Elbir T, Müezzinoğlu A, Bayram A (2000) Evaluation of some air pollution indicators in Turkey, Environment International, 26, 5–10.

EMEP/CORINAIR (1996) Atmospheric emission inventory guidebook, European Environmental Agency, First edition, Mc Innes, G. (ed.), Copenhagen.

GLOED (1991) User's Guide for the Global Emissions Development System (GLOED). PC-Software prepared for Office of Research & Development, US-Environmental Protection Agency, Radian Corporation. North Carolina, 1991.

Grell GA, Dudhia J, Stauffer DR (1993 A Description of the Fifth-Generation PENN STATE/NCAR Mesoscale Model (MM5), NCAR/TN-398+IA, 1993, 107

Grell GA, Emeis S, Stockwell WR, Schoenemeyer T, Forkel R, Michalakes J, Knoche R, Seidl W (2000) Application of a multiscale, coupled MM5/Chemistry Model to the complex terrain of the VOTALP Valley Campaign. Atmospheric Environment 34, 1435–1453

Guenther A, Hewitt CN, Erickson D, Fall R, Geron C, Graedel T, Harley P, Klinger L, Lerdau M, McKay WA, Pierce T, Scholes B, Steinbrecher R, Tallamraju R, Taylor J, Zimmerman P (1995) A global model of natural volatile organic compound emissions. Journal of Geophysical Research 100, D5: 8873-8892

Laing, R (1991) CAREAIR - Ein computergestütztes Modellsystem zur Analyse von Luftschadstoffemissionen. In: Informatik Fachberichte No. 296, M. Hälker, A. Jaeschke (Hrsg.), Informatik für den Umweltschutz (Computer Science for Environmental Protection). 6. Symposium, München, Proceedings

MEET (1999) Methodology for Calculating transport emissions and energy consumption, Transport Research, 4th Framework Programme, Strategic Research, DG VII, *ISBN 92-828-6785-4.*

Mensink C (2000) Validation of urban emission inventories, *Environ. Monitoring and Assessment*, **65**, 31–39.

Mensink C, Van Rensbergen J, Viaene P, De Vlieger I and Beirens F (1999) Temporal and spatial emission modelling for urban environments using emission measurement data in: Borell PM and Borell P (eds.) *Proceedings of EUROTRAC Symposium '98*, **2**, 711–714, WIT Press, Southampton.

Mensink C, De Vlieger I, Nys J (2000a). An urban transport emission model for the Antwerp area, *Atmos. Environ.*, **34**, 4595–4602.

Mensink C, Bomans B and Janssen L (2000b) An assessment of urban VOC emissions and concentrations by comparing model results and measurements, *Int. J. Environment and Pollution*, **16 (1-6)**, 345–356.

MINERGG (1995) Software according to the Guidelines for National Greenhouse Gas Inventories. Sieber and Raskin (eds.), Inter Governmental Panel on Climate Change (IPCC). Stockholm Environment Institute (SEI) and Organisation of Economic Cooperation and Development (OECD), Boston

Mücher CA, Steinnocher K, Kressler F, Heunks C (2000) Land cover characterization and change detection for environmental monitoring of pan-Europe. Int. J. Remote Sens. 21, 1159–1181

Müezzinoglu A, Bayram A, Yıldızhan A, Elbir T (1997a) Emission Inventory of Izmir and Surroundings for Use in Air Quality Modelling Studies, 10th Regional IUAPPA Conference - Air Quality Management at Urban, Regional and Global Scales, eds: Ekinci, E.; Yardım, F.; Bayram, A., Environmental Research Forum, Vol. 7-8, TransTec Publications, Zurich, 119-124.

Müezzinoglu A, Tırıs M, Bayram A, ElbirT (1997b) Emission Inventory of Major Pollutant Categories in Turkey. In: Tropospheric Modelling and Emission Estimation (Ebel, A., Friedrich, R., Rodhe, H., eds) EUROTRAC Vol.7, Springer, Heidelberg.

Müezzinoğlu A, Elbir T, Bayram A (1998) Inventory of Emissions from Major Air Pollutant Categories in Turkey, Environmental Engineering and Policy, Vol. 1, 109–116, Springer Verlag, Berlin.

Müezzinoglu A, Bayram A, Elbir T, Seyfioglu R (2000) Emission Inventory of Aegean Region, TÜBITAK-YDABÇAG-198Y094 Project, Izmir (2000) (in Turkish).

Müezzinoğlu A (2003) A study of volatile organic sulfur emissions causing urban odors. Chemosphere (online, in press) 51(4), 245-252. Scire, J.S., Strimaitis, D.G.,& Yamartino, R.J., 2000: A User's Guide for the CALPUFF Dispersion Model. Concord, MA: Earth Tech, Inc

Müller Th, Boysen B, Friedrich R, Voß A (1990) Ermittlung und Analyse des zeitlichen Verlaufs und der räumlichen Verteilung der derzeitigen und zukünftigen SO2- und NO_x-Emissionen in Baden-Württemberg. Projekt Europäisches Forschungszentrum für Maßnahmen zur Luftreinhaltung (PEF). KfK-PEF 71. Forschungszentrum Karlsruhe.

Ntziachristos L and Samaras Z (2000) COPERT-III (Computer Programme to Calculate Emissions from Road Transport) – Methodology and Emission Factors (version 2.1), Technical report N°49, European Environment Agency, http://vergina.eng.auth.gr/mech/lat/copert/copert.htm, 86

Obermeier A, Friedrich R, John C, Seier J, Vogel H, Fiedler F and Vogel B (1994) Effiziente Immissionsminderungsstrategien. Institut für Meteorologie und Klimaforschung (IMK), Universität Karlsruhe/Kernforschungszentrum Karlsruhe. Projekt Europäisches Forschungszentrum für Maßnahmen zur Luftreinhaltung (PEF). Karlsruhe

San José R, Cortés J, Prieto JF, Hernández JM and González RM (1996) A sensitivity analysis of high-resolution emission model under different emission scenarios over Madrid area. Air Pollution IV: Monitoring, simulation and control. Computational Mech. pp 707-716.

San José R, Sanz MA, Moreno B, Ramírez-Montesino A, Hernández J, Rodríguez L (1995) Anthropogenic and biogenic emission model for mesoscale urban areas by using LANDSAT satellite data: Madrid case study. European Symposium on Optics for Environmental and Public Safety. LASER'95. SPIE.- The International Society for Optical Engineering, SPIE Vol. 2506, pp 274-285.

Schmulius C, Flügel A, Frotscher K, Hochschild V, Müschen B (2000) The Shuttle Radar Topography Mission (SRTM) and Applications in Europe, Africa and Siberia, Photogrammetrie Fernerkundung Geoinformation (PFG) 5, 361–366

Smiatek G (2000) Scale effects and GIS. In: Proceedings of the SIMPAQ Symposium: Scale Interactions in models and Policies for Air Quality Management, Antwerp Apr. 13-14, 47–51

Smiatek G (2002) Application of GIS and RDBMS to Numerical Atmospheric Meteorology/Chemistry Models. Proceedigs of the EUROTRAC2 Symposium 2002

Steinbrecher R, Schoenemeyer T, Forkel R and Smiatek G (1997) Biogenic emissions in Germany and Europe: Current status of inventory activities. 1er GENEMIS Workshop, Stuttgart

Tayanç M (2000) An Assessment of Spatial and Temporal Variation of Sulfur Dioxide Levels over İstanbul, Turkey. GENEMIS Annual Report 1999, ISS Munich

U.S Environmental Protection Agency, 1995: Compilation of Air Pollutant Emission Factors - Stationary Point and Area Sources. Research Triangle Park, NC: AP-42 Team.

U.S.Geological Survey (2002a) GTOPO30 Documentation.
http://edcdaac.usgs.gov/gtopo30/README.html

U.S.Geological Survey (2002b) Global Land Cover Characteristics Data Base.
http://edcdaac.usgs.gov/glcc/globdoc2_0.html

Veldt C (1992) Updating and upgrading the PHOXA emission data base to 1990, TNO report, TNO Institute of Environmental and Energy Technology, Apeldoorn, March 1992.

Printing: Mercedes-Druck, Berlin
Binding: Stein+Lehmann, Berlin